Understanding
Fiber Optics
Second Edition

Understanding Fiber Optics

Second Edition

JEFF HECHT

Prentice Hall
Upper Saddle River, New Jersey 07458

Library of Congress Catalog Card Number 93–83599

All terms mentioned in this book that are known to be trademarks or service marks have been appropriately capitalized. Prentice Hall Publishing cannot attest to the accuracy of this information. Use of a term in this book should not be regarded as affecting the validity of any trademark or service mark.

Composed in AGaramond and MCPdigital by Prentice Hall Publishing.

 © 1993 by Prentice-Hall, Inc.
Simon & Schuster / A Viacom Company
Upper Saddle River, New Jersey 07458

Printed in the United States of America

10 9 8 7

ISBN 0-13-649070-0

Prentice-Hall International (UK) Limited, *London*
Prentice-Hall of Australia Pty. Limited, *Sydney*
Prentice-Hall Canada Inc., *Toronto*
Prentice-Hall Hispanoamericana, S.A., *Mexico*
Prentice-Hall of India Private Limited, *New Delhi*
Prentice-Hall of Japan, Inc., *Tokyo*
Simon & Schuster Asia Pte. Ltd., *Singapore*
Editora Prentice-Hall do Brasil, Ltda., *Rio de Janeiro*

Overview

Contents

Acknowledgments

Many people in the fiber-optics industry have given generously of their time to patiently answer my questions. I want to give special thanks to those who made helpful comments on earlier drafts of various chapters: Jim Hayes, Robert Gallawa, David Charlton, Lois Boynton, Tony Beam, Marvin Ashby, Jay Schiestle, Carol Hutchinson, Vince Martinelli, Jim Walyus, Marjorie Katz, Art Nelson, and Ron Doyle.

About the Author

Jeff Hecht has been writing about fiber optics since 1974. He has been managing editor of *Laser Focus,* was a co-founder of *Lasers & Optronics* and *Fiberoptic Product News,* and is now contributing editor to *Laser Focus World.* He received a B.S. in electronic engineering from the California Institute of Technology in 1969 and has written several other books on lasers and optics.

Introduction to Fiber Optics

A Personal View

Light is an old friend. I've been fascinated with light and optics ever since I can remember. I started playing with lenses and prisms when I was about 12, and though my box of optical toys has spent some time in the closet over the years, I've added some new playthings, and many of them involve fiber-optic technology.

At Home

The first optical fibers I saw were in decorative lamps. A group of fibers was tied together at one end and splayed out in a fan at the other. A bulb at the tied end illuminated them, and the light emerging from the loose ends made them glitter. The effect was pretty enough that when I was in college I bought one for my sister as a Christmas present, but useless enough that I wandered away to explore other things.

In the Community

When I next saw fiber optics, in the mid-1970s, the technology had come a long way. Fibers had been improved enough that telephone companies were looking at them for communications. Those were the days when phone companies were—with good reason—described as "traditionally conservative" in their use of technology. Cautiously, they probed and tested fiber optics, almost like a bomb squad examining a suspicious package. It was not until 1977 that, within a month, first GTE and then AT&T dared to venture down manholes and stick fiber-optic cables into telephone circuits carrying live traffic.

Looking back, that technology looks primitive. It was daring then, and it worked. Not only that, it worked flawlessly. The small armies of engineers monitoring those test beds came to countless technical meetings afterward repeating the same monotonous but thrilling conclusion: "It works. Nothing has gone wrong."

In the Business World

I was at the first fiber-optic trade show in the late 1970s and have watched the excitement spread since then. Each year the meetings have grown larger. For a few years, breakthroughs were almost routine. The first generation of systems was barely in the ground before a second generation was ready. Today the telephone industry is using a third generation of fiber-optic technology, and a fourth generation is being installed in some places. The rate of change seems slower now, but the accomplishments are incredible when measured on the scale of the first fiber-optic systems.

The Future

Looking back, it's been an incredible ride. I've watched a technology spring from the laboratory into the real world. Once I heard about fiber optics from research scientists; now I hear about fiber optics from telephone service people. But the fun isn't over yet. The fiber-optics revolution will continue until fiber comes all the way to homes. It won't come tomorrow, but when it does come, it will bring a wealth of new information services. First, the services will go to businesses; eventually they will go into homes. The visionaries who foresaw a wired city were wrong—we will have a fibered society instead. We can all watch it happen.

But that's enough of this visionary stuff. Let's get down to the nuts and bolts—and fiber.

About This Chapter

The idea of communication by light was around long before fiber optics, as were fibers of glass. It took many years for the ideas behind fiber optics to evolve from conventional optics. Even then, people were thinking more of making special optical devices than of optical communications. In this chapter you will see how fiber-optic technology evolved and how it can solve a wide variety of problems in communications.

How and Why Fiber Optics Evolved

Left alone, light will travel in straight lines. Even though lenses can bend light and mirrors can deflect it, light still travels in a straight line between optical devices. This is fine for most purposes. Cameras, binoculars, telescopes, and microscopes wouldn't form images properly if light didn't travel in a straight line.

However, there also are times when people want to look around corners or probe inside places that are not in a straight line from their eyes. Or they may just need to pipe light from place to place, for communicating, viewing, illuminating, or other purposes. That's when they need fiber optics.

●
Light normally travels in straight lines, but sometimes it is useful to make it go around corners.

Piping Light

The problem arose long before the solution was recognized. In 1880, a Concord, Massachusetts, engineer named William Wheeler patented a scheme for piping light through buildings. Evidently not believing that Thomas Edison's incandescent bulb would prove practical, Wheeler planned to use light from a bright electric arc to illuminate distant rooms. He devised a set of pipes with reflective linings and diffusing optics to carry light through a building, then diffuse it into other rooms, a concept shown in one of his patent drawings in Figure 1.1.

Although he was in his twenties when he received his patent, Wheeler had already helped found a Japanese engineering school. He went on to become a widely known hydraulic engineer. Nevertheless, light piping was not one of his successes. Incandescent bulbs proved so practical that they're still in use today. Even if they hadn't, Wheeler's light pipes probably wouldn't have reflected enough light to do the job. However, his idea of light piping reappeared again and again until it finally coalesced into the optical fiber.

Total Internal Reflection

Ironically, the fundamental concept underlying the optical fiber was known well before Wheeler's time. A phenomenon called total internal reflection, described in more detail in Chapter 2, can confine light inside glass or other transparent materials denser than air. If the light in the glass strikes the edge at a glancing angle, it cannot pass out of the material and is instead reflected back inside it. Glassblowers probably saw this effect long ago in bent glass rods, but it wasn't widely recognized until the mid-1800s, when British physicist John Tyndall used it in his popular lectures on science.

Tyndall's trick, shown in Figure 1.2, worked like this. He shone a bright light down a horizontal pipe leading out of a tank of water. When he turned the water on, the liquid flowed out, with the pull of gravity forming a parabolic arc. The light was trapped within the water by total internal reflection, first bouncing off the top surface of the jet, then off the lower surface, until the turbulence in the water broke up the beam. Wheeler may have seen Tyndall perform this trick when he lectured in Boston, but presumably it didn't seem practical to his engineering mind.

FIGURE 1.1.

Wheeler's vision of piping light (U.S. Patent 247,229).

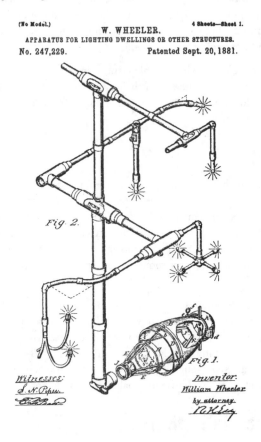

Optical Communication

An optical telegraph was invented in France in the 1790s and made obsolete by the electric telegraph.

Meanwhile, in Washington, a young scientist who already had an international reputation was working on what he considered his greatest invention—the Photophone. Earlier, Alexander Graham Bell had used electricity to carry voices in the telephone. However, Bell was intrigued by the idea of sending signals without wires. He thought of optical communication, an idea that probably goes back to signal fires on prehistoric hilltops. The first "telegraph," devised by French engineer Claude Chappe in the 1790s, was an optical telegraph. Operators in towers relayed signals from one hilltop to the next by moving semaphore arms. Samuel Morse's electric telegraph put the optical telegraph out of business, but it left behind countless Telegraph Hills.

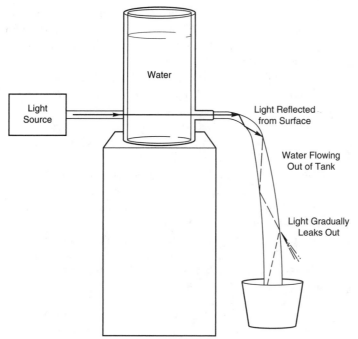

FIGURE 1.2.

Light guided down a water jet.

Light beam becomes more diffuse as it passes down the water jet, because turbulence breaks up surface.

In 1880, Bell demonstrated that light could carry voices through the air without wires. Bell's Photophone reproduced voices by detecting variations in the amount of sunlight or artificial light reaching a receiver. It was the first "wireless" voice communication. However, it never proved practical because too many things could get in the way of the beam.

Later, others used light to carry voices through the open air, something that is now done with a few laser systems but is not used widely. In the 1930s, another engineer, Norman R. French—ironically, an employee of the American Telephone & Telegraph Corp., the company built around Bell's telephone—patented the idea of communicating via light sent through pipes. Even more time passed before the optical fiber was invented.

The Clad Fiber

Although Tyndall could guide light in his stream of water, he couldn't do so very well. The rough boundary of the water broke up the light beam. Other effects also limited light guiding in bent glass rods. The problem is that light can leak out wherever the rod touches something other than air. Because the rod cannot hang unsupported in the air, it has to touch something.

The key development in making optical fibers usable was a cladding to keep the light from leaking out.

The solution to that problem seems obvious with 20/20 hindsight. In the 1950s, Brian O'Brien, Sr., in the United States and Harry Hopkins and Narinder Kapany in England started looking for ways to guide light. The key concept was making a two-layer fiber, shown in Figure 1.3. Light would be carried in the inner layer or core of the fiber. The outer layer, the cladding, would confine the light in the core, because its refractive index (like that of air) was below that of the core. It would also prevent the core from touching anything with a higher refractive index that might let the light leak out.

Imaging

●
Many optical fibers can be bundled together to transmit images.

O'Brien, who had been dean of the prestigious school of optics at the University of Rochester, brought the idea of the clad fiber with him when he became director of research at the American Optical Co. in Southbridge, Massachusetts. His interest was imaging. A single fiber would not transmit an image because the light from different parts of the image would be blurred together. O'Brien bundled many optical fibers together so that a pattern of light formed on one end of the bundle would be re-created on the other end. This requires that all the fibers be at precisely the same position relative to each other at each end of the bundle, as I will examine in more detail in Chapter 23.

FIGURE 1.3.

Light cannot leak out of clad fibers.

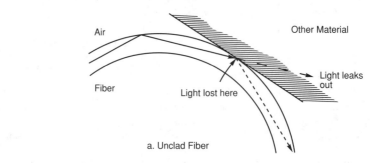

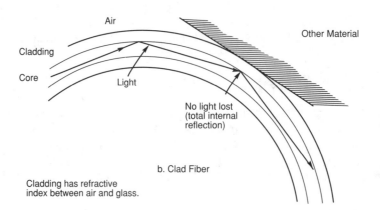

Expanded Applications

Engineers at American Optical and elsewhere soon began extending the technology. Fibers could be grouped in flexible as well as rigid bundles. The fibers need not be aligned with respect to each other if the only goal was to deliver light, as in a fiber-optic lamp. Fibers could be made of plastic as well as of glass. Applications began to emerge in fields such as medicine and inspection.

Reducing Fiber Loss

Recognizing Transmission Problems

One important fundamental limit remained: optical fibers absorbed too much light. That statement must be put in context. Even comparatively high-loss fibers are much more transparent than ordinary window glass. (Ordinary windows are very thin, so they absorb little light; most light lost at windows is reflected from the glass.) Those high fiber losses were acceptable in transmitting light a few feet or a couple meters, something impractical through window glass. However, only 10% of the light remained after traveling 10 m (32 ft), and after 20 m only 1% was left.

At first, no one seriously thought that fibers had a transmission problem, because they weren't trying to send light very far. Theodore H. Maiman's demonstration of the first laser in 1960 renewed interest in optical communications, but people were slow to give optical fibers serious thought. Scientists thought all transparent solids absorbed too much light to transmit optical signals long distances, so they started looking at new types of light pipes.

Purifying Glass Filters

Two engineers working at Standard Telecommunications Laboratories in England, Charles K. Kao and George Hockham, did take a careful look at the possibilities of optical-fiber telecommunications. Their theoretical analysis, published in 1966, indicated that fiber loss was high because of impurities, not because of the glass itself. By removing the impurities, they said, it should be possible to reduce loss to levels low enough that 10% of the light would remain after it had passed through 500 m (1600 ft) of fiber. Their prediction may have sounded fantastic then, but it proved too conservative.

Publication of Kao and Hockham's paper set off a worldwide race to make better fibers. The first to beat the theoretical prediction were Robert Maurer, Donald Keck, and Peter Schultz at the Corning Glass Works (now Corning Inc.) in 1970. Others soon followed, and losses were pushed down to even lower levels. In today's best optical fibers, 10% of the entering light remains after the light has passed through more than 50 kilometers (30 miles) of fiber. Losses are not quite that low in practical telecommunication systems, but

●
Invention of the laser stimulated interest in optical communications and led to efforts to reduce light absorption in fibers, which was essential for communications.

as you will see in Chapter 4, impressive progress has been made. Because of that progress, fiber optics have become the backbones of long-distance telephone networks around the world.

The Basics of Fiber Optics

Materials

●
Many types of optical fibers are made, and they have different characteristics.

Not all optical fibers are the same. Many distinct types are available, designed for specific applications. Many fibers may be bundled together, or fibers may be used individually.

Individual optical fibers are used in virtually all communications applications and for many other purposes. Each fiber is optically separate from other fibers, although many separate fibers may be housed in a common cable. Most fibers are made of glass, plastic, or plastic-clad glass; some special fibers are made of other materials, such as exotic fluoride compounds. Standard fibers are flexible but somewhat stiff; flexibility depends on fiber diameter. Optical fibers are often compared to human hairs, but whoever thought of that comparison must have had some very stiff hairs or very thin plastic fibers. Communication fibers are stiffer than even a man's coarse beard hair of the same length. A better comparison would be to monofilament fishing line. Unlike wires, fibers spring back to their original straight form after being bent.

Alternatively, fibers can be bundled together in either of two forms. A flexible bundle is made up of many separate fibers, assembled together, with the ends of the bundle fixed and the rest unattached to each other (although typically encased in some overall housing). A rigid bundle is made by melting many fibers together into a single rod, which typically is bent to the desired shape during manufacture. Such rigid or fused bundles cost less than flexible bundles, but their inflexibility makes them unsuitable for some applications.

Systems

●
Fiber-optic communication systems include transmitter, receiver, and a cable structure (to house the fiber) as well as the fiber itself.

You need more than just fiber to make a communication system. The basic elements of a system are shown in Figure 1.4. The signal originates from a modulated light source, which feeds it into a fiber, which delivers it to a receiver. The receiver decodes the optical signal and converts it to electronic form for use by equipment at the receiving end.

Real-world fiber-optic systems are more complex than indicated in this simple example. Most communication travels in two directions and requires a transmitter and a receiver at each end. Usually separate fibers carry signals in each direction. The fiber or fibers are housed in a cable to simplify handling and protect them from environmental stresses. Fibers must be precisely aligned with light sources to collect their output efficiently. Likewise, if light is transferred between fibers, the two ends must be precisely aligned. Because their diameters are very small, the mechanical tolerances for proper alignment are tight. Consequently, much more attention must be paid to connectors and splices than in electrical communication over wires.

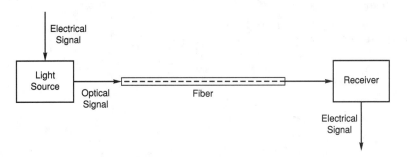

FIGURE 1.4.
Fiber-optic system components.

Auxiliary equipment is needed for many of these tasks. Unlike metal wires, optical fibers cannot be spliced with wire cutters and a soldering iron. More elaborate fiber splicers are needed, as you will see in Chapter 11. Connector installation is also a complex task, and often special measurement tools are needed to assess the quality of a fiber-optic link, as you will see in Chapter 14.

Why Fiber Optics?

Light Transmission

The crucial operating difference between a fiber-optic communication system and other types is that signals are transmitted as light. Conventional electronic communication relies on electrons passing through wires. Radio-frequency and microwave communications (including satellite links) rely on radio waves and microwaves traveling through open space. (I will ignore free-space optical communication systems, which send a laser beam through free space or air, because there are few such systems. To prevent confusion, I'll avoid the term "optical" communications.)

●
Because fiber optics is a unique transmission medium, it has some unique advantages for certain types of communications.

Different media are suited for different communication jobs. The choice depends on the job and the nature of the transmission medium. One important factor is how signals are to be distributed. If the same signal is to be sent from one point to many people in an area—as in broadcast television or radio—the best choice may be nondirectional radio transmission. Radio-frequency communication is the best way to avoid cables for cellular phones and to reach remote places like tropical islands or arctic bases. On the other hand, a cable system is preferable for making physical links among many fixed points, as in telephone and cable television networks. Cable is also useful for permanent connections between two fixed points. Some types of transmission are shown in Figure 1.5.

FIGURE 1.5.

Types of communication transmission.

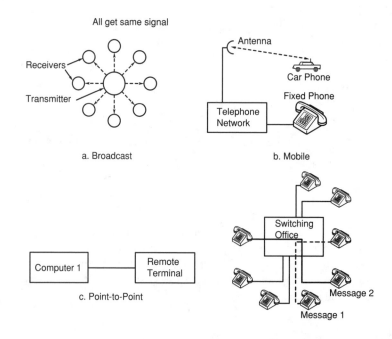

a. Broadcast

b. Mobile

c. Point-to-Point

d. Telephone Network

Capacity and Distance

> High transmission capacity and long transmission distance are two major advantages of fiber-optic cables.

When choosing the type of cable, key factors to consider are how far the signal must go and how much information it carries. Transmission distance depends on such effects as transmitter power, receiver sensitivity, and losses in the intervening medium. The amount of information carried is measured in the passage of signals through the system, much as we measure water volume flowing through a pipe. Some transmission media, such as simple pairs of wires, work fine for low-speed signals but cannot carry higher-speed signals very far. Others, such as coaxial cables, can carry higher-speed signals, but only over limited distances. A major attraction of optical fibers is that they can carry information at high speeds over long distances. (I'll attach some numbers to those terms in later chapters.)

Secondary Factors

Secondary factors also influence the choice of transmission medium. A typical example is electromagnetic interference (EMI), which, like static on an AM radio, can sometimes block signal transmission on wires. Optical fibers cannot pick up EMI because they carry signals as light.

Different factors can dictate the choice of fiber for other applications. For example, security of the transmitted signals is a vital concern in military facilities and financial institutions. When cables must be installed in existing buildings, cable size and rigidity determine ease of installation, which is a major cost issue. The related issues of weight and bulk are major practical concerns in communication systems designed to be portable, ranging from electronic news-gathering equipment to battlefield communication systems. Avoiding sparks is a must in systems being installed in refineries and chemical plants, where the atmosphere may contain explosive gases. As you will see in Chapter 3, fiber optics can solve many of these problems.

Fiber Optic Applications

The wide variety of fiber-optic systems that have come into use because of the advantages of optical fibers will be described in more detail in following chapters. They include the following:

● The advantages of fiber optics have led to many applications in long-haul and short-distance communications.

- Long-haul telecommunication systems on land and at sea to carry many simultaneous telephone calls (or other signals) over long distances. These include ocean-spanning submarine cables and national backbone networks for telephone and computer data transmission.

- Interoffice trunks that carry many telephone conversations simultaneously between local and regional switching facilities.

- Drop-offs for telephone lines operating above the normal speed of single voice lines, both to businesses and to groups of homes.

- Systems to carry cable television signals between microwave receivers and control facilities (called head-ends), and to distribute the video signals from head-ends to groups of homes.

- Connections between the telephone network and antennas for mobile telephone service.

- Cables for remote news-gathering equipment.

- Links among computers and high-resolution video terminals used for such purposes as computer-aided design.

- Local area networks operating at high speeds or over large areas, and backbone systems connecting slower local area networks.

- High-speed interconnections between computers and peripheral devices, or between computers, or even within segments of single large computers.

- Moderate-speed transmission of computer data in places where fiber is most economical to install.

- Transmission in difficult environments, especially those plagued with severe EMI.

- Portable communication equipment for battlefield use.

- Transmission of signals within ships and aircraft.

Meanwhile, fiber-optic technology has continued to expand into many areas outside of communications. These include fiber-optic bundles for illumination and imaging, endoscopes to view inside the body and treat diseases with light and without surgery, and optical sensors to measure rotation, pressure, sound waves, magnetic fields, and many other quantities.

These are today's fiber-optic applications. More are coming as fiber optics and other technologies develop. For example, fiber-optic systems are in development to provide a broad array of new and existing communication services to homes and businesses. Telephone companies and cable television providers are maneuvering to corner that potentially large market. For now, however, I will concentrate on the realities of present-day technology.

What Have You Learned?

1. Light rays want to go in straight lines, but optical fibers can guide them around corners.
2. Early optical communication systems sent light through the air.
3. Optical fibers must have a cladding layer to keep light from leaking out.
4. The first applications of fiber optics were outside of communications.
5. Loss of fibers was reduced dramatically to allow their use in communications.
6. Optical fibers are best for transmitting signals at high speeds or over long distances between fixed points.
7. Fiber optics have some special advantages, including immunity to electromagnetic interference, which can block transmission on wires.
8. Fiber optics are likely to bring a new generation of communication services to homes in the future.

What's Next?

In this chapter, we examined the background of fiber-optic technology. In Chapter 2, you'll learn some of the basic physics behind fiber optics before going into some specifics about fiber-optic hardware.

Quiz for Chapter 1

1. The first use of light for communication was
   a. Claude Chappe's optical telegraph.
   b. native American smoke signals.
   c. a testbed that GTE installed in 1977.
   d. prehistoric signal fires.

2. Light can be guided around corners most efficiently in
   a. reflective pipes.
   b. hollow pipes with gas lenses.
   c. clad optical fibers.
   d. bare glass fibers.

3. The first low-loss optical fibers were made
   a. at Corning Glass Works in 1970.
   b. at Standard Telecommunication Labs in 1966.
   c. at Bell Telephone Labs in 1960.

4. Today's best optical fibers transmit light so well that 10% of the input light remains after
   a. 0.5 km.
   b. 4 km.
   c. 20 km.
   d. 50 km.
   e. 100 km.

5. Optical fibers are made of
   a. glass coated with plastic.
   b. ultrapure glass.
   c. plastic.
   d. all of these.

6. Essential components of any fiber-optic communication system are
   a. light source, fiber, and receiver.
   b. light source and cable.
   c. fiber and receiver.
   d. fiber only.

7. The small size of optical fibers makes what necessary in any device connecting them?
   a. Special glue.
   b. Tight mechanical tolerances.
   c. Low optical absorption.
   d. Small overall size.

8. What are the major advantages of optical fibers for long-distance communications?
   a. Small fiber size.
   b. Nonmetallic.
   c. Low loss when carrying high-speed signals.
   d. Low loss only.
   e. High-speed signal capacity only.

9. Unlike wires, optical fibers are immune to

    a. electromagnetic interference.

    b. high-frequency transmission.

    c. signal losses.

10. Applications of fiber-optic communications include

    a. ocean-spanning submarine cables.

    b. long-distance telephone transmission on land.

    c. connecting telephone-company facilities.

    d. transmitting data to high-resolution video terminals.

    e. all of the above.

Fundamentals of Fiber Optics

About This Chapter

Fiber optics is a hybrid field that started as a branch of optics. The basic concept of a fiber is optical, and some optical fibers are used singly or as bundles as optical components. However, as fiber became a communication medium, the field borrowed concepts and terminology from electronic communications. Transmitters and receivers convert signals from electrical to optical format and back; they are part optics and part electronics. To understand fiber-optic communications, you need to know about three fields: optics, electronics, and communications.

This chapter is a starting point. In later chapters, I'll go into more detail on such topics as how light is guided in fibers and how optical fibers can serve as the basis of a communication system.

Basics of Optics

The workings of optical fibers depend on basic principles of optics and the interaction of light with matter. The first step in understanding fiber optics is to review the relevant parts of optics. The summary that follows does not cover all of optics, and some parts may seem basic, but you should read it to make sure you understand the fundamentals.

From a physical standpoint, light can be seen either as electromagnetic waves or as photons, quanta of electromagnetic energy. This is the famous wave-particle duality of modern physics. Both viewpoints are valid and valuable, but the most useful viewpoint for optics is often to consider light as rays traveling in straight lines between or within optical elements, which can reflect or refract (bend) light rays at their surfaces.

The Electromagnetic Spectrum

The light carried in fiber-optic communication systems can be viewed as either a wave or a particle.

What we call "light" is only a small part of the spectrum of electromagnetic radiation. The fundamental nature of all electromagnetic radiation is the same: it can be viewed as photons or waves and travels at the speed of light (c), which is approximately 300,000 kilometers per second (km/s) or 180,000 miles per second (mi/s). The difference between radiation in different parts of the electromagnetic spectrum is a quantity that can be measured in several ways: as the length of a wave, as the energy of a photon, or as the oscillation frequency of an electromagnetic field. These three views are compared in Figure 2.1.

Each measurement—wavelength, energy, or frequency—has its own characteristic unit.

FIGURE 2.1.

Electromagnetic spectrum.

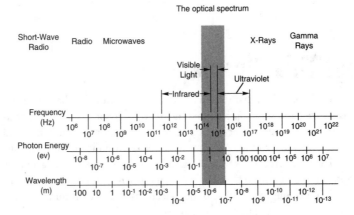

In some parts of the spectrum, frequency is used most; in others, photon energy or wavelength is. The optical world talks in wavelength, which is measured in metric units—meters, micrometers (μm or 10^{-6} m), and nanometers (nm or 10^{-9} m). Don't even think of wavelength in inches. (If you absolutely have to know, 1 μm is 1/25,000 of an inch.) Frequency is measured in cycles per second (cps) or hertz (Hz), with megahertz (MHz) meaning a million hertz and gigahertz (GHz) meaning a billion hertz. (The metric system uses the standard prefixes listed in Table 2.1 to provide different units of length, weight, frequency, and other quantities. The prefix makes a unit a multiple of a standard unit. For example, a millimeter is a thousandth [10^{-3}] of a meter, and a kilometer is a thousand [10^{3}] meters.) Photon energy can be measured in many ways, but the most convenient here is in electron volts (eV)—the energy that an electron gains in moving through a 1-volt (V) electric field.

Table 2.1. Metric unit prefixes and their meanings.

Prefix	Symbol	Multiple
tera	T	10^{12} (trillion)
giga	G	10^9 (billion)
mega	M	10^6 (million)
kilo	k	10^3 (thousand)
hecto	h	10^2 (hundred)
deca	da	10^1 (ten)
deci	d	10^{-1} (tenth)
centi	c	10^{-2} (hundredth)
milli	m	10^{-3} (thousandth)
micro	μ	10^{-6} (millionth)
nano	n	10^{-9} (billionth)
pico	p	10^{-12} (trillionth)
femto	f	10^{-15} (quadrillionth)

All the measurement units shown on the spectrum chart are actually different rulers that measure the same thing. There are simple ways to convert between them. Wavelength is inversely proportional to frequency, according to the following formula:

$$\text{WAVELENGTH} = c/\text{FREQUENCY}$$

or

$$\lambda = c/v$$

where c is the speed of light, λ is wavelength, and v is frequency. To get the right answer, all terms must be measured in the same units. Thus c must be in meters per second (m/s), λ must be in meters, and frequency must be in hertz (or cycles per second). Plugging in the number for c, we have a more useful formula for wavelength:

$$\lambda = 3 \times 10^8 \text{ m/s } /v$$

You can also turn this around to get the frequency if you know the wavelength:

$$v = 3 \times 10^8 \text{ m/s } /\lambda$$

Not many people talk about photon energy (E) in fiber optics, but a value can be gotten from Planck's law, which states:

$$E = hv$$

where h is Planck's constant (6.63×10^{-34} joule-second or 4.14 eV-second) and ν is the frequency. Because most interest in photon energy is in the part of the spectrum measured in wavelength, a more useful formula is

$$E(eV) = 1.2406/\lambda(\mu m)$$

which gives energy in electron volts when wavelength is measured in micrometers (μm).

●

Light waves that are 180° out of phase with each other can cancel each other out.

There is one practical consequence of the wave aspect of light's personality, which you will meet later on. Light waves can interfere with each other. Normally this does not show up because many different light waves are present and the effect averages out. Suppose, however, that only two identical light waves are present, as shown in Figure 2.2. The total amount of light detectable is the sum of the amplitudes of the light waves squared. If the light waves are neatly lined up, what is called "in phase," they add together and give a bright spot. However, if the two light waves are aligned so that the peaks of one coincide with the troughs of the other, they interfere destructively and cancel each other out. This happens when the two light waves are 180° out of phase with each other. If the two waves are out of phase by a different amount, they add together to give an intensity between the maximum and minimum possible.

FIGURE 2.2.

Constructive and destructive interference.

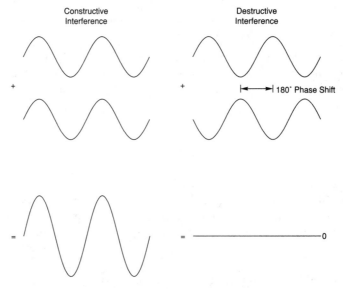

We are mainly interested in a small part of the spectrum shown in Figure 2.1—the optical region, where optical fibers and other optical devices work. That region includes light visible to the human eye at wavelengths of 400-700 nm, and nearby parts of the infrared and ultraviolet, which have similar properties. Roughly speaking, this means wavelengths of 200-20,000 nm (0.2-20 μm).

Most optical fibers used for communications transmit light in the near-infrared at wavelengths of 700-1600 nm (0.7-1.6 μm), where glasses made of silica (SiO_2) are most transparent. Plastic fibers transmit best at visible wavelengths, but they are not used as widely as silica glass fibers. Special fibers made of non-silica glasses transmit longer infrared wavelengths better. Special grades of silica can transmit some near-ultraviolet light.

Fiber-optic communication systems transmit near infrared light invisible to the human eye.

Refractive Index

The most important optical measurement for any transparent material is its refractive index (n). The refractive index is the ratio of the speed of light in a vacuum to the speed of light in the medium

$$n = c_{vac}/c_{mat}$$

Light always travels more slowly in a material than in a vacuum, so the refractive index is always greater than 1.0 in the optical part of the spectrum. In practice, the refractive index is measured by comparing the speed of light in the material to that in air rather than in a vacuum. The refractive index of air at atmospheric pressure and room temperature is 1.00029, so close to 1.0 that the difference is insignificant.

The refractive index of a material is the ratio of the speed of light in a vacuum to the speed of light in the material.

Although light rays travel in straight lines through optical materials, something different happens at the surface. Light is bent as it passes through a surface where the refractive index changes—for example, as it passes from air into glass, as shown in Figure 2.3. The amount of bending depends on the refractive indexes of the two media and the angle at which the light strikes the surface between them. The angles of incidence and refraction are measured not from the plane of the surface but from a line normal (perpendicular) to the surface. The relationship is known as Snell's law, which is written

$$n_i \sin I = n_r \sin R$$

Refraction occurs when light passes through a surface where the refractive index changes.

where n_i and n_r are the refractive indexes of the initial medium and the medium into which the light is refracted, and I and R are the angles of incidence and refraction, respectively, as shown in Figure 2.3.

Figure 2.3 shows the standard textbook example of light going from air into glass. The frequency of the wave does not change, but because it slows down, the wavelength gets shorter, causing the light wave to bend, whether the surface is flat or curved. However, if both front and rear surfaces are flat, light emerges at the same angle that it entered, and the net refraction is zero, as when you look through a flat window. If one or both surfaces are curved, the net effect is that of looking through a lens—which in fact you are doing with your eyes. That is, light rays emerge from the lens at a different angle than they entered. These overall refractive effects are shown in Figure 2.4.

Light refraction as
it enters glass.

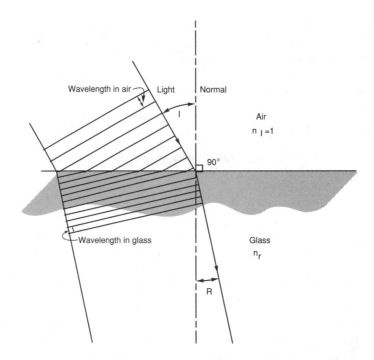

What does this have to do with fiber optics? Stop and consider what happens when light in a medium with a high refractive index (such as glass) comes to an interface with a medium having a lower refractive index (such as air). If the glass has a refractive index of 1.5 and the air an index of 1.0, the equation becomes

$$1.5 \sin I = 1 \sin R$$

That means that instead of being bent closer to the normal, as in Figure 2.3, the light is bent farther from it, as in Figure 2.5. This isn't a problem if the angle of incidence is small. For $I = 30°$, $\sin I = 0.5$, and $\sin R = 0.75$. But a problem does occur when the angle of incidence becomes too steep. For $I = 60°$, $\sin I = 0.866$, so Snell's law says that $\sin R = 1.299$. That angle can't exist because the sine can't be greater than 1.0!

If light hits a boundary with a material of lower refractive index at a steep enough (i.e., glancing) angle, it is reflected back into the high-index medium. This total internal reflection is the basic concept behind the optical fiber.

Snell's law indicates that refraction can't take place when the angle of incidence is too large, and that's true. Light cannot get out of the glass if the angle of incidence exceeds a value called the critical angle, where the sine of the angle of refraction would equal 1.0. (Recall from trigonometry that the maximum value of the sine is 1.0 at 90°.) Instead, total internal reflection bounces the light back into the glass, obeying the law that the angle of incidence equals the angle of reflection, as shown in Figure 2.5. It is this total internal reflection that keeps light confined in optical fibers, at least to a first approximation. As you will see in Chapter 4, the mechanism of light guiding can be more complex.

FIGURE 2.4.

*Light refraction
through a window
and a lens.*

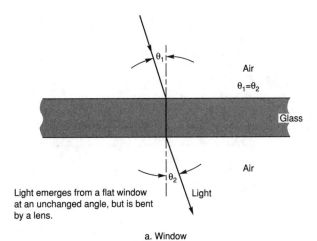

Light emerges from a flat window
at an unchanged angle, but is bent
by a lens.

a. Window

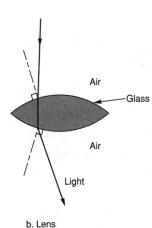

b. Lens

The critical angle above which total internal reflection takes place, θ_c, can be deduced by turning Snell's law around, to give

$$\theta_c = \arcsin \left(n_r / n_i \right)$$

For the example given, with light trying to emerge from glass into air, the critical angle is arcsin (1/1.5) or 41.8°.

FIGURE 2.5.

Refraction and total internal reflection.

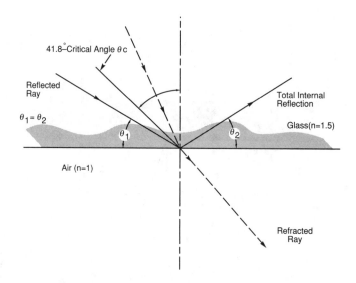

Light Guiding

Light is guided in the core of an optical fiber by total internal reflection at the boundary of the lower-index cladding.

The two key elements of an optical fiber—from an optical standpoint—are its core and cladding. The core is the inner part of the fiber, through which light is guided. The cladding surrounds it completely. The refractive index of the core is higher than that of the cladding, so light in the core that strikes the boundary with the cladding at a glancing angle is confined in the core by total internal reflection, as shown in Figure 2.6.

The difference in refractive index between core and cladding need not be large. In practice, it is only about 1%. This still allows light guiding in fibers. For $n_r/n_i = 0.99$, the critical angle θ_c is about 82°. Thus, light is confined in the core if it strikes the interface with the cladding at an angle of 8° or less to the surface. The upper limit can be considered the confinement angle in the fiber.

FIGURE 2.6.

Light guiding in an optical fiber.

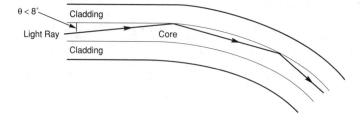

Another way to look at light guiding in a fiber is to measure the fiber's acceptance angle—the angle over which light rays entering the fiber will be guided along its core, shown in Figure 2.7. (Because the acceptance angle is measured in air outside the fiber, it differs from the confinement angle in the glass.) The acceptance angle normally is measured as numerical aperture (NA), which for light entering a fiber from air is approximately

$$NA = (n_0^2 - n_1^2)^{1/2}$$

where n_0 is the refractive index of the core and n_1 is the index of the cladding. For a fiber with core index of 1.50 and cladding index of 1.485 (a 1% difference), NA = 0.21. An alternate but equivalent definition is the sine of the half-angle over which the fiber can accept light rays, 12° in this example (θ in Figure 2.7). Another alternate definition is $NA = n_0 \sin \theta_c$, where θ_c is the confinement angle in the fiber core (8° in this example).

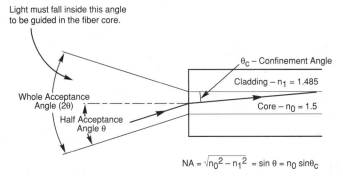

Note that the half acceptance angle is larger than the largest glancing angle at which light rays must strike the cladding interface to be reflected, which I said earlier was 8°. What does this mean? Go back and look at Snell's law of refraction again. The difference is the factor n_0, which is the refractive index of the core glass, or 1.5. As you can see in Figure 2.7, refraction bends a light ray entering the fiber so that it is at a smaller angle to the fiber axis than it was in the air. The sine of the angle inside the glass equals that of the angle outside the glass, divided by the refractive index of the core (n_0).

Light Collection

Numerical aperture and acceptance angle measure a crucial concern in practical fiber-optic systems: getting light into the fiber. A century ago, British physicist Charles Vernon Boys made fine fibers of glass comparable in size to today's optical fibers, but as far as we know he didn't try to transmit light along their lengths. Even when the first optical fibers were developed in the 1950s, no one seemed to think that single fibers could collect much light. Instead, they grouped fibers into bundles that together could collect reasonable amounts of light. Only when lasers made highly directional beams available did researchers seriously begin to consider using single optical fibers.

● The angle over which a fiber accepts light depends on the refractive indexes of the core and cladding glass.

FIGURE 2.7.
Measuring the acceptance angle.

● Practical applications for single fibers require ways to get light into their small cores.

Light Source and Fiber Size

As you saw earlier, optical fibers can accept light only over limited angles. Conventional optics can readily produce narrow beams; just look at a flashlight. The biggest problem is that individual fibers are small. The most common optical fibers used in communications are 0.25-0.5 mm in diameter, but that includes a plastic coating that protects the fiber from mechanical damage. The cladding—the outer part of the fiber proper—normally is only 125 μm across (0.125 mm), and the inner core that carries the light is only a fraction of that diameter (from 8 to 62.5 μm). Some fibers are larger, with cores 100-1000 μm (0.1-1 mm) in diameter. Even smaller fibers may be used in bundles; the problem of getting light into them is simplified by collecting them together.

A fiber will pick up some light from any source. Hold a single fiber so one end points at a light bulb. Now look into the other end, bending the fiber so you don't look directly at the light. You can see that the fiber has collected some light, but that is only a tiny fraction of the light from the bulb.

Efficient collection of light requires a light source about the same size as the fiber core. For small-core fibers, the best match is a semiconductor diode laser, which emits light from a region a fraction of a micrometer high and a few micrometers wide. Light-emitting diodes (LEDs), with larger emitting areas, work well with larger-core fibers. The main trade-offs are cost versus performance; diode lasers cost more than LEDs, but they can deliver higher powers into fibers and operate at higher speeds. I will say more about light-source trade-offs in Chapters 6 and 7.

Alignment

Joining the ends of optical fibers requires careful alignment and tight tolerances.

Transferring light between fibers requires careful alignment and tight tolerances. The greatest efficiency comes when the ends of two fibers are permanently joined in a splice (described in Chapter 11). Temporary junctions between two fiber ends, made by connectors (described in Chapter 10) typically have slightly higher losses but allow much greater flexibility in reconfiguring a fiber-optic network. Special devices called couplers (described in Chapter 12) are needed to join three or more fiber ends. One of the most important functional differences between fiber-optic and wire communications is that fiber couplers are much harder to make than their metal-wire counterparts.

Transfer losses must always be considered in fiber-optic communication systems.

Losses in transferring signals between wires are so small that they can normally be neglected. This is not so for fiber optics. As you will see in Chapter 16, system designers should account for coupling losses at each connector, coupler, splice, and light source.

Transmission and Attenuation

Transmission of light by optical fibers is not 100% efficient. Some light is lost, causing attenuation of the signal. Several mechanisms are involved, including absorption by materials within the fiber, scattering of light out of the fiber core, and leakage of light out of the core caused by environmental factors. The degree of attenuation depends on the wavelength of light transmitted, as you will see in Chapter 4. This makes operating wavelength an important feature of a fiber system.

Attenuation measures the reduction in signal strength by comparing output power with input power. Measurements are made in decibels (dB), a very useful unit, albeit a peculiar one. The decibel is a logarithmic unit measuring the ratio of output to input power. (It is actually a tenth of a unit called a "bel" after Alexander Graham Bell, but that base unit is virtually never used.) Loss in decibels is defined as

$$\text{dB LOSS} = -10 \times \log_{10} (\text{POWER OUT/POWER IN})$$

Thus, if output power is 0.001 of input power, the signal has experienced a 30 dB loss. The minus sign is added to avoid negative numbers in attenuation measurements. It is not used in systems where the signal level might increase, where the sign of the logarithm indicates if the signal has decreased (minus) or increased (plus).

Each optical fiber has a characteristic attenuation that is measured in decibels per unit length, normally decibels per kilometer. The total attenuation (in decibels) in the fiber equals the characteristic attenuation times the length. To understand why, consider a simple example, with a fiber having the relatively high attenuation of 10 dB/km. That is, only 10% of the light that enters the fiber emerges from a 1 km length. If that output light was sent through another kilometer of the same fiber, only 10% of it would emerge (or 1% of the original signal), for a total loss of 20 dB.

As you will see later in this chapter, the choice of operating wavelength depends not only on fiber loss but also on the available light sources and on other fiber properties. The loss of silica fibers in the near infrared is much lower than signal attenuation in other media with comparable transmission capability. Attenuation is very low at 1300 nanometers, and even lower at 1550 nm. At the 0.4 dB/km attenuation typical near 1300 nm, 1% of the light entering the fiber remains after 50 km, for a 20 dB loss. At the 0.25 dB/km attenuation common at 1550 nm, 1% of the input light remains after 80 km. This allows signals to go through 50 to 100 km of fiber without amplification, an important advantage in communications. Metal coaxial cables, which have comparable information capacity, would require dozens of amplifiers over such distances (depending on transmission rate).

Bandwidth and Dispersion

Low attenuation alone is not enough to make optical fibers invaluable for telecommunications. The thick wires used to transmit electrical power also have very low loss, but they cannot transmit information at high speeds. Optical fibers are attractive because they combine low loss with high bandwidth to support the transmission of high-speed signals over long distances (i.e., high information capacity).

● Some light is lost in transmission through a fiber. The amount of loss depends on wavelength.

● Attenuation of a fiber is the product of the length times the characteristic loss in decibels per kilometer.

● Optical fibers are unique in allowing high-speed signal transmission at low attenuation.

Information Capacity

Information capacity is very important in all types of communications but is measured differently in different types of systems. Where data is transmitted digitally (i.e., in digitized "bits" or units of information), transmission capacity is measured in bits per second. The more bits that can pass through a system in a given time, the more information it can carry. A 100-megabit-per-second (Mbit/s) system can carry as much information as a hundred 1 Mbit/s systems. As long as all the information is going between the same two points, it is much cheaper to build one 100 Mbit/s system than a hundred 1 Mbit/s systems. The same principle works for analog communications, like the telephone line to your home or most television signals, but we measure capacity by the frequency bandwidth in hertz.

In practice, the attenuation of wires increases with the frequency of the electrical signals they carry. Electrical power wires have low attenuation only at very low frequencies, including the 60-hertz variation of alternating current. Coaxial cables can transmit higher frequencies, but their loss increases sharply with frequency, as shown in Figure 2.8. However, the loss of optical fibers is essentially independent of signal frequency over their normal operating range. (The scale is measured in loss per kilometer of cable and does not take into account the transfer losses mentioned earlier.)

FIGURE 2.8.

Loss as a function of frequency.

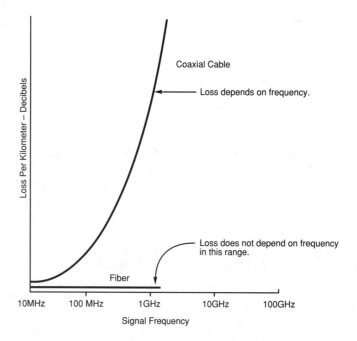

Transmission Speed Limits

Effects other than loss limit fiber transmission speed. They are best seen by looking at digital transmission, although they also occur for analog transmission. Consider another view of the basic optical fiber shown in Figure 2.9. As long as the fiber has a core diameter much larger than the wavelength of light (i.e., well over 10 μm), rays can enter the fiber at many different angles to its axis. A ray that bounces back and forth within the core many times will travel a slightly greater distance than one that goes straight through. For instance, if one ray traveled straight through a 1 km fiber and another bounced back and forth at a 5° angle through a 100 μm core through the same fiber, the second ray would travel 3.8 m farther.

You can see what this means for communications by thinking about what happens to light rays that start together in an instantaneous pulse. The one that goes down the center of the fiber will reach the output end first, 12.7 ns before the ray that bounced back and forth at a 5° angle. Thus an instantaneous pulse at the start of the fiber would spread out to 12.7 ns at the end.

This phenomenon is called pulse dispersion. This example gives a rather simplified view of how light travels through the fiber in what are called different modes. As you will see in Chapter 4, other effects can also cause various degrees of pulse dispersion. Dispersion is important because, as the pulses spread, they can overlap and interfere with each other, limiting data transmission speed. In this example, if each pulse went through 1 km of fiber, it would acquire a 12.7 ns tail. So the time between pulses would have to be at least that long, limiting transmission to 80 Mbit/s. Because actual input pulses are not instantaneous, the real maximum pulse rate would be even slower.

●
Pulse dispersion limits fiber transmission capacity.

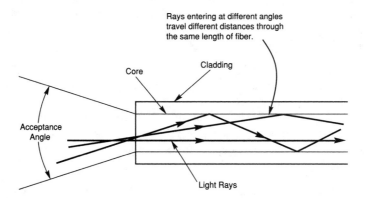

Rays entering at different angles
travel different distances through
the same length of fiber.

Cladding

Core

Acceptance
Angle

Light Rays

FIGURE 2.9.

Light rays take different paths in a fiber core.

Dispersion and Distance

Because pulses stretch out a certain amount for each kilometer of fiber, pulse dispersion—like attenuation measured in decibels—increases linearly with distance traveled. That is, pulse dispersion doubles if transmission distance doubles. Typical dispersion values for fibers that carry multiple modes are measured in nanoseconds (of dispersion) per kilometer of fiber. These can be translated into an analog bandwidth limit or a maximum data rate for digital transmission. Both bandwidth and data rate are the inverse of pulse dispersion, so they decrease as the distance of the fiber increases.

The bandwidths quoted so far are low by fiber-optic standards because they are for fibers that carry light in many modes. Dispersion is much smaller in fibers that carry light in only one mode, which will be described in more detail in Chapter 4. These single-mode fibers can carry billions of bits per second over tens of kilometers without amplification. The residual dispersion in them depends on the range of wavelengths emitted by the light source.

Light Sources

Fiber-optic communication systems require a light source to generate the signal that the fiber transmits. In practical systems, these light sources are almost always semiconductor diode lasers or LEDs, described in Chapter 6. Some inexpensive short-distance systems use LEDs that emit visible light, but most systems carry near-infrared wavelengths.

Many fiber systems use light sources of gallium arsenide (GaAs) and gallium aluminum arsenide (GaAlAs) emitting at 750-900 nm. GaAs/GaAlAs lasers and LEDs were the best sources available for the first fiber-optic systems, and they remain inexpensive. However, loss and dispersion are considerably lower at longer wavelengths, so GaAs and GaAlAs sources are used in short, low-speed systems.

Most long-distance systems now operate at 1300 nm, where fiber loss is 0.35-0.5 dB/km (about 1/5 that at 850 nm) and where pulse dispersion is naturally very low. Semiconductor lasers or LEDs that emit at 1300 nm are made of indium gallium arsenide phosphide (InGaAsP). Those devices are harder to make than GaAs devices and are thus much more expensive. A few systems use crystalline neodymium lasers that emit a continuous beam near 1300 nm, which can generate higher powers than semiconductor lasers but require external modulation, as described in the next section.

Some systems operate at 1550 nm, where fiber loss is only 0.2-0.3 dB/km and optical amplifiers can stretch transmission distances even further. However, systems operating at 1550 nm must be designed carefully to reduce effects that can cause much larger pulse dispersion than at 1300 nm. InGaAsP is used to make 1550 nm devices, but the task is harder than producing 1300 nm lasers and LEDs. Another possible light source is a length of optical fiber doped with the rare earth erbium so that it operates as a laser at 1550 nm. However, like neodymium lasers, erbium-doped fiber lasers require external modulation.

Modulation

An important practical advantage of semiconductor lasers and LEDs is that they can be modulated directly by the same electrical current that powers them. The signal is imposed on the drive current, causing light output to vary in proportion to the signal.

The relationships between drive current and optical output power for an LED and a semiconductor diode laser are shown in Figure 2.10. Changes in drive current cause changes in the optical output. Thus, turning the modulation current off and on with a digital signal produces a series of light pulses. Analog modulation makes input current vary continuously over a range, so light output also rises and falls. The current-output curves of optical devices are not perfectly linear, so direct modulation can induce some distortion in analog signals. (Digital transmission is largely immune to such effects, which is one reason it is so popular.)

Direct modulation sounds so logical that it may seem hard to see why anyone would choose a different approach. However, other lasers cannot be modulated in the same way at high speeds; they have to be powered with a steady source and modulated with an external device (called a modulator) that varies its transmission of light. As you will see later, external modulation offers some advantages both for high-power neodymium or erbium laser sources and for very high-speed semiconductor laser systems. However, direct modulation remains much more common.

●
Changing the current passing through a semiconductor laser or LED changes its light output, modulating it with a signal.

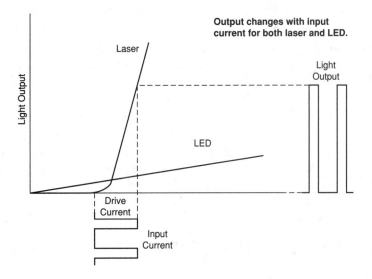

FIGURE 2.10.

Relationship between drive current and optical output.

Multiplexing

Many signals can be combined or multiplexed into a single transmission in an optical fiber system.

A fiber-optic system may carry a single simple signal, such as one video channel. If communications are two-way, separate fibers usually carry the signals going in opposite directions. However, many systems carry several separate signals that have been "multiplexed" or merged together to go through a single fiber. The idea is shown in Figure 2.11. It is particularly useful in fiber optics because of fiber's high transmission capacity. Hundreds or thousands of individual telephone circuits can be multiplexed onto a single pair of fibers, as you will see in Chapter 17.

FIGURE 2.11.

Multiplexing signals.

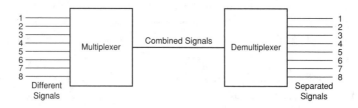

Detection

Optical signals emerging from an optical fiber are detected by a receiver that converts them back to electronic form.

The final element of a fiber-optic communication system is the receiver, which detects the optical signal and converts it to electronic form for further transmission or processing. Typically, a receiver consists of a semiconductor photodetector and amplifying electronics. (Details will be described in Chapter 8.)

How well the detector can do its job depends on its sensitivity and the signal level reaching it. Sensitivity, in turn, depends on operating conditions, detector design, and how the material used in the detector responds to different wavelengths. Silicon photodiodes, for example, work well at 750 to 900 nm, but are not usable at 1300 or 1550 nm, where detectors made of indium gallium arsenide or other semiconductors must be used.

The receiver electronics process the electrical signal generated by the detector to replicate the original signal sent through the fiber. In essence, this is a cleaning-up process. The signal itself is amplified. Digital pulse edges, which had grown blurred as the signal passed through the system, are sharpened by detecting when the signal strength passes a threshold level. Timing of a series of pulses is checked and corrected.

Amplifiers, Repeaters, and Regenerators

Although optical fibers can carry signals long distances because of their low loss, the signal would eventually become too faint to detect if it is not amplified. There are two basic approaches: electro-optical repeaters or regenerators, and optical amplifiers. I describe them briefly here and provide more detail in Chapter 9.

Electro-optical repeaters combine a receiver and a transmitter. The receiver detects the optical signal and converts it into an electrical signal. The electrical signal is then amplified to drive a transmitter that produces an optical signal which goes through the next length of optical fiber. Signal-processing electronics between the receiver and transmitter may clean up or "regenerate" the signal, reducing noise and adjusting the timing of pulses.

●
Insertion of amplifiers or repeaters in a communication system allows signals to be transmitted over long distances.

Although repeaters can extend transmission distances, they are costly, complex, and prone to failure. A much simpler alternative for fiber-optic systems is the use of optical amplifiers, which directly amplify optical signals without converting them into electrical form. The best developed optical amplifiers at this writing are lengths of special optical fiber doped with the rare earth erbium to amplify 1550 nm light. Semiconductor laser amplifiers and fiber amplifiers for 1300 nm are also in development. Optical amplifiers do not regenerate signals—they amplify noise along with the signal—but because they are simpler and less costly, they can be spaced more closely than regenerators, reducing noise and allowing transmission over thousands of kilometers.

System Considerations

Fiber-optic communications has found a vast array of applications, as you will see in Chapter 3. System design depends largely on the application. For long-distance telecommunications—including the most demanding applications in transoceanic submarine cables—crucial factors include maximizing transmission speed and distance and minimizing fiber and splice loss. By contrast, connector loss becomes vital in local area networks that distribute signals within buildings. In long-haul systems, it is important to minimize the cost of cable, while the goal is to reduce the cost of terminal equipment for local area networks (because there are many separate terminals, rather than the two required for point-to-point communications).

●
Design of fiber-optic systems and choice of hardware depend upon the application.

These system considerations make design and construction of practical fiber-optic systems a multifaceted task. Guidelines appropriate for one type of system should not be followed blindly for others, because they often lead directly to the wrong answers. Applications are diverse enough that many different components are offered.

What Have You Learned?

1. Light is guided through optical fibers by total internal reflection of light that enters them within an acceptance angle, measured directly or as the numerical aperture (NA).

2. Refractive index is a crucial property of optical materials.

3. The core of an optical fiber must have a higher refractive index than the cladding surrounding it.

4. Fiber-optic communication systems transmit near-infrared light at 750-900, 1300, or 1550 nm.

5. The small dimensions of optical fibers make tolerances tight in transferring light into fibers.

6. A major attraction of fiber optics is its capability to send high-speed signals over long distances.

7. Attenuation of optical fibers (measured in decibels) is proportional to a characteristic value for the fiber (measured in decibels per kilometer) and the length of the fiber.

8. Transmission capacity of optical fibers depends on the dispersion of light signals sent through them, which is proportional to a characteristic value (measured in nanoseconds per kilometer) times the length of the fiber.

9. Attenuation and dispersion in an optical fiber depend on wavelength.

10. Light sources in fiber-optic systems are semiconductor lasers, LEDs, and crystalline or glass lasers.

What's Next?

In Chapter 3, we will look at the major applications for fiber optics, particularly in communication systems.

Quiz for Chapter 2

1. If light passes from air to glass, it is
   a. reflected.
   b. refracted.
   c. absorbed.
   d. scattered.

2. Light is confined within the core of a simple optical fiber by
   a. refraction.
   b. total internal reflection at the outer edge of the cladding.
   c. total internal reflection at the core-cladding boundary.
   d. reflection from the fiber's plastic coating.

3. An optical fiber has a core with refractive index of 1.52 and a cladding with index of 1.45. Its numerical aperture is
   a. 0.15.
   b. 0.20.
   c. 0.35.
   d. 0.46.
   e. 0.70.

4. The input power to a fiber-optic cable is 1 milliwatt (mW). The cable's loss is 20 dB. What is the output power, assuming there are no other losses?
   a. 0.10 mW.
   b. 0.05 mW.
   c. 0.01 mW.

    d. 0.001 mW.

    e. None of the above.

5. The output of a 20 km fiber-optic cable is measured at 0.005 mW. The fiber loss is 0.5 dB/km. What is the input power to the fiber?

    a. 1 mW.

    b. 0.5 mW.

    c. 0.05 mW.

    d. 0.01 mW.

    e. None of the above.

6. Optical fiber attenuation is lowest at

    a. 800 nm.

    b. 900 nm.

    c. 1300 nm.

    d. 1400 nm.

    e. 1500 nm.

7. Optical fiber attenuation can be as low as

    a. 0.1 dB/km.

    b. 0.2 dB/km.

    c. 0.4 dB/km.

    d. 0.5 dB/km.

    e. 1.0 dB/km.

8. A pulse sent through a 20 km optical fiber is 300 ns long when it emerges. What is the fiber dispersion? Assume that the pulse was instantaneous when it entered the fiber.

    a. 35 ns/km.

    b. 30 ns/km.

    c. 25 ns/km.

    d. 20 ns/km.

    e. 15 ns/km.

9. The most common wavelength for long-distance communications is

    a. 800 nm.

    b. 900 nm.

    c. 1300 nm.

    d. 1400 nm.

    e. 1550 nm.

10. What can be used to extend transmission distance of fiber-optic systems?

    a. Electro-optical repeaters.

    b. Optical fiber amplifiers.

    c. Coaxial cable transmission.

    d. a and b.

    e. None of the above.

Applications of Fiber Optics

About This Chapter

Fiber optics is best known for its communication applications, particularly in the telephone industry. That vision reflects reality, because the telephone industry is the largest user of fiber optics. Nonetheless, optical fibers are used in many other types of communications, including short-distance data transmission, cable television and video systems, and local area networks. More applications are coming in sensing and military systems. And some important applications, such as imaging, inspection, sensing, and delivery of laser power, are entirely separate from communications.

This chapter will introduce how and why fiber optics are used. I will go into more detail on applications later, but a general understanding of how fibers are used will help you better appreciate important features of fiber technology that will be described in later chapters.

Types of Communications

Communication systems come in many sizes, shapes, and forms. They may send signals between two locations, link together many points, or distribute the same signals to many individual subscribers. They may carry different kinds of signals at high or low speeds. Many link only fixed points, but others can serve mobile users (like cellular telephones). Fiber optics are used to transmit signals in many such systems, but details can differ greatly.

UNDERSTANDING

●
Fiber optics can be
used in many—but
not all—types of
communications, but
system designs vary
widely.

At first glance, it is easy to overlook many differences among communication systems. For example, traditional telephone networks and cable television systems both distribute signals to individual homes, but their structures differ in important ways. The telephone network contains switches that make temporary two-way connections between phones in different places, whereas a traditional cable television network distributes the same signals to all homes (although not all homes have the equipment needed to decode all the signals). This approach worked fine for many years, because people were satisfied to get the same video signals—but not the same phone calls. However, it is changing rapidly as cable systems add switching and other services, and phone companies try to add video. (You will learn more about these new services in Chapters 18 and 19.)

Other differences are more obvious. The telephone network spans the globe, so you can call Australia from your home phone (although you may wish you hadn't after you get the bill). In contrast, computer data communication is often over short distances within buildings or among separate buildings on a campus. Local area networks link many computers at data rates much higher than a single phone line, but lower than the long-distance lines that carry many combined telephone signals.

Some communication systems are dedicated to a specific job, like a cable between a closed-circuit video camera and a monitor used for security in a store or bank. Others—notably the telephone system—are "common carriers" that carry many sorts of signals for a variety of users.

One common trend is a steady increase in the demand for transmission capacity, for more voice and facsimile telephone lines, more video channels, and higher data-transfer rates between computers. This has helped push the spread of high-capacity fiber-optic systems. Fiber optics can also solve special problems, such as avoiding electromagnetic interference with signal transmission.

In the rest of this chapter, you will look at the important types of communications and how well fiber optics meet their needs. First, however, you need to look at the crucial distinction between analog and digital communications.

●
Signals can be
transmitted in
analog or digital
formats. Each has its
advantages, and
both are compatible
with fiber optics.

Analog and Digital Communications

Communication signals can be transmitted in two fundamentally different forms, as shown in Figure 3.1: continuous analog signals and discrete digital signals. The level of an analog signal varies continuously. A digital signal, on the other hand, can be at only certain levels. The most common number of levels, as shown in Figure 3.1, is two, with signals coded in binary format, either off or on.

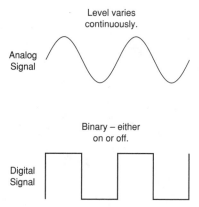

FIGURE 3.1.

Analog and digital signals.

Each format has its advantages. The older analog technology is more compatible with people and much existing equipment. Our ears, for instance, detect continuous variations in the level of sound, not just the presence or absence of sound. Our eyes likewise detect levels of brightness, not simply the presence or absence of light. For that reason, audio and video communications have traditionally been in analog form. Telephone wires deliver a continuously varying signal to your telephone handset, which converts those electronic signals into continuously varying sound waves. The video signals that drive standard televisions are also analog.

On the other hand, digital signals are easier to process with electronics and fiber optics. It is much simpler and cheaper to design a circuit to detect whether a signal is at a high or a low level (off or on) than to design and build one to accurately replicate a continuously varying signal. Digital signals are also much less prone to distortion, as shown in Figure 3.2. When an analog signal goes through a system that doesn't reproduce it exactly, the result is a garbled signal that can be unintelligible. That is exactly what happens when you get a distorted voice on the phone. However, when a digital signal is not reproduced exactly, it is still possible to tell the on from the off state, so the signal is clearer. That is one reason digital compact discs reproduce sound much better than analog cassette tapes or phonograph records.

Converting Analog and Digital Signals

If people need analog signals, but transmission works best with digital, what about converting between the two? That is often done for audio. Compact disc players use a laser to read sound digitized as spots on a rapidly spinning disc, then use internal electronics to convert the digitized sound back to analog form. The telephone network converts the analog signals from a telephone handset into digital code for long-distance transmission, then translates the digital code back to analog form on the other end.

FIGURE 3.2.

Distortion of analog and digital signals.

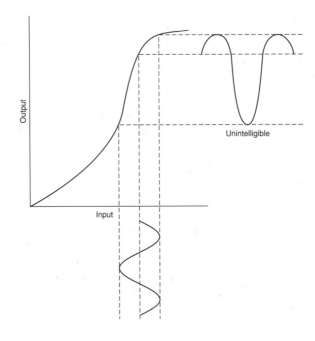

a. Analog

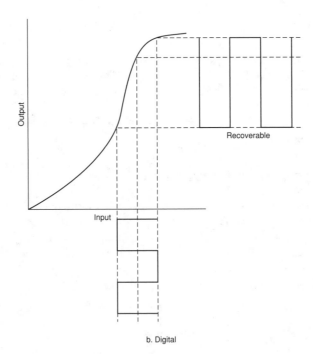

b. Digital

The idea of digitization is simple, as shown in Figure 3.3. A circuit called an analog-to-digital converter samples an analog waveform to measure its amplitude. The samples are taken at uniform intervals (8,000 times per second in a telephone circuit). The converter assigns the signal amplitude to one of a predetermined number of possible levels. For a telephone circuit, that number is 128 (exactly the number of levels that can be encoded by 7 bits). This converts the 4 kHz analog telephone signal into a digital stream of sets of 7 bits sent 8,000 times a second (56,000 kbits/s).

● Analog telephone signals are converted to digital format by sampling them 8,000 times a second.

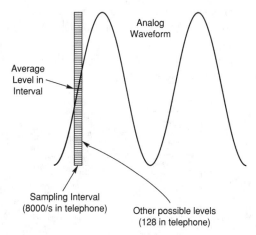

FIGURE 3.3.

Digitization of an analog signal.

The figures in the last paragraph show one disadvantage of digital transmission. Accurate reproduction of an analog signal requires sampling at a rate faster than the highest frequency to be reproduced. In the telephone example, the sampling rate is twice the highest frequency to be reproduced (4,000 Hz). Many bits (7 for telephony) have to be sent per sampling interval. This requires a large transmission bandwidth. There is no precise equivalence between analog and digital transmission capacity, but the two are comparable: a transmission line capable of handling 10 Mbit/s has an analog capacity of around 10 MHz. That means an analog signal takes only about a tenth of the transmission capacity that it needs in digital form. That is not a problem in telephony, but it led cable television carriers to stay with analog transmission for many years. That is changing with new technology that can compress digital video signals so that they use many fewer bits to carry the same information.

● Signals require more transmission capacity in digital than in analog form.

Fiber optics work well for digital signals and were initially used mainly for digital systems. They have the high transmission capacity needed for digital transmission, and many light sources suffer from nonlinearities that induce distortion in analog signals at high frequencies. However, developers have also succeeded in making highly linear analog fiber systems, which are widely used to distribute signals for cable television.

● Fiber-optic systems handle both digital and analog transmission.

Long-Distance Telecommunications

Telephone Network Structure

The telephone network can be loosely divided into a hierarchy of systems, shown in simplified form in Figure 3.4. Your home or business telephone is part of the subscriber loop or local loop, the part between individual subscribers and telephone-company switching offices (called central offices in the telephone industry). Trunk lines run between central offices, for example, carrying telephone calls from one suburb to another or from suburbs to central cities. Long-distance telephone carriers operate long-haul backbone systems between regions, and these backbone systems connect with international communication systems like transatlantic cables and satellites.

The telephone network includes subscriber loops, trunk lines, and backbone systems. Fiber optics is widely used for trunk and backbone systems.

FIGURE 3.4.

Parts of the telephone network.

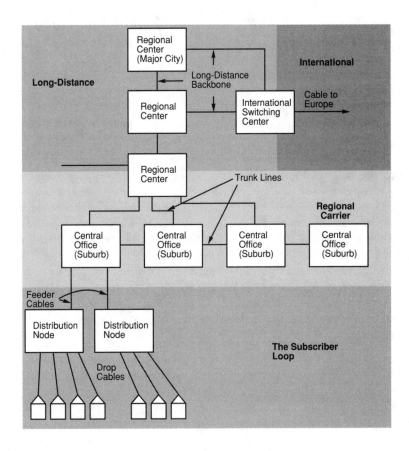

In practice, the subscriber loop may be broken down into two segments—a feeder cable from the central office to a distribution node or drop-off point and individual wires from there to homes. A typical central office may serve thousands of homes, but each drop-off serves between several and several dozen. Regional telephone operating companies and long-distance carriers each have networks that reach throughout the regions where they operate, making it hard to define a boundary between trunk and backbone network. Optical fiber is common in all parts of the telephone network except connections to individual homes.

Telephone System Hardware

The telephone network is rapidly changing from analog to digital technology. That change is not visible to most telephone users because it has yet to reach most homes; most digital systems go no further than the distribution node or to business users. However, the telephone industry plans to bring digital technology all the way to home subscribers eventually.

● The telephone system is changing from analog to digital transmission.

The telephone network makes extensive use of a technique called multiplexing so that cables can carry many phone circuits simultaneously. Only in the local loop, between homes and central offices or remote drop-off points, is one conversation routed over each pair of wires. At the central office or a distribution node, signals from many telephone lines are combined (multiplexed) for simultaneous transmission through a single cable. Successive levels of multiplexing raise transmission speeds and number of voice circuits carried on a single cable until the level reaches thousands in backbone systems, as you will see in Chapters 17 and 18. Initially, signals were multiplexed on analog lines by assigning them different transmission frequencies. Now multiplexers combine many signals digitized at slow speeds (e.g., 56,000 bits per second) to produce a single faster digital signal.

● The telephone network makes extensive use of multiplexing.

As you saw in Chapter 2, the advantages of fiber optics grow with transmission speed and distance. Thus, many early uses of fiber optics were in systems that operated at what were then high speeds. Long-distance telephone carriers (e.g., the American Telephone & Telegraph Co., Sprint, and MCI) use fiber-optic cables for their high-speed "backbone" systems that carry thousands of voice signals between regions of the country at overall speeds to 3.4 Gbit/s. Some long-distance traffic still goes via satellite or microwave transmission on the ground, but optical fibers are taking much of the load on land.

● Backbone communications systems rely on fiber optics to carry thousands of telephone circuits per fiber pair.

The telephone industry's first widespread use of fiber optics was for 45 Mbit/s trunks between central offices. Most urban and suburban central offices are several kilometers apart, and even the first commercial fiber systems, operating at 800-900 nm, could transmit that far without repeaters. When a second generation of fiber technology became available, transmitting at 1300 nm, longer and faster fiber systems came into use. Now a third generation of fiber systems is the standard choice for trunk cables at a wide range of speeds below those of backbone systems. Repeaters can be placed tens of kilometers apart, allowing transmission over long distances.

In recent years, fibers have spread beyond the central office, to connect business users and provide drop-off points in the subscriber loop. Some operate at fairly low speeds, such as 1.5 Mbit/s in North America (standard transmission rates are different in Europe). However, others operate at higher speeds made possible by steady drops in the cost of high-speed electronic and fiber equipment.

●

By replacing one thick metal cable with four fiber cables that can fit in the same duct, telephone companies can save much money on installation costs.

Telephone companies have found that fiber optics have many advantages besides their high transmission capacity and low signal attenuation. Fiber-optic cables are much smaller than old metal cables, so four fiber cables can fit into an underground duct that could hold only one wire cable. That means important savings because installing new ducts in urban areas costs far more than replacing old cables with new. What's more, the cables can be installed with extra fibers to leave room for future expansion—at minimal extra cost and without the need for major construction or excavation.

Fibers have other advantages that can be important for particular applications. Fiber-optic cables can be made with no electrical conductors—so they won't carry dangerous current pulses injected by lightning strikes. Signals carried in fibers cannot pick up electromagnetic interference from power lines, generating plants, or other sources. The high capacity of fibers also leaves plenty of room for future expansion—an important consideration for telephone-industry planners looking for continued steady growth in communication demand.

Submarine Fiber Cables versus Satellites

●

Undersea fiber-optic cables offer important advantages over satellites for intercontinental communications.

The most dramatic example of the role fiber optics are playing in long-distance communications is the development of ocean-spanning cable systems that have brought some satellite communications back to earth. Satellite links and submarine cables have competed for intercontinental telephone traffic since the 1960s. Satellites steadily gained ground on coaxial cables, which came to the end of their rope with a transatlantic cable laid in 1983. Fiber-optic cables completely outclass coaxial cables, with much longer repeater spacing and vastly greater transmission capacity. The first transatlantic fiber cable, installed in 1988, carries almost 10 times more voice circuits than the last coaxial cable. Transatlantic conversations sound crystal clear. Newer cables have even higher capacity, and the generation of fiber systems planned to begin operation in the mid-1990s will have more than 10 times the capacity of the first submarine fiber cables. By that time, submarine fiber cables should circle the globe, crossing the Atlantic, Pacific, and Indian Oceans.

Submarine fiber cables have an importance that goes beyond their considerable role in the global communications network. Their needs for long repeater spacing and high capacity have pushed new generations of fiber-optic technology that have been adapted for other applications. Design of the first transatlantic fiber cable pushed development of the single-mode fiber technology now standard in most telephone applications. Optical amplifiers were perfected for the new generation of submarine fiber systems being installed during the 1990s, but they are starting to find applications on land.

Computer Data Communications

Fiber optics have been gaining acceptance in computer data communications, but wire links remain much more common because most computer data communications are not very demanding. Most personal computers send data only short distances, such as the few feet separating the keyboard, central processor, display screen, and printer. Network connections typically are at modest speeds and over modest distances. Standard twisted-wire telephone lines often suffice for communications with the external world via modems, although noise can be a problem on poor-quality lines at data rates above 2400 bit/s.

● Fibers are used for some computer data communications, mostly at high speeds.

Fibers were first used for data communications in places where wire links would not work properly because of problems such as electromagnetic interference, which can block wire or radio transmission. Fibers are immune to electromagnetic interference, so they can run alongside power lines to carry data to monitor electric power utilities or to carry data up and down an elevator shaft, which is a convenient place to string cables in a high-rise building. Fiber cables, which lack conductors, are immune to power surges from lightning strikes, which can damage sensitive electronic equipment. The small size of fiber cables makes them a good choice for installation in tight quarters or through already-crowded cable ducts. Because they are difficult to tap and do not radiate electromagnetic signals, fibers provide more secure data transmission than wires, important for financial institutions and military agencies.

Growing needs for data transmission at higher speeds and over longer distances have increased the demand for fiber-optic systems in many other areas. Graphic computer interfaces require much more data than text-based systems, leading to demands for faster transmission. Steady decreases in costs of memory and storage, coupled with the growing complexity of programs, push information demands ever upwards. Many personal computers are attached to networks, which carry signals at ever-higher speeds. Designers of local area networks several years ago turned to fiber as the basis for the 100 Mbit/s Fiber Distributed Data Interface (FDDI), although recently FDDI has been implemented (over short distances) on copper wires. Fiber networks operating at higher speeds are in development.

Specialists in data transmission make a subtle but important distinction between digital data links and networks. Digital data links connect two points, whereas networks interconnect a multitude of points, as shown in Figure 3.5. In other words, a point-to-point data link transfers information between two points at opposite ends of a transmission line. Local area networks allow the interchange of data among many points within the same area (e.g., within a department of a large company); other networks may cover larger areas.

● The need to interconnect many terminals complicates building local area networks with fibers.

The distinction between networks and data links is crucial. One reason is that optical signals cannot be divided as readily as electrical ones. Electrical signals are delivered as voltage, so a signal can be tapped from a wire while drawing minimal current. Optical signals are delivered as power, which must be divided among the terminals, limiting the number of terminals a single optical source can drive. In addition, wires can pick up signals directly from electronic devices, but fiber systems require an interface to convert the electronic signal to optical form. The more terminals are needed, the more converters are needed—and the more expensive the fiber network becomes.

FIGURE 3.5.

Types of data transmission systems.

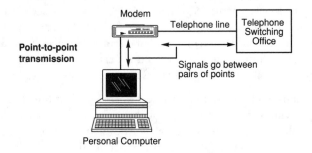

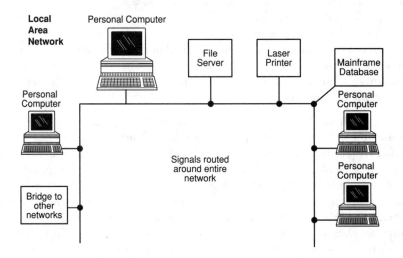

Video Transmission

The design of the typical cable television system looks at first glance like the distribution of signals from a central office in Figure 3.4. However, fundamental differences exist. Cable (often called CATV, from Community Antenna TeleVision) was designed to distribute the same video signals to many subscribers. The telephone network and computer networks make temporary connections that route voice or data between subscribers. Although the cable and phone systems are evolving to compete with each other, these differences have shaped their design.

Video signals require more bandwidth than sound, so the cable network has much more signal capacity than phone lines. However, cable signals go only one way, because video originates in studios, not in homes. All homes get the same video signals, over cables that spread out in a "tree" architecture from a central point, called the "head end." (Extra-cost subscription services are sent to all homes in a form that requires special decoders supplied by the cable company.) Video transmission is analog, matching the input requirements of current television sets while maximizing transmission capacity and minimizing conversion costs.

As in the telephone network, fiber optics are spreading out from the centers of cable television systems toward subscribers. Fiber cables have lower noise than coaxial cables, providing better signal quality to subscribers. They also greatly reduce the need for amplifiers. Although coax can carry many video channels, it requires many closely spaced repeaters to do the job—and failure of any one repeater can knock out the entire chain. Fiber is spreading steadily to serve smaller and smaller groups of subscribers; typical sizes now are a few hundred to a few thousand. Cable networks are adding switches to route signals (including services other than video), and developers are working on a new generation of digital cable systems to carry hundreds of channels.

> ⬤ Design of the cable television network differs from that of the telephone network.

Ships, Automobiles, and Airplanes

The small size, light weight, and immunity to electromagnetic interference of optical fibers has attracted designers of ships, automobiles, and aircraft. The electronic content of ships, cars, and planes has increased steadily, and with it the need for communications. Such systems are quite different from those intended for long-distance communication, because of their small overall size and the need for very low-cost components for cars.

Automakers have studied the use of fibers to carry control signals from a central point—perhaps in the center of the steering wheel—to remote points, where they would switch on accessories such as radios and power windows. Their goals are to simplify electrical wiring harnesses and avoid EMI, which can scramble signals in electronic control systems. However, automakers have been very slow to take advantage of fibers, apparently for lack of fibers that meet their strict requirements on performance and serviceability.

> ⬤ Key attractions of fiber optics for use in ships and planes are its light weight and immunity to electromagnetic interference.

Electromagnetic interference is an even more serious problem in aircraft and aboard ships. The complex avionics of modern ships and planes require much greater information-carrying capacity than automobile electronics. In military aircraft, EMI is a crucial concern because an enemy might try to use it, in the form of electronic countermeasures, to disable the plane's sophisticated weapon systems. The light weight of fibers is also important because of the performance advantages that can come from shaving weight from a plane.

Military Systems

Many military systems are beginning to use fiber-optic communications.

Fiber optics can also play important communications roles in many other military systems. As new technology is brought into military service, command, control, communications, and intelligence—C^{3}I in Pentagon acronymese—become crucial on the battlefield. The Army is using optical fibers to replace bulky 26-pair wire cables in portable battlefield communication systems. The fiber cable promises to be more reliable than the metal cable, which often suffers from broken wires. Military planners are also working on fiber-optic systems for guiding battlefield missiles to their targets.

Noncommunication Fiber Optics

Optical fibers can be used for transmitting light and images and for sensing.

Most of this book is about fiber-optic communications, but fiber optics have a variety of other applications, as you will see in Chapters 22 and 23. Many of them rely on fiber-optic bundles; others depend on the properties of single fibers.

Illumination

A light source at one end of a bundle of optical fibers illuminates whatever is at the other end. This can make a decorative lamp, a flexible illuminator for hard-to-reach places, or a practical display, such as the one shown in Figure 3.6. Short lengths of optical fibers, pressed together to make a flat plate, can concentrate light from a bulb toward a specific angle, a property that makes them useful in traffic signals that can be viewed only from certain lanes of a road. (An example is a left-turn signal visible only from the left-turn lane.)

Single fibers or small bundles can deliver laser energy to hard-to-reach targets. For example, a fiber catheter can be threaded through the ureter to carry laser pulses powerful enough to shatter kidney stones inside the body—avoiding the need for surgery. Fiber bundles can deliver laser light for cutting or drilling objects.

FIGURE 3.6.
Fiber-optic display sign. (Courtesy of Mitsubishi Rayon)

Imaging and Inspection

If the fibers in a bundle are arranged in the same pattern at the ends, they can transmit images. One of the most important applications for fiber-optic image transmission is endoscopy, which lets physicians view an otherwise inaccessible part of the body. For example, an endoscope can be passed down the esophagus so a physician can see into the stomach. Endoscopes can also carry laser light to treat certain conditions, such as bleeding ulcers.

Imaging fiber bundles also permits inspection of the insides of otherwise-inaccessible objects, such as the interiors of engines. Some fibers in the bundle can carry light to dark regions, and the remaining fibers collect the light and deliver it to the inspector.

Sensors

Optical fibers can also be used as sensors. Some fiber sensors use the fiber only to pick up light and bring it to a place where it can be detected. Others rely on fibers to carry light to and from a sensing element. In still others, the fiber itself is the sensing element. Examples of the latter include the sensitive acoustic detectors that the Navy hopes will keep track of the whereabouts of potentially hostile submarines and loops of optical fiber used to sense rotation, which can serve as gyroscopes in guidance systems for missiles and planes—and perhaps for land vehicles as well.

What Have You Learned?

1. Fiber optics can transmit signals in both digital and analog format. Analog transmission requires linear light transmitters.

2. The telephone system is composed of a backbone system, trunk lines, and the subscriber loop. Fiber is used mostly in the backbone system, trunk lines, and in parts of the subscriber loop where individual fiber cables carry multiplexed signals.

3. High-capacity fiber-optic cables are the backbones of national and international telecommunication networks. Submarine fiber cables compete with satellite channels for telephone communication across the oceans.

4. Fibers are finding growing use in data communications and local area networks, but they are not used for short, low-speed links between personal computers and peripherals.

5. Cable television systems use fiber to deliver analog video signals to remote distribution points, where signals are split among many subscribers.

6. Cable television and telephone networks are converging toward similar architectures using fiber optics.

7. Data networks in ships and planes may use fiber.

8. Optical fibers can be used to sense, illuminate, deliver laser power, display, and image, as well as to communicate.

What's Next?

In Chapter 4, you will go back and take a closer look at the characteristics of optical fibers. Later on, in Chapters 17-23, you will look more closely at the variety of fiber-optic applications.

Quiz for Chapter 3

1. Which of the following are true for analog signals?

   a. They vary continuously in intensity.

   b. They are transmitted in parts of the telephone network.

   c. They are compatible with human senses.

   d. They can be processed electronically.

   e. All of the above.

2. Which of the following are true for digital signals?

   a. They are used in parts of the telephone network.

   b. Their intensity can be at only certain discrete levels.

   c. They can be processed electronically.

   d. They can encode analog signals.

   e. All of the above.

3. You digitize a 10 kHz signal by sampling it at twice the highest frequency (e.g., 20,000 times a second) and encoding the intensity in 7 bits. What is the resulting data rate?

   a. 20 kHz.

   b. 56 kHz.

   c. 128 kHz.

   d. 140 kHz.

   e. 1.28 MHz.

4. What part of the telephone network is connected directly to your home telephone?

   a. Subscriber loop.

   b. Feeder cable.

   c. Trunk line.

   d. Backbone system.

5. What part of the telephone network carries the highest-speed signals?

   a. Subscriber loop.

   b. Feeder cable.

   c. Trunk cable.

   d. Backbone system.

6. Which of the following characteristics of fiber optics is least important for telephone transmission?

   a. Will not propagate current pulses from lightning strikes.

   b. High-quality analog transmission.

   c. Small size lets four fiber cables fit into ducts that could hold only one metal cable.

   d. Immunity to electromagnetic interference.

   e. Long-distance high-speed transmission.

7. When are fiber optics used for point-to-point computer data transmission?

  a. When personal computers need to be connected.

  b. When there is concern about electromagnetic interference.

  c. When new technology is being tested.

  d. When the data comes from a fiber-optic telephone system.

8. Fiber optics are specified as a standard transmission medium for

  a. the 100 Mbit/s FDDI local area network.

  b. 10 Mbit/s Ethernet local area networks.

  c. connection of printers to personal computers.

  d. short-distance transmission within buildings.

9. Which is not a reason fiber optics are used in cable television distribution today?

  a. Fibers can carry digital signals over extremely long distances.

  b. Low fiber loss eliminates the need for most repeaters.

  c. Fiber transmission has lower analog noise than coaxial cable systems.

10. Which is not a present or potential application of fiber optics?

  a. Sensing.

  b. Power transmission to homes.

  c. Image transmission.

  d. Signal transmission in planes.

  e. Delivering laser energy inside the human body.

Types of Fibers

About This Chapter

All optical fibers are not alike. There are several major types, which are made differently, operate in different ways, have different characteristics, and serve different functions. Some differences that seem subtle can lead to large functional differences.

This chapter describes the basic types of fibers and how they work. You will learn the differences between various types of fibers and what those mean for the applications of such fibers. This is an essential groundwork to understanding the range of applications of fiber optics.

Functional Requirements and Types

The major types of optical fibers have evolved over many years, with some types going back a decade or two. Functional requirements have evolved along with fiber technology, so interest in some types has grown and declined. For example, the first low-loss fibers were small-core single-mode types. Concern that getting light into their tiny cores would be impractical led to the development of larger-core graded-index multimode fibers with higher transmission capacity than the simple step-index multimode fibers you examined in Chapter 2. Later, advances in light-transfer technology and a desire for even higher-speed transmission renewed interest in single-mode fibers, which today dominate the telecommunications market. Meanwhile, graded-index multimode fibers have become common for short-distance intrabuilding communications. Step-index multimode fibers still find some uses, but they are limited.

● Types of fibers have evolved in response to user needs, which themselves have changed, but many fiber types have found uses other than originally expected.

● Different fiber applications often have conflicting requirements.

The main reason so many fiber types are needed is that fiber applications are diverse. Different applications often have conflicting requirements. For example, the low-loss high-speed transmission essential for long-distance telecommunications is best provided by a small-core single-mode fiber. On the other hand, the needs of local area networks for inexpensive terminal components and easy coupling are best met by a larger-core graded-index fiber.

The major factors that can dictate the choice of certain fibers for specific applications include the following:

- Low attenuation to maximize spacing of repeaters or amplifiers or to avoid them altogether
- Maximization of transmission bandwidth or speed
- Ease of collecting light from inexpensive large-area emitters
- Ease of making splices or attaching connectors in the field
- Large tolerances to allow inexpensive connectors
- Cost of fiber
- Transmission wavelength
- Tolerance of high temperatures or other environmental conditions
- Strength and flexibility of the fiber

The rest of this chapter describes the types of fibers now available, including some in development, and how they can meet these and other criteria. Figure 4.1 shows important types of single fibers, with relative dimensions in cross section, showing variation of the key parameter of refractive index. Only core and cladding are shown for simplicity; actual fibers are coated with an outer plastic layer. I will discuss them in a rough but somewhat arbitrary sequence of complexity, after first outlining how fibers are made. Basic fiber-optic principles are included in the discussions of the various fiber types.

Fiber Manufacturing

● Glass fibers are drawn from a preform of highly purified glass, which is heated so the glass flows.

The manufacture of optical fibers is a precise and highly demanding process requiring special equipment. The first step in making glass fibers is to make a rod or "preform" of highly purified glass, with a core and cladding structure. Then the preform is heated and drawn out into a thin fiber, which is coated with a protective plastic layer as it is drawn. There are many variations in the ways preforms are made, but the other steps are similar in most processes. (Plastic fibers, however, are produced in somewhat different ways, which will not be discussed here.)

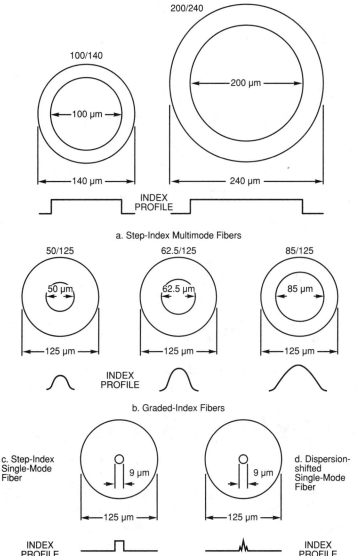

FIGURE 4.1.

Types of glass optical fiber.

Preforms can be made in several different ways. One method, shown in Figure 4.2, is to burn ultrapure vapors of silicon tetrachloride and germanium tetrachloride in a flame, producing a fine soot of silica (SiO_2) and germania (GeO_2). The soot is deposited on a rotating rod or mandrel. First the core glass is deposited, then the composition is changed to deposit the cladding. Finally, the rod is removed, and the remaining glass is heated and collapsed into a dense glass. This approach, called outside vapor deposition, produces an extremely pure preform because all the material is deposited from the vapor phase.

FIGURE 4.2.

Outside vapor deposition to make a preform.

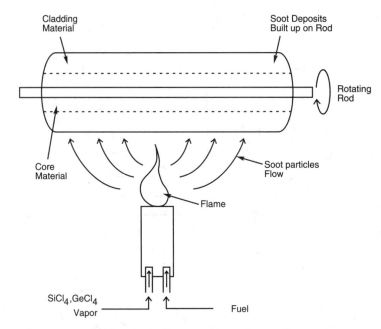

Later, as shown in Figure 4.3, the preform is mounted vertically in a furnace and heated until molten glass can be pulled from it in a fine fiber. The fiber diameter is measured and the fiber is coated with a protective plastic layer as it is pulled from the preform. The equipment used for drawing fibers is called a drawing tower, and tower it does. It is easily a couple of stories high, and it towers over everything else on the floor of a fiber factory.

Step-Index Multimode Fiber

Step-index multimode fibers are conceptually the simplest fibers and were the first to find practical uses. These fibers work in the simplified way described in Chapter 2. The fiber core has a refractive index slightly higher than the cladding material, confining the light by total internal reflection to the core. The term "step-index" comes from the abrupt change in refractive index of the fiber material at the core-cladding boundary, the interface that confines light within the core. The amount of the index difference depends on the fiber design and material, but is typically small—less than 1% usually gives adequate light guiding in glass fibers.

> Step-index multimode fibers confine light to the core by total internal reflection from the lower-index cladding.

> Step-index multimode fibers collect light easily but have a limited bandwidth.

The major attraction of step-index multimode fibers is the ease with which they can collect light. Their core diameters start at 100 μm, and some all-plastic fibers have cores larger than a millimeter. Typical numerical apertures are 0.2 to 0.4 for silica-core fibers and 0.4 to 0.75 for all-plastic fibers. This combination of large core and high numerical aperture allows the use of inexpensive large-area light sources and avoids the need for extremely precise connectors. Indeed, tolerances are so large for 1 mm core all-plastic fibers that signals can be transmitted even when the fibers are partly out of the connector. However, these advantages come at the cost of limited transmission bandwidth and higher losses—much higher for plastic fibers.

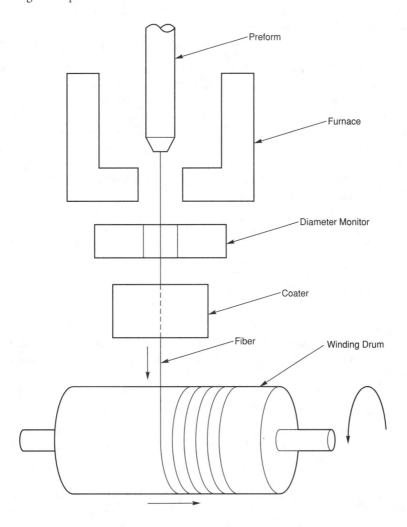

FIGURE 4.3.
Drawing glass fibers from preforms.
(Courtesy of Corning Inc.)

Modes and How They Work

The term "multimode" indicates that light can travel in many ways through such fibers. The easiest way to visualize these multiple paths is by drawing light rays that are reflected back and forth from the core-cladding interface, as you saw in Chapter 2. Unfortunately, reality—as usual—is considerably more complicated. Instead of talking about light rays, I should be talking about transmission modes in an optical waveguide.

What is a mode? A mode is a stable propagation state in an optical fiber. Dig into mode-propagation theory and you will find that it is an effect caused by the wave nature of light. If light travels through an optical fiber along certain paths, the electromagnetic fields in the light waves reinforce each other to form a field distribution that is stable as it travels down the fiber. These stable operating points (standing waves) are modes. If the light tries to travel other paths, a stable wave will not propagate down the fiber—thus no mode.

The details of mode-propagation theory are far too complex to discuss here and have little relevance to the day-to-day concerns of fiber-optic users. I will take the shortcut of treating modes as bundles of light rays entering the fiber at the same angle. However, some results of mode theory are worth noting. One is that the light actually penetrates slightly into the fiber cladding layer, even though it nominally undergoes total internal reflection, as shown in Figure 4.4. This means that attenuation in the cladding layer, although not crucial, cannot be ignored altogether. In addition, some modes may propagate partly in the cladding, where losses tend to be high because light can leak out of the fiber, as well as suffer absorption in the cladding.

FIGURE 4.4.

Light's path in a step-index multimode fiber.

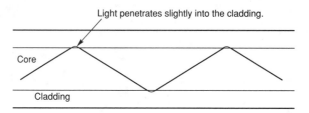

Propagation modes are standing waves that travel through the fiber. To see some consequences of multimode transmission, think of a mode as a family of rays all traveling down the fiber at the same angle to the fiber axis. (Because the cone of rays is circularly symmetric about the fiber axis, adding them together cancels out the parts of the rays going in and out, leaving only the mode traveling down the fiber.)

Each mode has a characteristic number N. A mode N is associated with all rays traveling at an angle θ_N with respect to the fiber axis, which is approximately

$$\theta_N \approx \lambda((N+1)/2Dn) \text{ radians}$$

where N is the mode number (the lowest-order mode is 0), λ is the wavelength, D is core diameter, and n is the refractive index of the core. Note that even the rays associated with the zero-order mode do not travel straight along the fiber core. For a fiber with a refractive index of 1.5 and a 100 μm core, the angle θ_0 (for the lowest-order mode) is approximately 0.5° for a wavelength of 0.85 μm.

The number of modes that can propagate in a fiber depends on the fiber's numerical aperture (or acceptance angle) as well as on its core diameter and the wavelength of the light. For a step-index multimode fiber, the number of such modes N_m is defined by

$$\text{MODES} = (\text{CORE DIAMETER} \times \text{NA} \times \pi/\text{WAVELENGTH})^2/2$$

or

$$N_m = (D \times NA \times \pi/\lambda)^2/2$$

where λ is the wavelength and D is the core diameter. To plug in some representative numbers, a 100 μm core step-index multimode fiber with NA = 0.29 (a typical value) would transmit 5744 modes at 850 nm.

Modal Dispersion Effects

Looking at the ray propagation angle θ_N gives another indication of how modal dispersion works. Remember that the mode can be viewed as a family of rays traveling at an angle of θ_N to the fiber axis. Light traveling straight along the fiber axis would move at a speed of c/n—that is, the speed of light in a vacuum divided by the refractive index of the core material. Because the ray family for each mode has a different propagation angle, their path lengths in the fiber are different. The differences are proportional to the cosines of the propagation angles. Because a pulse of light travels through a step-index multimode fiber in many different modes, it spreads out as it travels along the fiber, causing the modal dispersion described in Chapter 2. Precise calculations are beyond the scope of this book, but typical bandwidths of 100 μm silica-core step-index multimode fiber are about 20 MHz-km. (A few have bandwidths of 100 MHz-km.)

Modal dispersion limits pulse rate in multimode fibers because it causes successive pulses to overlap and interfere with each other. (It also distorts analog waveforms.) Total dispersion is roughly the product of the dispersion characteristic of the fiber, D_0 (measured in nanoseconds per kilometer) times the length L (measured in kilometers). To be strictly accurate, the total dispersion of multimode fibers should be calculated by the formula

$$D = D_0 \times L^\gamma$$

where γ is a factor close to 1, dependent on the fiber type. For most practical purposes, however, assume $\gamma = 1$, especially because multimode fiber is normally used only over short distances.

If we neglect the effects of the range of wavelengths in the signal light, fiber dispersion can be converted to transmission bandwidth by the approximate formula

$$\text{BANDWIDTH (MHz-km)} = 350/\text{DISPERSION (ns/km)}$$

● The number of modes a fiber can transmit depends on its numerical aperture and core diameter, as well as the wavelength.

● Modal dispersion comes from differences in the propagation angles of the ray families associated with individual modes.

● Dispersion and bandwidth depend on the fiber's internal characteristics and its length.

(The bandwidth in this formula—and elsewhere in the book—is defined as the frequency where output has dropped by one-half [3 dB] from its usual level.) As with dispersion, bandwidth of a fiber segment depends on length L and the fiber's characteristic bandwidth BW_0, in this case according to the formula

$$BW = BW_0 / (L^\gamma)$$

where the γ factor is the same as described previously and can usually be assumed to equal 1, except for long-distance transmission. Thus a 5 km length of 20 MHz-km fiber would have a net bandwidth of 4 MHz.

Leaky Modes

Some modes can propagate short distances in the cladding of a multimode fiber.

Measurements of numerical aperture of multimode fibers show a peculiar phenomenon: the NA appears highest for a short segment of fiber. Thus a 2 m length may have NA of 0.37, while a 1 km length has 0.30, which is close to the theoretical prediction. There is no room for this in the formula for NA given in Chapter 2, so what's happening? Modes that are just slightly beyond the threshold for propagating in a multimode fiber can travel for short distances in the fiber cladding. Because this extra light appears to have been transmitted by the fiber, it increases the measured NA.

Similarly, the highest-order modes that meet the conditions for propagation are just within the threshold. If conditions alter just a tiny bit—for example, if the fiber is bent—they might leak out. These are called leaky modes, and you will see some of their effects in the following section.

Bending Effects

Bending can increase fiber losses by letting high-order modes leak out of the core.

So far I've assumed that the fiber is straight, but in any real application, it will bend around corners. In practice, fiber bends are gradual relative to the diameter of the fiber, with curvature of a few centimeters or more compared to the 100 μm diameter of a typical step-index fiber core. Larger-core fibers are more rigid and have larger minimum bend radii.

To see how a bend can change transmission, recall the simple ray model of transmission and look at Figure 4.5. When light rays strike a bend in the fiber, those in higher-order modes can leak out if they hit the side of the fiber at an angle beyond the critical angle θ_c. That increases the loss in the fiber. Lower-order modes are not likely to leak out, but they can be transformed into higher-order modes, which can leak out further along the fiber at the next bend. The bends need not be large to cause losses in the fiber. Indeed, the most serious bending losses in multimode fibers come from microbending, which causes tiny kinks.

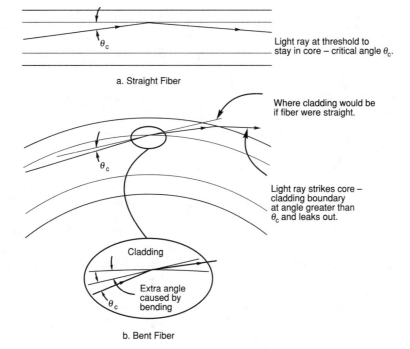

FIGURE 4.5.

Light can leak out of bent fibers.

Light ray at threshold to stay in core – critical angle θ_c.

a. Straight Fiber

Where cladding would be if fiber were straight.

Light ray strikes core – cladding boundary at angle greater than θ_c and leaks out.

Cladding

Extra angle caused by bending

b. Bent Fiber

Types of Step-Index Multimode Fiber

All step-index multimode fibers share many characteristics, but not all step-index multimode fibers are alike. Two key sets of parameters can vary: composition and diameter of both the core and cladding. The variation of these parameters can lead to major differences in crucial functional characteristics such as bandwidth and attenuation. The most common core diameter is 100 μm, but glass cores come as large as 1 millimeter. Some examples are listed in Table 4.1, along with representative values from data sheets.

Differences in composition and diameter of core and cladding lead to differences in properties of step-index multimode fibers.

Table 4.1. Types of silica-core step-index multimode fiber.

Fiber Type	Diameter, Core/Clad	Attenuation @850 nm	Bandwidth, MHz-km	NA
All-glass	100/140	4 dB/km	100 MHz-km	0.29
All-glass	100/130	5 dB/km	25 MHz-km	0.22
Hard-clad silica	110/125	12 dB/km	17 MHz-km	0.37
All-glass	200/230	6 dB/km	20 MHz-km	0.39
Plastic-clad silica	200/?	10 dB/km	20 MHz-km	—

continues

Table 4.1. continued

Fiber Type	Diameter, Core/Clad	Attenuation @850 nm	Bandwidth, MHz-km	NA
Hard-clad silica	200/230	6 dB/km	17 MHz-km	0.37
Hard-clad silica	600/630	8 dB/km	9 MHz-km	0.37
Hard-clad silica	1000/1035	10 dB/km	—	0.37

The most common step-index multimode fiber has a 100 μm core and a 140 μm cladding.

There are three basic types of step-index multimode silica fibers. In the glass-on-glass fiber, both the core and cladding are silica-based glasses, with dopants added to adjust the difference in refractive index to the desired level. In plastic-clad silica, the core is a nominally pure silica and the cladding layer a soft plastic. Hard-clad silica has a cladding of hard rather than soft plastic and is made with a slightly larger core. Claddings are typically fairly thin, as shown in Table 4.1. In practice, all these fibers are coated with one or more plastic buffer layers to protect them from degradation and to simplify handling. A typical attenuation curve for hard-clad silica fiber is shown in Figure 4.6.

FIGURE 4.6.

Typical spectral attenuation of hard-clad silica step-index fiber. (Courtesy of Ensign-Bickford)

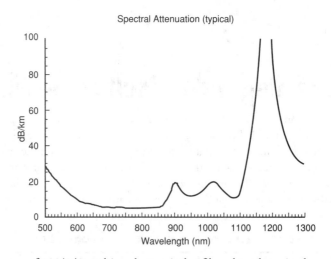

The main uses of 100/140 multimode step-index fibers have been in short data links at low speeds. The large cores can accept light from inexpensive LEDs and allow the use of inexpensive connectors with large tolerances. Losses are higher than in most single-mode or multimode graded-index fibers, but this is not a major concern for short distances. Nonetheless, most new communication systems use graded-index multimode fibers, which offer nearly the same advantages plus higher transmission capacity for intrabuilding use, and have a standard cladding size of 125 μm.

Some larger-core step-index multimode fibers are used for communications, but more often they are used for laser power delivery, sensing, measurement, or illumination. The larger cores offer larger collecting areas, and permit the fibers to handle higher laser powers, but make the fibers stiffer and more costly. The larger cores also reduce transmission bandwidth, but that is not important for illumination, sensing, or beam delivery.

Some step-index multimode fibers are designed for special applications, particularly to survive exposure to nuclear radiation, which can cause temporary or permanent darkening of conventional fibers. The main uses of these fibers are in military systems. Their transmission characteristics are otherwise similar to those of standard fibers, but they tend to contain more hydroxyl (OH) ions, so they have somewhat higher absorption at certain wavelengths.

Ultraviolet Fibers

Extrapolation from Figure 4.6 indicates that fiber loss should be high in the ultraviolet region. Although intrinsic losses of silica are much higher in the ultraviolet region than in the near infrared, special fibers are made to transmit reasonable amounts of ultraviolet light over short distances. Figure 4.7 plots attenuation of one such ultraviolet-transmitting fiber. Special ultraviolet-transmitting fibers can transmit some light through a 1 m length at wavelengths to 180 nm. Obviously these transmission figures cannot compare with fibers operating at longer wavelengths. However, they are adequate to carry ultraviolet light over short distances, for illumination or measurement, which otherwise can be difficult.

●
Special fibers are made for ultraviolet transmission at wavelengths as short as 200 nm.

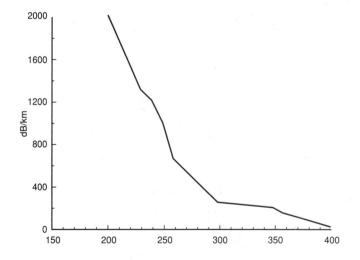

FIGURE 4.7.

Absorption in an ultraviolet-transmitting fiber. (Courtesy of Ensign-Bickford)

Plastic Fibers

High-loss, large-core multimode step-index fibers can be made entirely of plastic.

Multimode step-index fibers can be made entirely of plastic as long as they contain a core with a refractive index higher than the cladding. Many have a core of polymethyl methacrylate (PMMA), which can be surrounded by lower-index materials, such as fluorine-containing polymers. This can lead to large core-cladding index differences. For example, in one commercial fiber the core has 1.495 refractive index and the cladding 1.402, corresponding to an NA of 0.50 and a full acceptance angle of 60°. (Single-mode plastic fibers are not made.)

All-plastic fibers have some important attractions, including low cost, better flexibility, and ease of handling, and they have been used for many years in applications such as light-outage indicators in cars and in many types of fiber-optic bundles. Flexibility and low cost are important in large-core fibers because silica types tend to be stiff and expensive. On the negative side, all-plastic fibers have much higher attenuation and less resistance to high temperatures than glass types. The attenuation curve in Figure 4.8 shows minimum loss measured in hundreds of decibels per kilometer, limiting transmission to short distances. Note that losses are lower at visible wavelengths than in the near-infrared, where glass fibers have lowest attenuation. Low glass transition temperatures limit most plastics to temperatures below about 85° C. New types can operate up to 125° C but have higher losses. These temperature limits have been an important problem in automobile applications.

FIGURE 4.8.

Attenuation of a plastic fiber.
(Courtesy of Mitsubishi Rayon)

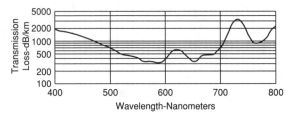

Plastic fibers have core diameters from about 85 to 3,000 µm. Cladding thicknesses differ considerably; some are very thin, others are 20% or more of fiber diameter. Smaller fibers are normally used as components in flexible bundles, but the larger fibers are used separately. As with other multimode step-index fibers, the numerical aperture and attenuation do not depend directly on fiber diameter.

All-plastic fibers have some potential communication applications in areas like automobiles, where distances are small and bandwidth requirements are modest. Signal transmission usually is with red LEDs emitting near 650 nm. However, most uses of plastic fibers are likely to remain in light piping and image transmission through bundles—where the goal is to carry light no more than a few feet and bandwidth (in the communications sense) is meaningless. Typically a 2 m bundle transmits about 60% of the incoming light, although both higher and lower transmission are possible.

Attenuation considerations for imaging fibers differ greatly from those for communication fibers, where only one wavelength matters. Imaging usually requires that the fiber transmit the entire visible spectrum. For instance, physicians use color in diagnosis, so endoscopes should accurately transmit the colors of tissue inside the body. Special plastic fibers are made for that purpose.

● Uniform transmission of light throughout the visible spectrum is important in imaging.

Graded-Index Multimode Fibers

Demonstration of the first low-loss fibers did not solve all problems of long-distance fiber-optic communications. The first low-loss fibers were single-mode types, with light-carrying cores only several micrometers in diameter. Such cores seemed much too small to collect light from then-available sources. The other major type of fiber—the step-index multimode fiber—had a different problem; its modal dispersion was too high to allow high-speed transmission over distances of kilometers. In the early 1970s, developers came up with a compromise solution—the graded-index multimode fiber.

● Graded-index multimode fibers collect light better than small-core single-mode fibers and have broader bandwidth than step-index multimode fibers.

Graded-index fibers get their name from the way the refractive index changes from core to cladding—gradually. For step-index fibers, the boundary between the core and cladding is considered abrupt. That is only an approximation, because any transition between two materials—particularly two connected as intimately as in glass optical fibers—takes place over a finite distance. However, it's a good enough approximation for all practical purposes.

In graded-index fibers, the transition is deliberately made gradual. In theory, the refractive index drops smoothly from the center of the fiber to the edge of the cladding, as shown in Figure 4.9. In practice, a good approximation of that smooth curve is made by depositing up to a couple hundred layers of glass with gradually changing composition in the early stages of making the preform. Heating and collapsing the preform and drawing it into fiber smooth the distribution.

FIGURE 4.9.

Graded-index fiber refractive index profile.

Graded-index multimode fibers have smaller cores than step-index multimode fibers—typically 50-85 μm in a fiber with 125 μm cladding versus 100 μm and up for step-index multimode fibers. Those cores are large enough to ease coupling tolerances, but they can also carry many modes. How does grading the refractive index help? It makes the path that light rays travel through the fiber dependent on refraction rather than total internal reflection, so light rays entering the fiber at different angles travel essentially the same distances through the fiber.

Light guiding in a graded-index multimode fiber is shown in Figure 4.10, which can be compared with the picture of light guiding in a step-index multimode fiber in Figure 4.4. In the step-index fiber, the light rays zigzag between the core-cladding boundary on each side of the fiber axis. In graded-index fiber, the gradient in the refractive index gradually bends the rays back toward the axis.

FIGURE 4.10.

Paths of light rays in a graded-index multimode fiber.

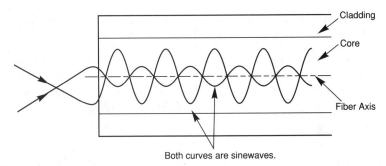

The mathematical details are far beyond the scope of this book, but they turn out to depend on how the refractive index varies with distance from the fiber axis, the index profile. The best way to limit dispersion in a graded-index fiber is to make the difference between core and cladding index vary with the square of the distance from the fiber axis. This results in the parabolic index curve shown in Figure 4.9 and the sine-wave light paths shown in Figure 4.10.

Modal Dispersion

Index grading equalizes the times that different rays take to pass through the fiber, minimizing modal dispersion.

How does index graduation help reduce modal dispersion? It doesn't really equalize the physical distances that light rays travel through the fiber. What it does instead is minimize the difference in time they take to pass through the fiber. This can be done because the speed at which light travels through the glass in the fiber is the speed of light in a vacuum divided by the refractive index, c/n. Thus the higher the refractive index, the slower light travels. The light rays that enter the fiber at a steeper angle still have to go farther through the glass. However, in the outer parts of the core, they travel through glass with a lower refractive index than the rays that go a shorter distance closer to the fiber axis. Thus, what the higher-order rays lose in speed by traveling a greater distance, they make up by going faster in the lower-index glass.

The graded index multimode design does not eliminate modal dispersion, but it does reduce it greatly. Typical graded-index multimode fibers have bandwidths of 100-1,000 MHz-km or more at their normal operating wavelengths of 850 or 1300 nm, normally with higher bandwidth at the longer wavelength.

Material Dispersion and Waveguide Dispersion

Although simple modal dispersion dominates the total dispersion picture for step-index multimode fibers, matters are more complex for graded-index multimode fibers, and wavelength becomes an important factor. The refractive index of glass is a function of wavelength, so the mode-equalizing effects of the fiber refractive index also depend upon wavelength.

There are two other kinds of dispersion: material dispersion and waveguide dispersion. Material dispersion occurs because a pulse of light in the fiber includes more than one wavelength. Thanks to the differences in refractive index with wavelength, different wavelengths travel through the fiber at different speeds. The range of wavelengths in a pulse also affects waveguide dispersion, which arises because of the way that light is divided between core and cladding (an effect more important for single-mode fibers).

The result shown in Figure 4.11 is a bandwidth-versus-wavelength curve with a pronounced peak. Fiber designers can pick this peak by their choice of manufacturing parameters and dopants. If the fiber is to be used at only a single wavelength, typically 1300 nm, the peak might be placed close to that wavelength. But if operation was to be at two or more wavelengths, the fiber might be designed with peak bandwidth at an intermediate wavelength. The figure shows two different peaks, one in fiber designed for 1300 nm, the other in fiber for use at shorter wavelengths. The curve labeled "laser material limit" indicates how material dispersion limits bandwidth. In practice, most graded-index fiber is made for use at both 850 and 1300 nm and is used only over modest distances where bandwidth is not critical. Single-mode fiber is used for higher bandwidths or longer distances.

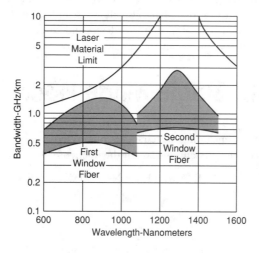

FIGURE 4.11.

Range of bandwidth in graded-index multimode fibers. (Courtesy of Corning Inc.)

Attenuation

Attenuation of graded-index multimode fibers is typically at least as low as the best step-index multimode fibers, and often better. Most graded-index fibers are designed to operate in one (or both) of two transmission windows—at 850 or 1300 nm. Attenuation at the shorter wavelength is higher (in the 3 dB/km range), and the bandwidth is normally lower. However, light sources and detectors for that wavelength cost much less than those for 1300 nm. For the longer wavelength, typical attenuation is 0.5 to 1 dB/km, allowing transmission over longer distances, but only at a cost of more expensive terminal components. Figure 4.12 plots typical attenuation of graded-index fibers.

FIGURE 4.12.

Attenuation of graded-index fibers. (Courtesy of Corning Inc.)

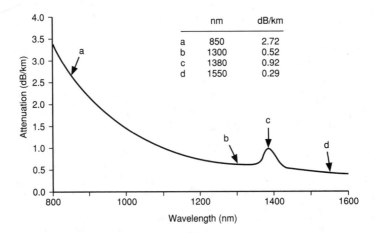

nm	dB/km
a	850	2.72
b	1300	0.52
c	1380	0.92
d	1550	0.29

Graded-index multimode fibers have dropped from favor for long-distance communication but are used widely in shorter systems, data links, and local area networks. Fibers with 62.5 μm cores are most widely used and are specified by some network standards, but 50 and 85 μm core fibers remain available. Although multimode graded-index fibers cannot match the bandwidth of single-mode fibers, they offer adequate capacity for most systems no more than a few kilometers long. They also allow use of less-costly terminal components, such as LEDs rather than semiconductor lasers and connectors with looser tolerances. In applications that require many terminals, such costs are crucial concerns.

Step-Index Single-Mode Fibers

If a fiber core is made small enough, it will carry only one mode.

Earlier in this chapter, you saw how the number of modes that a step-index fiber could carry depended on factors including the refractive-index difference between core and cladding, the operating wavelength, and the size of the fiber core. The relationship is

$$\text{MODES} = (\text{CORE DIAMETER} \times \text{NA} \times \pi/\text{WAVELENGTH})^2/2$$

or

$$N_m = (D \times NA \times \pi/\lambda)^2/2$$

where N_m is the number of modes, D is the core diameter, NA is the numerical aperture, and λ is the wavelength. This shows that there are three ways to reduce the number of modes to get a single-mode fiber: reduce core diameter, reduce numerical aperture, or increase wavelength.

Core Size

Operating wavelength is generally fixed by considerations such as attenuation. Reducing numerical aperture too much can make it almost impossible for the fiber to collect light. That leaves reducing the core diameter as the only reasonable way to reduce the number of modes. It is not possible just to turn the equation around and solve for the D needed to limit transmission to a single mode. Proper derivation of the answer requires Bessel functions, which are the sort of things taught in third-year advanced calculus courses that flunk out engineering majors. Let's skip that step and go instead to the formula that gives conditions for single-mode operation of a step-index fiber:

$$D < 2.4\lambda/(\pi \times NA)$$

For an NA of about 0.15, the core diameter must be no more than about five times the wavelength. If the NA is 0.1, the core could be up to 7.6 times the wavelength. If the NA were raised to 0.2, the core would have to be smaller than 3.8 times the wavelength. These considerations lead to core diameters of around 8 or 9 μm for single-mode fibers used in 1.3 μm communications. (The mode-field diameter, the actual region that carries light, is slightly larger than the core, about 9.3 μm versus 8.3 μm for one widely used commercial fiber, because part of the mode travels in the cladding.)

The small core size puts tight requirements on light collection by the fiber. For handling reasons, a single-mode fiber cladding should be at least 125 μm in diameter, a dozen times (or more) the core diameter. Light sources should have output areas measured in micrometers, to match the fiber core. Connection tolerances also must be tight. If two fibers with 10 μm cores are misaligned by just 1 μm, the overlap area is reduced by 12.7%. Such problems made early developers wary of single-mode fibers, but current technology has them under control.

⬤ The small size of a single-mode fiber core makes light transfer into the fiber difficult.

Transmission Capacity and Dispersion

On the other hand, fibers that can carry only a single mode banish the whole problem of modal dispersion—the main limitation on the bandwidth of multimode fibers. What remains is called chromatic dispersion (because it is dependent on the range of wavelengths being transmitted by the fiber). It is the sum of material dispersion and waveguide dispersion. These are plotted roughly in Figure 4.13 for a standard single-mode fiber. Here, for once, some good luck intervenes. Dispersion can be positive or negative because it

⬤ Step-index single-mode fibers have zero chromatic dispersion around 1300 nm.

measures the change in the refractive index with wavelength. That change can be an increase or decrease (i.e., positive or negative). In other words, material dispersion and waveguide dispersion can be opposite in sign. Better yet, they cancel out for step-index single-mode fibers at a wavelength close to 1300 nm, where optical fibers also have low attenuation.

FIGURE 4.13.

Chromatic dispersion in single-mode step-index fibers.

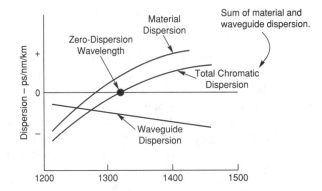

This does not mean that optical fibers have infinite transmission capacity at 1300 nm or even true zero dispersion. (Nature is not that kind to engineers.) Chromatic dispersion equals zero only at one wavelength, but all light sources emit a range of wavelengths. Other, smaller effects also keep dispersion from precisely equaling zero. But dispersion is close enough to zero that it is not a practical concern for present long-distance telephone applications using semiconductor laser sources until data rates become very high. (Lasers are needed because the range of wavelengths emitted by LEDs is large enough that dispersion becomes significant. Also, as you shall see later, LEDs cannot couple enough power into single-mode fibers for long-distance transmission.)

Attenuation

The main limitation on 1300 nm transmission is loss as low as 0.34 dB/km in mass-produced fibers.

The main limitation on 1300 nm systems thus becomes attenuation, plotted in Figure 4.14 for step-index single-mode fiber, which is a little lower than in multimode fibers. One reason is eminently practical: the light-carrying cores that must have ultra-low loss are much smaller, simplifying production. Other factors also enter in, allowing attenuation in the best commercial single-mode fibers to reach 0.34-0.5 dB/km at 1300 nm. Somewhat lower losses have been demonstrated in the laboratory but are not specified in mass-produced fibers. In practice, cabled fibers have slightly higher losses, because they include splices and because of cabling losses described in the next chapter.

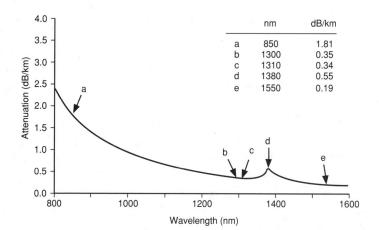

	nm	dB/km
a	850	1.81
b	1300	0.35
c	1310	0.34
d	1380	0.55
e	1550	0.19

FIGURE 4.14.

Attenuation of step-index single-mode fiber. (Courtesy of Corning Inc.)

Long-Distance Transmission

Today's single-mode fiber systems can transmit 2.4 Gbit/s over distances of 30 to 50 km without amplification or repeaters at 1300 nm. Developmental systems promise transmission of 5 to 10 Gbit/s over similar distances in the near future. That performance is good enough that single-mode fibers are now virtually the only type the telephone industry uses outside the subscriber loop. But for some purposes, such as submarine cables, developers would like to send signals still farther without amplification.

The simplest way to extend transmission distance might seem to be turning up the optical power. Unfortunately, that creates other problems. Semiconductor laser lifetimes decrease with operating power, but more fundamentally, the fiber itself has limited power transmission capacity. Moving to powers above tens of milliwatts could trigger nonlinear effects that would distort the optical signal traveling through the fiber.

At lower powers, the interaction between light and optical materials depends directly on the electric field E, but as power levels increase, higher-order terms like E^2 become important. These higher-order terms can distort the signal. Tens of milliwatts might not seem like much, but the critical factor is power density, and that is high because the power is concentrated in a fiber core only 9 μm across. What's more, the light travels many kilometers through the fiber, so weak interactions accumulate. Therefore, increasing optical power beyond a certain level causes more problems than it solves.

1550 nm Transmission

What about moving to 1550 nm, where attenuation is considerably lower? Laboratory researchers have pushed attenuation below 0.16 dB/km—an impressive feat and very close to the theoretical limit. Commercial fibers have losses of 0.20 to 0.25 dB/km. Such low losses could really stretch repeater spacing.

● Step-index single-mode fiber systems can now transmit 2.4 Gbit/s 30-50 km without repeaters.

Reaching such low losses is no minor task. Intrinsic scattering and absorption inherent in silicate glasses set the ultimate limits on lowest fiber loss. However, absorption by impurities in the glass keeps attenuation above those minimums. Some impurities are inevitable in most fiber designs because dopants must be added to pure silica to raise the refractive index of the core glass above that of the cladding material. Recently, losses have been reduced by using pure silica as a core glass and doping the cladding glass with fluorine to reduce its refractive index below that of the core glass.

Unfortunately, dispersion, the other key parameter in single-mode fiber performance, enters the picture at 1550 nm. Recall that in Figure 4.13 dispersion is zero only at 1300 nm. At longer wavelengths, chromatic dispersion is well above zero, and at 1550 nm it is large enough to become a problem.

One possible solution to the dispersion problem can be seen by looking closer at chromatic dispersion. The units of measurement are picoseconds of pulse spreading per kilometer of fiber per nanometer of source bandwidth. Note the effect of source bandwidth. If inherent dispersion is large, the total pulse spreading could be reduced by reducing source bandwidth well below the few nanometers typical of conventional semiconductor lasers. What is needed is a laser stabilized to emit only a single frequency. This is not easy, as described in more detail in Chapter 6, but it can be done, and such lasers are used to transmit up to 2.4 Gbit/s at 1550 nm.

> One way to reduce dispersion at 1550 nm is to use a single-frequency laser source.

The other approach is to reduce the fiber dispersion. As described next, this requires a different fiber design.

Dispersion-Shifted Single-Mode Fibers

You saw that chromatic dispersion of a single-mode fiber is the sum of material dispersion and waveguide dispersion. Material dispersion depends on glass composition, and little can be done to alter it without harming fiber transmission. Waveguide dispersion occurs because light moves faster in the low-index cladding than in the higher-index core. (The difference, like material dispersion, depends on wavelength.) The degree of waveguide dispersion depends not only on materials but also on how the light is divided between core and cladding. That is a consequence of fiber design, which can be altered. Thus, changing the waveguide dispersion offers a way to shift the zero dispersion wavelength—where waveguide dispersion and material dispersion cancel—to 1550 nm.

> Changing waveguide dispersion by changing the fiber design allows shifting zero dispersion wavelength to 1550 nm.

As you have seen, the interface between core and cladding in conventional single-mode fibers is a refractive-index step, where glass composition changes abruptly. This approach was originally chosen because such fibers are simple to make, but it has proven a workhorse. Changing the waveguide dispersion to make dispersion-shifted fibers requires a more elaborate design that divides light differently between core and cladding. Figure 4.15 shows one approach used in commercial dispersion-shifted fiber, where the core is divided into two layers. The inner core has a refractive index graded with a triangular (gradient $\alpha=1$) profile. It is surrounded with a lower-index silica layer, which in turn is surrounded by a layer with higher refractive index (but lower than the inner core), then an outer low-index cladding. This design makes waveguide dispersion at 1550 nm equal in magnitude but opposite in sign to material dispersion at that wavelength, so that the chromatic dispersion equals zero.

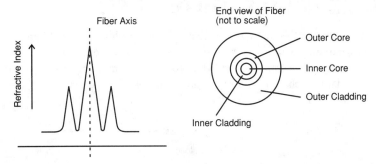

FIGURE 4.15.

Dispersion-shifted fiber with segmented core.

Shifting the dispersion does not come without cost. The addition of germanium to the fiber core as a dopant increases fiber attenuation slightly above that of step-index single-mode fibers, but you have to look very carefully to see the difference in the plots. With care, the change can be kept within a few hundredths of a decibel per kilometer of that of step-index single-mode fiber. In addition, the dispersion-shifted design is inherently more complex and thus harder and more costly to produce than step-index single-mode fibers. The lower cost and large installed base of single-mode fibers with zero dispersion at 1300 nm is likely to keep them in use, but dispersion-shifted fibers offer important advantages for high-performance systems where long repeater spacing is critical, such as undersea cables. (Some types of systems also require dispersion-shifted fiber, as you will see later.)

Dispersion-Flattened Single-Mode Fibers

Both step-index and dispersion-shifted single-mode fibers share one transmission-limiting characteristic—zero dispersion is limited to one wavelength. Look back at Figure 4.13, and you can see that the graph of chromatic dispersion has a significant slope. This means that dispersion in any such fiber is zero only at a single wavelength, and low only at a narrow range of wavelengths around the zero-dispersion wavelength. Typical specifications for a dispersion-shifted fiber call for zero-dispersion slope (S_0) under 0.085 ps/(nm²-km) and dispersion below 2.7 ps/(nm-km) between 1525 and 1575 nm. That is not a problem if high-speed signals are transmitted only at closely spaced wavelengths in that range, but it makes use of both 1300 and 1550 nm windows difficult.

To get around this problem, developers have studied ways to make material dispersion and waveguide dispersion add to zero over a broader range. "Dispersion-flattened" fibers

● Dispersion-flattened fibers might offer low dispersion at a range of wavelengths.

have been tested in the laboratory, but they are difficult to make because fiber designers have only a limited number of degrees of freedom to work with. They have yet to prove practical.

Polarization-Sensitive Fibers

So far, I have oversimplified the way in which light propagates in optical fibers by totally ignoring polarization. Polarization is a consequence of the nature of electromagnetic waves. An electromagnetic wave contains two fields—one electric and one magnetic—oscillating perpendicular to each other and propagating in a direction perpendicular to both, as shown in Figure 4.16.

FIGURE 4.16.

An electromagnetic wave.

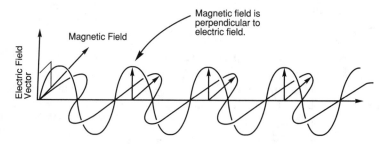

Ordinary unpolarized light is made up of many waves with their electric and magnetic fields oriented randomly (although always perpendicular to each other for each wave). If all the electric fields (and hence the magnetic fields as well) were aligned parallel to one another, the light would be linearly polarized, which is the simplest type of polarization. The two polarization directions are, logically enough, called horizontal and vertical. (Light can also be polarized circularly and elliptically, depending on the way electric and magnetic fields oscillate with respect to each other's phase, a matter beyond the scope of this chapter.)

Special fibers maintain polarization of input light or transmit only one polarization.

Normal optical fibers are essentially insensitive to polarization because their cores are circularly symmetrical. Thus, a single-mode fiber carries light in both horizontal and vertical polarization modes. That doesn't matter for most applications, including ordinary communication systems. However, it can be a problem for some fiber-optic sensors and for advanced communication schemes under development. To deal with such concerns, two types of polarization-sensitive single-mode fibers have been developed.

The difference between the two types is subtle. True single-polarization (or polarizing) fiber has different attenuations for light in the two linear polarizations. One polarization is transmitted well, but the other is strongly attenuated, so it is lost after transmission through the fiber. Thus only one polarization remains at the end.

Polarization-maintaining fiber maintains the polarization of light that originally entered the fiber by isolating the two orthogonal polarizations from each other even while they travel down the same single-mode fiber. Light in the two polarization modes experiences about the same attenuation. However, the fiber manufacturing process deliberately introduces strain, which makes the refractive index differ for the two polarizations, a condition called birefringence. This means that the two polarizations travel through the fiber at different speeds. If linearly polarized light enters a polarization-maintaining fiber with polarization parallel to one of its polarizing axes, the polarization will remain unchanged along the length of the fiber. (Otherwise, the polarization axis will rotate along the fiber because the light is broken into two parts that travel at unequal speeds through the fiber.)

Polarization-sensitive fibers are similar in some ways to conventional single-mode fibers. They are made by similar processes and have only slightly higher attenuation. However, they have the desired polarization characteristics over only a limited range (for example, from 780 to 850 nm or from 1270 to 1330 nm).

Both single-polarization and polarization-maintaining fibers are available today, but they are not used in practical communication systems. Their primary uses are in fiber-optic sensing, particularly rotation sensors or gyroscopes, which will be described in Chapter 22. They are also used as pigtails for polarization-sensitive optical components.

Sensing and Special-Purpose Fibers

Polarization-sensitive fibers are one example of special-purpose fibers made for applications other than general-purpose communications. There are a variety of others, some made for specific types of communications, and others designed for noncommunication uses.

Radiation-hardened fibers are made for military systems. As I mentioned earlier, exposure to nuclear radiation increases the loss of standard optical fibers. Excess radiation losses can be reduced by optimizing fiber design for that purpose. This affects other fiber parameters slightly, and also increases cost. Fibers also can be designed to control other parameters, such as cutoff wavelength for single-mode transmission, or to minimize the extra losses caused by bending the fibers. (Bend-insensitive fibers may be needed if the fiber will be bent strongly during operation, such as being used on a tightly wound reel.)

Special fibers are made for use in fiber devices such as couplers, modulators, switches, and filters. These fibers typically need characteristics that differ from ordinary communication fibers. For example, couplers need to transfer light between adjacent fibers, and some modulators work in the same way, but the fraction of light transferred varies with the signal applied. Researchers are working on new types of fiber devices formed by writing patterns onto the core-cladding interface with ultraviolet light, creating, for example, gratings that can make the fiber serve as a filter. For the UV exposure to create a permanent grating, special dopants must be added to the fiber.

Fibers can also be designed to be particularly sensitive to the surrounding environment, so they can serve as sensors. Standard communication fibers are quite insensitive to the environment (an important practical advantage), but sensing fibers may need to respond strongly to changes in temperature, microbending, pressure, and other ambient conditions. You'll learn more about fiber sensors in Chapter 22.

Fiber Amplifiers

One of the hottest new fiber technologies is the fiber amplifier, which directly amplifies an optical signal without converting it into electronic form. I will cover them in more detail as light sources in Chapter 6 and as amplifiers in Chapter 9. I mention them here because fiber amplifiers are based on special fibers in which the core is doped with a rare earth element. Light from an external source excites the rare-earth atoms so they can emit light, amplifying the optical signal.

Bundled Fibers

The bundling together of many optical fibers was mentioned earlier in connection with plastic fibers. However, the technology is general and can also be used with glass and silica fibers.

Fiber bundles can be loosely divided into two categories, based on physical characteristics and manufacture. Some are rigid, with the fibers fused together in a solid bundle. Others are flexible, with the fibers physically discrete, so the bundle can bend. Each approach has its advantages.

A second way to view fiber bundles is by application. For imaging, the fibers are arranged in the same way at each end of the bundle. The output ends of such coherent fiber bundles replicate the pattern of light at the input ends. Alternatively, for illumination or power delivery, the fibers may be mixed up or randomized (sometimes with great care to make sure the pattern is truly random). Both rigid and flexible bundles can be made coherent or randomized.

One important goal in all fiber bundles is to make fiber cores cover as much of the surface area as possible. The cores transmit light through the bundle, but light coupled into the cladding is generally lost. Thus, bundled fibers have thin claddings to maximize the packing fraction, the part of the surface occupied by fiber cores. A good fiber bundle might have an 85% packing fraction. Fiber bundle technology and its uses will be described in more detail in Chapter 23.

Mid-Infrared Fibers

After years of steady reductions in fiber loss, developers approached the theoretical minimum attenuations for silica glass fibers in the early 1980s. Careful analysis showed that several factors contributed to the attenuation. Basic light scattering processes common to all transparent materials set a lower limit at short wavelengths, but that effect drops off sharply as wavelength increases. (It is proportional to $1/\lambda^4$, so the decline is rapid, but it causes most of the fiber attenuation at 1550 nm.) Silica absorbs longer infrared wavelengths, so the loss of silica fibers increases sharply at longer wavelengths. However, other materials are transparent at longer wavelengths in the mid-infrared, and theorists hoped that some of them might be useful at wavelengths where lower scattering losses would allow much lower attenuation—perhaps as low as 0.001 dB/km. Such extremely low losses would allow incredibly long distances between amplifiers.

Unfortunately, very low-loss infrared fibers have proven exceedingly difficult to make. Purification of the materials is difficult. The raw materials are far more expensive than those for silicate glasses. (Contrary to occasional jokes about Saudi Arabia cornering the raw materials market for optical fibers, conventional optical fibers can't be made from raw sand like some glass products. Fiber materials require extensive purification and processing.) Infrared materials are also harder to pull into fibers because they are much less viscous when molten than silicate glass. The fibers that can be produced are not as strong mechanically as silica glass fibers, and they suffer other environmental limitations. Other approaches appear far more promising for achieving extremely long transmission distances.

On the other hand, non-silica glass fibers can be used for transmitting infrared wavelengths that cannot pass through silica fibers. Two types are available commercially.

Fluorozirconate fibers transmit between 0.8 and 3.8 μm, as shown in Figure 4.17. They are made primarily of zirconium fluoride (ZrF_4) and barium fluoride (BaF_2), with some other components added to the glass. The lowest losses specified for commercial fluorozirconate fibers are about 25 dB/km at 2.6 μm, but losses as low as about 1 dB/km have been reported in the laboratory. Fluoride fibers are vulnerable to excess humidity but can be shielded from the atmosphere.

Figure 4.17 also shows that chalcogenide fibers transmit over a broader range, between about 3.3 and 11 μm, but overall loss is much higher. The lowest loss quoted for commercial fibers is 0.7 decibels per meter at 5.5 μm. Chalcogenide fibers are made of sulfur and selenium compounds, which resist water and acids but are attacked by strong bases.

Fiber Coatings and Strength

So far, this chapter has talked about fibers as if they were made of only two components, the light-carrying core and the cladding surrounding it. In practice, fibers are coated with one or more additional layers, typically plastic, as shown in Figure 4.18. This typically raises overall diameter of fibers with 125 μm cladding to 250 or 500 μm, making them easier to handle.

FIGURE 4.17.

Attenuation of commercial mid-infrared fibers. (Reprinted from *Laser Focus World,* June 1991, p. 149, by permission of PennWell Publishing Co. Copyright 1991 PennWell Publishing Co. Also courtesy of Galileo Electro-Optics Corp.)

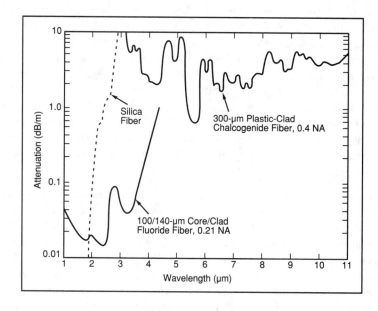

FIGURE 4.18.

Optical fiber with plastic coating layers.

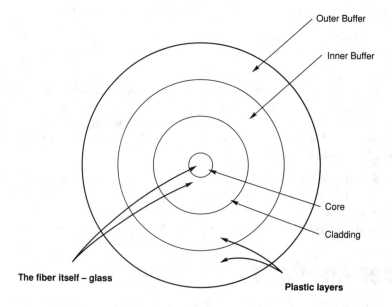

The thickness and precise composition of these coating layers depend on the type of fiber and the application for which it is designed. Their basic functions are to protect the fiber from physical and environmental damage and degradation. Researchers have found that weak points where fibers break develop from tiny microcracks in the surface of the fiber cladding. Coatings protect the fiber from damage that can produce microcracks and provide extra physical strength for the fiber. They also protect against damage from moisture or other environmental factors. They can ease handling, particularly as the fiber is stranded into the cables needed for most applications, which are described in more detail in Chapter 5.

Fibers are coated immediately after they are drawn. Thus, what you see and feel when you handle optical fibers is mostly this outer coating layer, not the cladding that is an optical part of the fiber.

Composition of the outer layers of the fiber also affects strength. Doping with certain materials, notably titanium, can increase resistance to formation of the small cracks that propagate to cause fiber failure.

What Have You Learned?

1. Optical fibers are made by depositing purified materials in a cylindrical preform, which is heated and drawn into fiber.

2. Step-index multimode fibers are simple types with cores 100-3,000 μm in diameter. They collect light efficiently from large-area sources, but modal dispersion usually limits bandwidth to about 20 MHz-km.

3. The number of modes a step-index fiber can transmit depends on its numerical aperture and core diameter as well as the wavelength. Single-mode types have core diameters of about 8-10 μm.

4. Bending can increase fiber losses by letting high-order modes leak out of the core.

5. Step-index multimode fibers come in all-glass, plastic-clad silica, and all-plastic versions of various sizes.

6. Plastic fibers have much higher attenuation than glass fibers.

7. Graded-index multimode fibers offer easier coupling than small-core single-mode fibers and better bandwidth than step-index multimode fibers. The gradation of their refractive index reduces modal dispersion to low levels, giving them a bandwidth of 100-1,000 MHz-km. Core diameters typically are 50-85 μm.

8. Single-mode step-index fibers, with cores about 8-10 μm across, suffer no modal dispersion. The two components of their chromatic dispersion—material and waveguide dispersion—add to zero around 1300 nm, giving them extremely high bandwidth with loss around 0.34-0.5 dB/km.

9. Fiber loss at 1550 nm is much less than at 1300 nm, but dispersion is much higher in standard single-mode fiber. Overall dispersion can be reduced either by using single-wavelength lasers or by making special fibers with dispersion shifted to 1550 nm.

10. Zero-dispersion wavelength of a single-mode fiber is the wavelength where material dispersion and waveguide dispersion add to zero. It is 1300 nm in step-index single-mode fibers but can be shifted to 1550 nm by redesigning the fiber core.

11. Standard single-mode fibers do not distinguish between the two polarizations of input light. Polarization-maintaining fibers preserve the input polarization. Single-polarization fibers transmit only one polarization. Such fibers are used mostly for rotation sensors.

12. Infrared fibers have higher loss than silica fibers at shorter wavelengths.

13. The part of fibers you see is actually a plastic coating, not the cladding of the fiber itself.

What's Next?

In Chapter 5, you will see what happens to fibers when they are put into cables.

Quiz for Chapter 4

1. How many modes would a step-index fiber with a core 100 μm in diameter and a numerical aperture of 0.29 transmit at 850 nm?

   a. 1.

   b. 2.

   c. About 50.

   d. Hundreds.

   e. Thousands.

2. Which of the following is not true for plastic fibers?

   a. They have lowest loss at visible wavelengths.

   b. Both single-mode and multimode forms are available.

   c. They are more flexible than glass optical fibers.

   d. They have much higher attenuation than glass fibers.

3. What diameter are the cores of multimode step-index fiber?

   a. 100 μm.

   b. 200 μm.

   c. 400 μm.

   d. 1000 μm.

   e. All of the above.

4. Light is guided in multimode graded-index fibers by

   a. total internal reflection.

   b. mode confinement in the cladding.

   c. refraction in the region where the core refractive index changes.

   d. the optics that couple light into the fiber.

5. Modal dispersion is highest in which type of fiber?

   a. Step-index multimode.

   b. Graded-index multimode.

   c. Step-index single-mode.

   d. Graded-index single-mode.

6. Which of the following will not reduce the number of modes that an optical fiber can carry?

   a. Reducing core diameter.

   b. Reducing numerical aperture.

   c. Increasing wavelength.

   d. Reducing attenuation.

7. If a fiber has numerical aperture of 0.1, what must its core diameter be less than for it to transmit only a single mode at 1.3 μm wavelength?

   a. 1.55 μm.

   b. 6.5 μm.

   c. 10 μm.

   d. 50 μm.

   e. 100 μm.

8. What makes dispersion zero at 1300 nm in step-index single-mode fibers?

   a. Waveguide and material dispersion cancel each other out.

   b. Chromatic dispersion cancels out modal dispersion.

   c. Waveguide dispersion equals the sum of material and modal dispersion.

   d. Dispersion is zero in all single-mode fibers.

9. Which of the following is needed for high-speed transmission at 1550 nm?

   a. Special fibers with zero dispersion at that wavelength.

   b. Special lasers with extremely narrow linewidth so they experience little dispersion.

   c. New technology to produce fibers with lower attenuation at 1550 nm.

   d. a and b.

   e. a and c.

10. Bandwidths of multimode graded-index fibers are

    a. 20-100 MHz-km.

    b. 100-1,000 MHz-km.

    c. 1-10 GHz-km.

    d. over 10 GHz-km.

    e. Cover entire range.

Cabling

About This Chapter

Cabling is not glamorous, but it is a necessity for virtually all communication uses of fiber optics. A cable structure protects optical fibers from mechanical damage and environmental degradation, eases handling of the small fibers, and isolates them from mechanical stresses that could occur in installation or operation. The cable makes the critical difference in determining whether optical fibers can transmit signals under the ocean or just within the confines of an environmentally controlled office building.

This chapter discusses the major types of fiber-optic cable you are likely to encounter. You will see what cables do, where and why different types are installed, what cables look like on the inside, how cables are installed, and what happens to fibers in cables.

Cabling Basics

Fiber-optic cables resemble conventional metal cables externally, and they use some materials and jacketing technology borrowed from copper wire cables. Polyvinyl chloride (PVC) sheaths are common on both fiber-optic cables and coaxial cables used inside buildings, but fiber cables are typically brightly colored. Polyethylene (PE) is used on both metal and fiber cables to protect against the environmental rigors of underground burial or aerial installation.

UNDERSTANDING

● Fiber-optic cables resemble metal-wire cables but differ because signals are transmitted as light, not electricity.

Some important differences can be subtle. Because optical fibers are not conductive, they do not require electrical insulation to isolate circuits from each other. Optical cables can be made nonconductive by avoiding use of metals in their construction, which produces all-dielectric cables that are immune to ground-loop problems and resistant to lightning strikes. Fiber-optic cables tend to be smaller because one fiber has the same capacity as many wire pairs and because fibers themselves are small.

Some major differences in cable design are necessary because glass fibers react differently to tension than copper wires. Pull on a fiber and it will stretch slightly, then spring back to its original length. Pull a fiber hard enough and it will break (starting at a weak point or surface flaw) after stretching about 5%. Pull a copper wire, applying less stress than you did to break the fiber, and it will stretch by up to about 30% and not spring back to its original length. In mechanical engineering terminology, fiber is elastic (because it contracts back to its original length), and copper is inelastic (because it stays stretched out).

● Fibers must be isolated from tension, which can cause breakage or long-term reliability problems.

Fibers are strong. They can withstand tensions of hundreds of thousands of pounds per square inch of cross-sectional area, or over a giganewton per square meter. (The usual units are thousands of pounds per square inch, kpsi.) Theoretically, fiber strength should reach 2,000 kpsi, stronger than steel, but in practice fibers break at lower tensions, with cracks starting from small surface defects. Applying tension to a fiber can cause formation of tiny surface defects that later lead to fiber failure. Although the strength per unit area is high for fibers, you should keep in mind that a strong tug applies a large force per unit area across the small diameter of a fiber. A standard 125 μm fiber has a cross-sectional area of only 0.000019 square inch, so a 10-pound force applied to the fiber alone corresponds to 500 kpsi.

The fact that fibers tend to break at surface flaws has an important consequence. The longer the fiber, the more likely it is to contain a flaw that will cause breakage at a certain tension. Fiber manufacturers have a simple test to weed out the weakest points in the fiber; they apply a weight to the fiber, which applies a certain tension (called the proof test) along its length. The fiber breaks if it contains any flaws that cannot withstand that tension. Normally all cabled fibers have undergone proof testing.

Cables are designed to isolate the fibers from tension, both during installation and afterwards (such as when they are hanging from poles). Most do so by applying tension to strength members that run the length of the cable, either at the center or in another layer. The strength members may be metallic or nonmetallic, as described later.

Reasons for Cabling

Cabling is the packaging of optical fibers for easier handling and protection. Uncabled fibers work fine in the laboratory and in certain applications such as sensors and the fiber-optic system for guiding missiles, which I will describe later. However, like wires, fibers must be cabled for most communications uses.

Ease of Handling

One reason for cabling fibers is to make them easier to handle. Physically, single glass optical fibers resemble monofilament fishing line, except the fibers are stiffer. Protective plastic coatings raise the outer diameter of fibers to 250-900 μm, but they are still so small that they are hard to handle. They are also transparent enough to be hard to see on many surfaces. Try to pick up one loose fiber with your fingers and you'll soon appreciate one of the virtues of cable.

Cabling also makes multiple fibers easy to handle. Most communication systems require at least two fibers, one carrying signals in each direction. Some require many fibers, and some cables contain hundreds of fibers. Cabling puts the fibers in a single easy-to-see and easy-to-handle structure.

Cables also serve as mounting points for connectors and other equipment used to interconnect fibers. If you take that function too much for granted, try butting two bare fibers together with your hands and finding some way to hold them together permanently.

> Cables make fibers easier to handle.

Protection

STRESS ALONG FIBERS

Another goal of cabling is to prevent physical damage to the fiber during installation and use. The most severe stresses along the length of cables normally come when they are pulled or laid in place. Aerial cables always experience some static stress after installation because they hang from supports. Dynamic stresses applied for short periods can be the most severe, and the most damaging to cables. The worst problems come from contractors with backhoes and other earth-moving equipment, who dig up buried cables, applying sharp forces and snapping the cables. Falling branches can break aerial cables. Cables can isolate the fibers from static stresses by applying the force to strength members. As you saw earlier, fibers are much more vulnerable to excess force than are copper wires, so strength members in fiber cables must resist stretching. However, the cable designer cannot provide absolute protection against careless contractors or heavy falling branches.

> Cables prevent physical damage to fibers during installation and use.

CRUSH-RESISTANCE

Cables must also withstand force applied from the sides. Requirements for crush-resistance differ greatly. Ordinary intrabuilding cables are not made to be walked on, but a few are made for installation under carpets, and military field cables must survive being driven over, requiring that the fibers be embedded in sturdy materials. Figure 5.1 shows the difference in those cable structures. Submarine cables must be capable of withstanding high static pressures underwater—and deep-sea cables must be capable of withstanding the pressure of several kilometers of seawater. Cables buried in the ground must withstand a different type of crushing force applied in a small area: the teeth of chewing gophers, whose front teeth—like those of other rodents—grow constantly. That is one case where the small size of fiber cables can be undesirable: small cables may be just bite-sized for a gopher.

Cables provide the rigidity needed to keep fibers from being bent too tightly. They also help protect fibers from developing tiny microcracks, caused by surface nicks, which can lead to fiber breakage.

FIGURE 5.1.

Cross sections of light-duty intrabuilding cable (top); under-carpet cable (middle); and heavy-duty military field cable (bottom). (Lower two courtesy of Optical Cable Corp., Roanoke, VA, maker of tight buffered fiber cables, and AT&T, respectively)

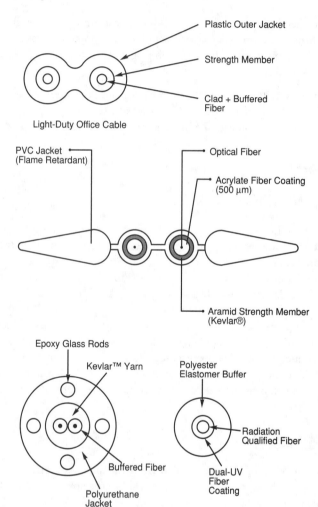

Plastic Outer Jacket

Strength Member

Clad + Buffered Fiber

Light-Duty Office Cable

PVC Jacket (Flame Retardant)

Optical Fiber

Acrylate Fiber Coating (500 µm)

Aramid Strength Member (Kevlar®)

Epoxy Glass Rods

Kevlar™ Yarn

Polyester Elastomer Buffer

Radiation Qualified Fiber

Buffered Fiber

Dual-UV Fiber Coating

Polyurethane Jacket

DEGRADATION

Cabling also protects fibers from more gradual degradation. Long-term exposure to moisture can degrade fiber strength and optical properties. Most cables designed for use in uncontrolled (i.e., outdoor or underground) environments include barriers to keep moisture out. Aerial cables must withstand extremes of temperature—from heating to high temperatures on a hot, sunny day in the summer to freezing in the winter. The combination of cold and moisture presents an added danger—freezing of moisture in the cable. Because water expands when it freezes, it could apply forces on the fiber that produce microbends and increase losses. Cables are designed to prevent the types of degradation important in the environments where they are used; for example, water-blocking materials are used to prevent water from entering loose-tube cables.

A significant long-term concern in some fibers that transmit at 1300 or (particularly) 1550 nm is the possible influx of molecular hydrogen into the fiber. If a fiber is kept in an atmosphere with a large hydrogen content, the tiny hydrogen molecules diffuse throughout the fiber, adding significantly to losses at long wavelengths. Hydrogen is rare in open air, but it can accumulate in some cable structures, for example, by diffusion from or decomposition of certain plastics or by electrolytic breakdown of moisture by electrical currents in the cable (e.g., the power delivered to repeaters in undersea cables). Hydrogen effects were not discovered until many fiber-optic systems had gone into use, and some early cables contained materials that could generate significant quantities of hydrogen. However, new materials have come into use, and heavy-duty cables intended for long-term use are now designed to avoid internal hydrogen build-up. In addition, researchers have learned that fibers containing phosphorous dopants are most vulnerable to hydrogen effects, so fiber vulnerability can be reduced by lowering phosphorous levels.

● Cabling helps protect fibers against degradation caused by moisture.

Types of Cable

The same optical fiber may be used in many different environments, but this is not so for cable. Cables are designed to withstand particular conditions and to provide a controlled environment for the fibers they contain. Thus, choice of a cable design depends on the environment where it is to be installed. To see what is involved, consider where cables may be installed.

● Cables are designed for particular environments.

Types of Environments

The major types of environments for optical cable can be loosely classified as follows:

- Inside devices (e.g., inside a telephone switching system or computer).
- Intraoffice or horizontal (e.g., across a room or under a raised floor in a computer room; often to individual terminals or work groups).
- Intrabuilding (e.g., between walls or above suspended ceilings between offices in a structure; typically between distribution nodes in a building).

- Plenum installations (i.e., through air spaces in a building; must meet building codes).
- Interbuilding or campus links (short exterior connections; link distribution nodes in separate buildings).
- Temporary light-duty cables (e.g., remote news gathering).
- Temporary heavy-duty cables (e.g., military battlefield communications).
- Aerial cables (e.g., strung from utility poles outdoors). May be supported by lashing to support wires or other cables.
- All-dielectric self-supporting cables.
- Cables installed in plastic ducts buried underground.
- Direct-burial cables (i.e., laid directly in a trench or plowed into the ground).
- Submarine cables (i.e., submerged in ocean water or sometimes fresh water).
- Instrumentation cables, which may have to meet special requirements (e.g., withstand high temperatures, corrosive vapors, or nuclear radiation).
- Composite cables, which include fibers and copper wires that carry signals (used in buildings). Note the differences from hybrid cables, below.
- Hybrid power-fiber cables, which carry electric power (or serve as the ground wire for an electric power system) as well as optical signals.

These categories are not exhaustive or exclusive, and some are deliberately broad and vague. Instrumentation, for example, covers cables used to log data collected while drilling to explore for oil or other minerals. Special cables are needed to withstand the high temperatures and severe physical stresses experienced within deep wells. There is some overlap among categories; composite cables, for example, may also be classed as intraoffice cables.

Cable Design Considerations

Each environment has special requirements leading to the design of many types of cables. The basic considerations differ considerably and are summarized below.

- Intradevice cables should be small, simple, and low-cost, because the device protects the cable.

Cables used within buildings must meet fire and electrical codes.

- Intraoffice and intrabuilding cables must meet the appropriate fire and electrical codes. The 1993 National Electrical Code covers fiber and hybrid fiber-copper cables for use within buildings. Underwriters Laboratories' tests cover three grades of intrabuilding cable: general-purpose (UL 1581), risers (UL 1666), and plenum cable (UL 910), described in more detail later. Unlisted cables designed for use outdoors may run for up to 50 feet (15 meters) indoors; if they run greater distances, they must be enclosed in metallic tubing.

● Fiber count depends on the number of terminals served. Individual terminals may be served by a two-fiber duplex cable that looks like the zip cord used for electric lamps. Other types are round or oval in cross-section and may contain up to hundreds of fibers. Multifiber cables often terminate at patch panels or communications "closets" where they connect to cables serving individual terminals.

● Breakout or fanout cables are intrabuilding cables in which the fibers are packaged as single- or multifiber subcables. This allows users to divide the cable to serve users with individual fibers, without the need for patch panels. Figure 5.2 shows an example.

●
Breakout cables are
intrabuilding cables
with fibers packaged
into subcables.

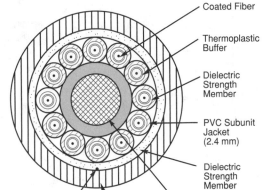

Coated Fiber

Thermoplastic
Buffer

Dielectric
Strength
Member

PVC Subunit
Jacket
(2.4 mm)

Dielectric
Strength
Member

Ripcord   PVC Outer
Jacket

Dielectric
Central
Member

FIGURE 5.2.
Breakout cable.
(Courtesy of Siecor
Corp., Hickory, NC)

● Plenum cables are special intrabuilding cables made of materials that retard the spread of flame and produce little smoke. In this context, the word "plenum" means air-handling spaces, including the space above suspended ceilings and below elevated floors, as well as heating and ventilation ducts. If they are listed with Underwriters Laboratories as meeting the UL 910 specification, they can be run through air spaces without special conduits to contain them. The need for special materials makes the cables relatively expensive, but installation savings offset that extra cost. Their construction is shown in Figure 5.3.

FIGURE 5.3.

Single- and dual-fiber plenum cable. (Courtesy of Math Associates Inc.)

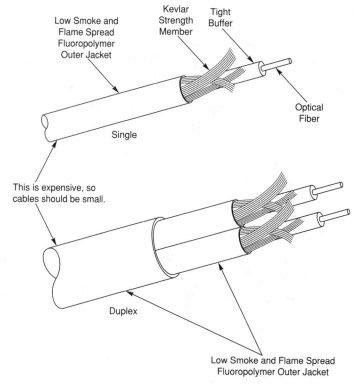

- Composite cables include both fibers and copper wires to deliver different communication services to the same point. For example, the fiber may connect a workstation to a local area network while the wires carry voice telephone service to the same user. Figure 5.4 shows an example.

FIGURE 5.4.

Composite cable contains both copper wires and fibers. (Courtesy of Siecor Corp., Hickory, NC)

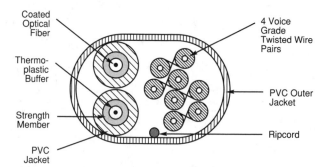

- Temporary light-duty cables are portable and rugged enough to withstand reasonable wear and tear. They may contain only a single fiber (e.g., a video feed from a camera) and should be durable enough to be laid and reused a few times.

● Temporary military cables are made rugged for military use in a field camp, where they must survive considerable abuse, such as being walked and driven across while laid in mud. They are special-purpose cables made to withstand both hostile conditions and unskilled users. Because they are unusual, little attention will be given to them.

● All-dielectric cables are made to be strung aerially from poles outdoors and typically can also be installed in underground ducts. They normally contain multiple fibers, and the internal arrangements of fibers can become elaborate. Figure 5.5 shows two types of aerial installations, which typically use different types of cables. One is the classical suspension of the cable between poles, depending on internal strength members or on a parallel strength member packaged with the fiber in what is sometimes called a "figure-8" cable. The other requires lashing the cable to a messenger wire that runs between poles, or winding the two together. Lashing or winding supports the cable at more frequent intervals and reduces stress applied along its length, which can be large if the only support points are at the poles. Many fiber cables are designed only for lashing, not to withstand the high stress applied by suspension between poles. All such cables have strength members and structures that isolate the fibers from stress. The outer plastic jacket is a material such as polyethylene, which can withstand temperature extremes and intense sunlight. The cable's internal structure is designed to keep moisture out. The all-dielectric construction prevents lightning surges and ground-loop problems.

Outdoor cables can withstand harsher environments than intrabuilding cables but do not meet the same fire and building codes.

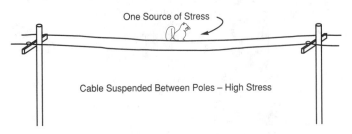

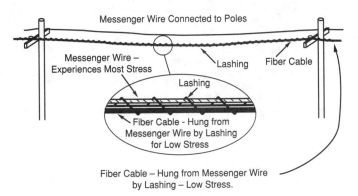

FIGURE 5.5.

Aerial cable installations.

● Armored cables are similar to all-dielectric cables but include an outer armor layer for mechanical protection and to prevent rodent damage. Steel or all-dielectric central members may be used. They can be installed in ducts or aerially, or directly buried underground (which requires extra protection against the demanding environment of dirt). Normally, the armor is surrounded inside and out with polyethylene layers that protect it from corrosion and cushion the inside from bending damage.

● Submarine cables can operate while submerged in fresh or salt water. Those intended to operate over relatively short distances—no more than a few kilometers—are essentially ruggedized and waterproof versions of direct-burial cables. Cables for long-distance submarine use are much more elaborate, as I will describe in Chapter 17. Some parts of submarine cables are buried under the floor of the river, lake, or ocean, largely to protect them from damage by fishermen and boat anchors. The multilayer design used by AT&T, shown in Figure 5.6, can withstand ocean floor pressures.

FIGURE 5.6.

Fiber-optic cable design for transatlantic cable. (Courtesy of AT&T and Simplex Wire & Cable Co.)

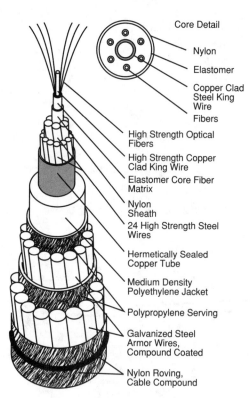

Core Detail

Nylon

Elastomer

Copper Clad Steel King Wire

Fibers

High Strength Optical Fibers

High Strength Copper Clad King Wire

Elastomer Core Fiber Matrix

Nylon Sheath

24 High Strength Steel Wires

Hermetically Sealed Copper Tube

Medium Density Polyethylene Jacket

Polypropylene Serving

Galvanized Steel Armor Wires, Compound Coated

Nylon Roving, Cable Compound

Outside Diameter 51mm (2.010″)

Elements of Cable Structure

As indicated previously, fiber-optic cables are diverse in nature, reflecting the diverse environments cables encounter. However, the same basic elements are used in those different cables.

●
All fiber-optic cables are made up of common elements.

Fiber Housing

One critical concern is the structure that houses individual fibers. Two basic approaches are the loose-tube structure and the tightly jacketed structure, both shown in Figure 5.7. Each has distinct advantages and has earned its own niche—in general, the loose-tube cables outdoors and the tightly buffered design indoors. A third approach is encasing parallel fibers in a plastic ribbon, shown in Figure 5.8. All three structures are made from fibers with protective plastic coatings, applied by the manufacturer.

●
Fibers can be housed in a loose tube, tightly jacketed with a plastic material, or placed in a plastic ribbon structure.

FIGURE 5.7.
Fiber housings.

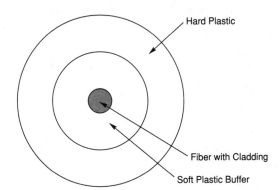

a. **Tightly Buffered Fiber – Fiber Encased in Plastic**

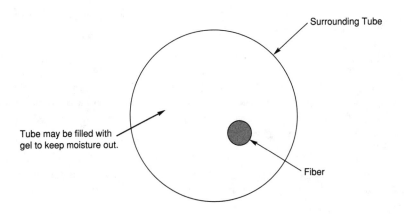

b. **Loose-Tube – Fiber Inside a Hollow Tube**

LOOSE-TUBE CABLE

In the simplest loose-tube design, a single plastic-coated fiber is contained in a long tube, with inner diameter much larger than the fiber diameter. The fiber is installed in a loose helix inside the tube, so it can move freely with respect to the tube walls. This design protects the fiber from stresses applied to the cable in installation or service, including effects of changing temperature. Such stresses can cause bending losses as well as damage the fiber.

There are several variations on the loose-tube approach. Multiple fibers can run through the same tube. The tube does not have to be a physically distinct cylinder running the length of the cable. It can be formed by running grooves along the length of a solid cylindrical structure encased in a larger tube, or by pressing corrugated structures together and running fibers through the interstices. The end result is the same: the fiber is isolated from stresses applied to the surrounding cable structure.

Loose tubes can be used without any filling. However, if they are to be used outdoors, they are normally filled with a jelly-like material. The gel acts as a buffer, keeping out moisture and letting the fibers move in the tube.

Loose-tube cables are filled with a gel for use outdoors.

Loose-tube cables are used outdoors because they effectively isolate the fibers from external stresses such as changes in temperature, preventing damage and resulting in lower fiber loss. They can be installed from poles, in ducts, or by direct burial. A single tube can contain up to a dozen fibers, making it possible to achieve high fiber densities in a compact cable. The cables can be made of flame-retardant materials to meet codes for indoor use, especially where high fiber counts are needed.

TIGHTLY BUFFERED FIBER

A tightly buffered fiber is encased (after coating) in a plastic layer. The coating is a soft plastic that allows deformation and reduces forces applied to the fiber. The surrounding buffer is a harder plastic, to provide physical protection. Tightly buffered fibers may be stranded in conventional cables.

Tightly buffered fibers are typically used indoors.

Tight buffering tolerances assure that the fibers are in precisely predictable positions, making it easier to install connectors. The tight-buffer structure creates subunits that can be divided among many terminals, without using patch panels. Tight-buffer cables are smaller for small fiber counts than loose-tube cables, but the ability to pack many fibers into a single loose tube makes that advantage disappear as the fiber count increases. Above about 36 fibers, loose-tube cables are smaller.

A major advantage of tight-buffered cable for indoor use is its compatibility with fire and electrical codes. (Loose-tube cables may require enclosure in metal tubing.) Although losses are somewhat higher than in loose-tube cables, indoor transmission distances are short enough that it's not a problem.

RIBBON CABLE

The ribbon cable shown in Figure 5.8 is in some ways a variation on the tight-buffered cable. A group of coated fibers is arranged in parallel, then coated with plastic to form a multifiber ribbon. This differs from tightly buffered cables in that one plastic layer encases many parallel fibers. The flat ribbon looks something like flat 4-wire cables used for household telephones. Typical ribbons contain 5 to 12 fibers. Up to 12 ribbons can be stacked together to form the core of a cable.

The simple structure makes a ribbon cable easy to splice in the field; a single splice can connect multiple fibers. Multifiber connectors can also be installed readily. Ribbon cables offer very dense packing of fibers, important for some applications. However, installation can cause uneven strain on different fibers in the ribbon, leading to unequal losses and potential problems with some fibers. Like loose-tube and tight-buffered cables, ribbon cables have their advocates but are not the solution to every fiber-optic cabling problem.

●
Many parallel fibers can be encased in plastic to form a ribbon, around which cables can be built.

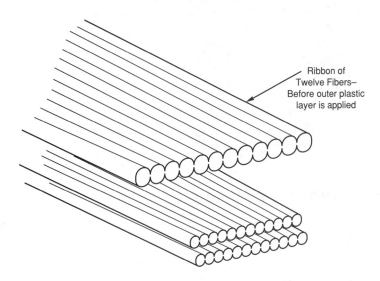

Ribbon of
Twelve Fibers–
Before outer plastic
layer is applied

FIGURE 5.8.
Core of a ribbon cable.

Fiber Arrangements in Cable

Fibers can be arranged in a cable in many different ways. The simplest cables are round with a single fiber at their center. Duplex (two-fiber) cables may either be circular or oval in cross-section or be made like electrical zip cord, with two single-fiber structures bonded together along their length, as in Figure 5.1.

The more fibers in the cable, the more complex the structure. One common cable structure has six buffered fibers wound loosely around a central member. The buffered fibers are wound so that they don't experience torsion in the cable. In loose-tube cables, the fiber count can be raised by putting multiple fibers in each tube. Groups of 8 or 12 fibers may also be wound around a central member.

Cables with more fibers are built up of modular structures. For example, a 36-fiber cable can be made from six loose-tube modules containing six fibers each, or from three 12-fiber ribbons. A dozen 12-fiber ribbons make a 144-fiber cable. Putting 12 fibers in each loose tube and adding a second ring of 12 loose tubes gives a 216-fiber cable, as shown in Figure 5.9. Design details depend on the manufacturer.

High fiber counts were rare in early installations, but they are becoming popular as fibers are used to distribute signals to many subscribers in metropolitan areas and large buildings or campuses. Cables with several hundred fibers are in use, and developers are working on cables containing thousands of fibers.

● 216-fiber cables are in use, and cables with many more fibers are in development.

FIGURE 5.9.

Modular cable containing 216 fibers, with 12 in each of 18 loose tubes. (Courtesy of Siecor Corp., Hickory, NC)

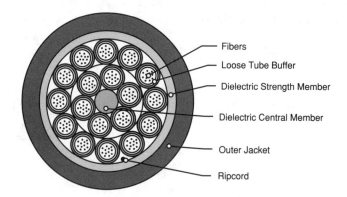

- Fibers
- Loose Tube Buffer
- Dielectric Strength Member
- Dielectric Central Member
- Outer Jacket
- Ripcord

Other Structural Elements

Fibers and their buffers are not the only structural elements of cables. Many—but not all—fiber-optic cables include other components to provide strength and rigidity.

● Many cables contain central members to make them rigid and strength members to withstand tensile forces.

Many cables are built around central members made of steel or fiberglass. These run along the center of the cable and provide the rigidity needed to keep it from buckling, as well as a core to build the cable structure around. It may be overcoated with plastic or other material to match cable size requirements and to prevent friction with other parts of the cable. Small indoor cables containing few fibers generally lack central members, but they are common in outdoor cables, and in indoor cables with high fiber counts.

Strength members in general are distinct from central members. Their role is to provide tensile strength along the length of the cable during and after installation. The usual strength members are strands of dielectric aramid yarn (better known under the trademark Kevlar) wound around the core of the cable. In some cases, tight-buffered fibers may be wound around them to form subunits of the cable. When a cable is pulled into a duct, the tension is applied directly to the strength member.

The structure containing the fibers normally surrounds the central member and, in turn, is surrounded by the strength member and one or more outer jacketing layers. For light-duty cables, the typical jacketing material is polyvinyl chloride (PVC) or polyurethane; polyethylene is often used in heavier-duty and outdoor cables because of its better durability. Electrical codes require fire-retardant materials in indoor cables.

Underwater and buried cables are among the types that require one or more layers of protecting armor. Typically for buried cables, steel is wound around an inner plastic sheath. An outer plastic sheath is then applied over the armor to prevent corrosion. The metal armor helps protect against crushing damage, rocks, and rodents. Underwater cables in shallow waters may have multiple layers to protect against damage from shipping and fishing operations, as shown in Figure 5.6.

● Buried and underwater cables require armor.

Cable Installation

Cable installation is not as simple as it sounds, but reliable methods developed for conventional metal cables have been successfully adapted for optical cables. The methods chosen depend on the type of installation.

● Special techniques are used to install different types of cable.

- Submarine cables are laid from special ships built for that purpose.

- Buried cables are normally installed by digging a deep, narrow trench with a cable plow, laying the cable in the trench, and filling the trench with dirt, or by plowing the cable directly into the ground.

- Cables are installed in ducts by threading a pull line through the duct, attaching it to the cable, then pulling the cable through the ductwork. Manholes or other access points are normally available along the duct route, so the cable need not be pulled all at once through a long route.

- Self-supporting aerial cables may be suspended directly from overhead poles. Other aerial cables can be suspended from messenger wires, strong steel wires strung between poles. If a messenger wire is used, the cable is lashed to it with a special lashing wire running around both the cable and the messenger wire, or sometimes wound around it. This is a common installation for many overhead fiber cables because it minimizes strength requirements.

- Plenum cables are strung through interior air spaces.

- Interior cables may be installed within walls, through cable risers, or elsewhere in buildings. Only special cables designed for installation under carpets should be laid on the floor where people walk.

- Temporary light-duty cables are laid by people carrying mobile equipment that requires a broad-band (typically video) connection to a fixed installation.

- Temporary military cables may be laid by helicopters from the air or by soldiers on the ground, during field exercises, in preparation for engagements, or in actual battle. Typically they are unreeled from cable spools.

Changes in Cabled Fiber

Microbending can cause fiber loss to change after cabling.

Ideally, characteristics of optical fibers should not change when they are cabled, but in practice some changes do occur, particularly in attenuation. A major cause of these changes is microbending, which depends on the fiber's local environment and the stresses applied to it. Tight buffering affects single-mode fiber more than multimode fiber, but in properly made cables the increase in loss is small.

Cabled fibers rarely suffer physical damage unless the entire cable is damaged. Fiber manufacturers apply a stress test to fibers before they leave the factory. The test is a simple one in which a series of pulleys and wheels apply a given stress to the fiber. If the fiber fails the test, it breaks and, thus, cannot be used. These proof tests assure levels of fiber strength that meet normal cable requirements.

Causes of Cable Failure

Most cable failures are a result of physical abuse.

Telephone companies were very cautious before beginning their massive switchover to fiber-optic cables and conducted extensive tests and field trials to evaluate the reliability of optical cables. These studies have shown that optical fibers have excellent reliability. Proof testing of fibers before they are cabled assures that mechanical strength is high, and proper splicing produces splices about as strong as unspliced fiber. Poor installation can damage fibers, but failures of properly installed fiber-optic cables are extremely rare.

Virtually all fiber-optic cable failures are due to physical abuse. The archetypical problem is a backhoe digging up and breaking a buried cable. Aerial cables are broken by falling branches or errant cranes. High-capacity cables have been cut mistakenly. Fibers in light-duty indoor cables could be broken by slamming doors or windows on the cable, although the cable might not show serious damage. Applying a sharp stress along a short indoor cable (e.g., tripping over it) is not likely to break the cable. However, it could jerk the cable out of a connector at one end. In general, connector junctions are the physical weak points of short-distance cables.

If the cable itself breaks, the fibers in it will also break. Because fibers tend to break at weak points, they may not break at precisely the point of the cable break but should break close to it. However, fibers can remain intact despite evident physical damage that does not completely sever the cable.

What Have You Learned?

1. Fibers can break at inherent flaws and develop microcracks under tension, so they must be protected from stretching forces.

2. Cabling packages fibers for protection and easier handling.

3. Cables must resist crushing as well as isolate the fiber from tension along its length.

4. Cable design depends on the environment where the cable will be used.

5. Elements of cable structures include housing for the fiber, strength members, jacketing, and armor.

6. Fibers can be enclosed in a loose tube, a tight plastic buffer, or a ribbon.

7. Physical arrangement of fibers in the cable depends on the number of fibers.

8. Most damage to cables is by application of sudden stresses, such as digging up buried cable.

What's Next?

In Chapter 6, we will examine the light sources used with fiber-optic cables.

Quiz for Chapter 5

1. What happens if an optical fiber is pulled along its length?

    a. It stretches out and does not return to its original length.

    b. It can stretch by about 5% before breaking.

    c. Its length is unchanged until it breaks.

    d. It breaks if any force is applied to it.

2. Cables cannot protect fibers effectively against

    a. gnawing rodents.

    b. stresses during cable installation.

    c. careless excavation.

    d. static stresses.

    e. crushing.

3. Light-duty cables are intended for use

    a. within office buildings.

    b. in underground ducts.

    c. deep underground where safe from contractors.

    d. on aerial poles where temperatures are not extreme.

4. The special advantages of plenum cables are what?

    a. They are small enough to fit in air ducts.

    b. They meet stringent fire codes for running through air spaces.

    c. They are crush-resistant and can run under carpets.

    d. They have special armor to keep rodents from damaging them.

5. Aerial cables are not used in which of the following situations?

   a. Suspended overhead between telephone poles.

   b. Tied to a separate messenger wire suspended between overhead poles.

   c. Inside air space in office buildings.

   d. Pulled through underground ducts.

6. A loose-tube cable is

   a. a cable in which fibers are housed in hollow tubes in the cable structure.

   b. a cable for installation in hollow tubes (ducts) underground.

   c. cable for installation in indoor air ducts.

   d. none of the above.

7. Which of the following are usually present in direct-burial cables but not in aerial cables?

   a. Strength members.

   b. Outer jacket.

   c. Armor.

   d. Fiber housing.

8. Which type of cable installation requires pulling the cable into place?

   a. Direct burial.

   b. Underground duct.

   c. Military field systems.

   d. Submarine cable.

9. The main cause of differences in properties of a fiber before and after cabling is

   a. microbending.

   b. temperature within the cable.

   c. application of forces to the fiber.

   d. damage during cabling.

10. The major reason for failure of cabled fiber is

   a. hydrogen-induced increases in attenuation.

   b. corrosion of the fiber by moisture trapped within the cable.

   c. severe microbending losses.

   d. physical damage to the cable.

Light Sources

About This Chapter

Many types of light sources are used in fiber-optic systems, from cheap LEDs directly driven by signal sources to sophisticated narrow-line lasers with external modulators. Some operate at telephone-like speeds and bandwidths over several meters; others send hundreds or thousands of megabits per second through tens of kilometers of fiber. All these light sources generate the signals transmitted through the fiber.

Light sources are actually parts of transmitters, but to approach the subject systematically I am dividing the two. This chapter covers the various light sources used in fiber-optic transmitters, which are covered in more detail in Chapter 7. In Chapter 8, you will look at receivers, which convert the optical signals back into electronic form at the other end of the fiber.

Light Source Considerations

Several important factors enter into selecting a light source for a fiber-optic system. The light must be at a wavelength transmitted effectively by the optical fiber, usually the 850, 1300, or 1550 nm windows for glass fibers. The range of wavelengths is also important, because the larger the range, the larger the potential for dispersion problems. The light source must also generate adequate power to send the signal through the fiber, but not so much power that it causes nonlinear effects or distortion in the fiber or receiver. The output light must be modulated so that it carries the signal. The light source must also transfer its output effectively into the fiber.

● Light source wavelength, modulation, spectral width, size, and power are all important for fiber-optic systems.

The main light sources used with fiber-optic systems are semiconductor devices—light-emitting diodes (LEDs) and semiconductor lasers (often called diode lasers). As semiconductor devices, they fit well with standard electronic circuitry used in communication systems. Certain other compact lasers can also be used in special cases. I'll look more closely at the light sources after a quick overview of key operational characteristics important for fiber-optic communications.

Operating Wavelength

● Source wavelength affects signal attenuation and pulse dispersion.

Source wavelength affects both the attenuation and pulse dispersion that signal experience in fibers. As you saw in Chapter 4, both attenuation and pulse dispersion are functions of wavelength. The usual transmission windows are 780-850, 1300, and 1550 nm in silica fibers and around 660 nm in plastic fibers, although other wavelengths can be used.

The spectral width, or range of wavelengths emitted, also affects pulse dispersion, which increases with the wavelength range. Spectral width is one of the major differences between LEDs and the narrower-line laser sources, as shown in Figure 6.1.

FIGURE 6.1.

Comparison of LED and laser spectral widths.

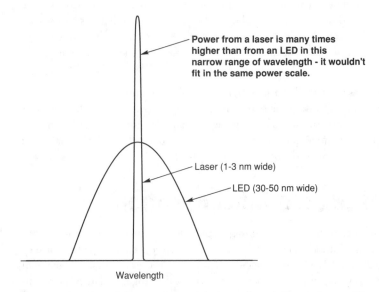

Power from a laser is many times higher than from an LED in this narrow range of wavelength - it wouldn't fit in the same power scale.

Laser (1-3 nm wide)

LED (30-50 nm wide)

Wavelength

The emitted wavelength and the spectral width depend on different factors. The emitted wavelength depends on the semiconductor material from which the light source is made, as described later. The spectral width depends on device structure. A laser and LED made of the same material have the same center wavelength, but the LED has a much broader spectral width. Two lasers with the same structure made of different materials have comparable spectral widths but different center wavelengths.

Output Power and Light Coupling

Power from communication light sources can range from more than 100 milliwatts for certain lasers to tens of microwatts for LEDs. Not all that power is useful. For fiber-optic system applications, the relevant value is the power delivered into an optical fiber. That power depends on the angle over which light is emitted, the size of the light-emitting area, the alignment of the source and fiber, and the light collection characteristics of the fiber, as shown in Figure 6.2. The light intensity is not uniform over the entire angle at which light is emitted but rather falls off with distance from the center. Typical semiconductor lasers emit light that spreads at an angle of 10-20°; the light from LEDs spreads out at larger angles.

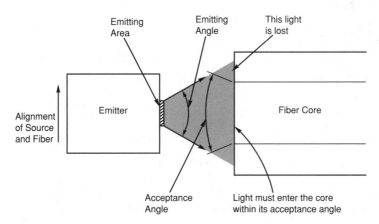

FIGURE 6.2.

Light transfer from an emitter into a fiber.

Losses of many decibels can easily occur in coupling light from an emitter into a fiber, especially for LEDs with broad emitting areas and wide emitting angles. This makes it important to be sure you know what power is specified.

Modulation

One important advantage of semiconductor lasers and LEDs is that their output power varies directly with the input current. This is called direct modulation. Other lasers require external modulators, which change their transparency in response to an applied signal.

●
Diode lasers and LEDs are modulated directly.

Several variables are important in modulation. One is speed; lasers are faster than LEDs. Another is how modulation affects the light signal; as you will see later, direct modulation can increase the spectral width of diode lasers. The linearity of the modulation is also important, especially for analog communications. Depth of modulation and modulation format are also important. Although I will cover some of these topics later in this chapter, many details of modulation belong in the realm of transmitters in Chapter 7.

Cost/Performance Trade-offs

As any student of engineering reality would expect, light sources with the most desirable characteristics cost the most. The cheapest light sources are LEDs with slow rise times, large emitting areas, and relatively low output power. Diode lasers that emit the 1300 and 1550 nm wavelengths where optical fibers have their lowest losses are the most expensive. The higher-power and narrower-line emission of lasers comes at a marked price premium. The only real performance advantage of LEDs is generally longer lifetime than lasers, which are driven harder during operation.

LED Sources

An LED emits light when a current flows through it.

LEDs that emit invisible near-infrared light are common light sources for short fiber systems. The basic concept of a light-emitting diode is shown in Figure 6.3. A small voltage is applied across a semiconductor diode, causing a current to flow across the junction. The diode is made up of two regions, each doped with impurities to give it the desired electrical characteristics. The p region is doped with impurities having fewer electrons than atoms they replace in the semiconductor compound, which creates "holes" where there is room for electrons in the crystalline lattice. The n region is doped with impurities that donate electrons, so extra electrons are left floating in the crystalline matrix. Applying a positive voltage to the p region and a negative voltage to the n region causes the electrons and holes to flow toward the junction of the two regions, where they combine (the process is actually called recombination). As long as the voltage is applied, electrons keep flowing through the diode and recombination continues at the junction.

In many semiconductors, notably silicon and germanium, the released energy is dissipated as heat—vibrations of the crystalline lattice. (The recent discovery of light emission from porous silicon appears to be a special case that depends on the microstructure of the silicon crystal; it has yet to find practical use.) However, in other materials, usable in LEDs, the recombination energy is released as a photon of light, which can emerge from the semiconductor material. The most important of these semiconductors, gallium arsenide and related materials, are made up of elements from the IIIa and Va columns of the periodic table:

IIIa	Va
Aluminum (Al)	Nitrogen (N)
Gallium (Ga)	Phosphorous (P)
Indium (In)	Arsenic (As)
	Antimony (Sb)

The wavelength emitted depends on the semiconductor's internal energy levels. In a pure semiconductor at low temperature, all the electrons are bonded within the crystalline lattice. As temperature rises, some electrons in this valence band jump to a higher-energy conduction level, where they are free to move about in the crystal. The valence and conductor bands are separated by a void where no energy levels exist—the band gap that gives semiconductors many of their special properties.

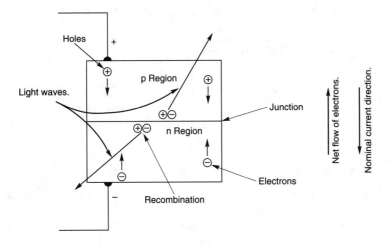

FIGURE 6.3.

LED operation.

a. **Basic Idea of LED**

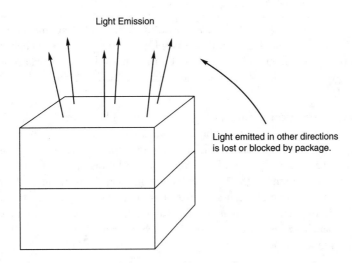

b. **LED as Packaged (surface-emitting)**

Conduction-band electrons leave behind a hole in the valence band, which is considered to have a positive charge. This hole can move about, as electrons from other spots in the crystalline lattice move to fill in the hole and leave behind their own hole (i.e., the hole moves from where the electron came to where the electron was). Impurity doping of semi-conductors can also generate free electrons and holes. When an electron drops from the conduction level to the valence level (i.e., when it recombines with a hole), it releases the difference in energy between the two levels, as shown in Figure 6.4.

FIGURE 6.4.

Semiconductor energy levels.

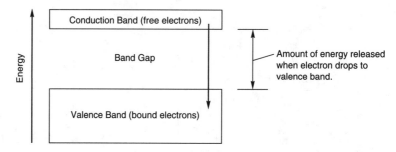

The band-gap difference between the energy levels—and hence the amount of energy released and the emitted wavelength—depends on the composition of the semiconductor. The usual LEDs used in fiber-optic systems are made of gallium aluminum arsenide or gallium arsenide. Gallium arsenide LEDs emit near 930 nm. Adding aluminum decreases the threshold current to improve lifetime and can also increase the energy gap and shift emission to shorter wavelengths of 750 to 900 nm. The usual wavelengths for fiber-optic applications are 820 or 850 nm. At room temperature, the typical 3 dB bandwidth of an 820 nm LED is about 40 nm.

> ●
> The usual LED wavelengths for glass fibers are 820 or 850 nm.

Other semiconductor compounds can be used to make LEDs that emit different wavelengths. Gallium-arsenide-phosphide (GaAsP) LEDs emitting visible red light at 665 nm are used with plastic fibers, which are most transparent in the red and transmit poorly at GaAlAs wavelengths. GaAsP LEDs cost less than GaAlAs LEDs but are lower in performance.

> ●
> Visible LEDs are used with plastic fibers, which transmit poorly in the infrared.

The most important compound for high-performance fiber optics is InGaAsP, made of indium, gallium, arsenic, and phosphorous mixed so the number of indium plus gallium atoms equals the number of arsenic plus phosphorous atoms. The resulting compound is written as $In_xGa_{1-x}As_yP_{1-y}$, where x is the fraction of indium and y the fraction of arsenic. These so-called "quaternary" (four-element) compounds are harder to make than "ternary" (three-element) compounds such as GaAlAs but are needed to produce output at 1300 and 1550 nm. In practice, LEDs are often used for short systems at 1300 nm, where conventional fibers have low chromatic dispersion, but are rarely used at 1550 nm, where dispersion is much higher.

> ●
> InGaAsP LEDs emit at 1300 and 1550 nm.

Other LED characteristics depend on device geometry and internal structure. The description of LEDs so far hasn't indicated in which direction they emit light. In fact, simple LEDs emit light in all directions, as shown in Figure 6.3, and are packaged so most emission comes from their surfaces. The light is emitted in a broad cone, with intensity falling off roughly with the cosine of the angle from the normal to the semiconductor junction. (This is called a Lambertian distribution.)

More complex internal structures can concentrate output of surface-emitting LEDs in a narrower angle, by means such as confining drive current to a small region of the LED. Such designs typically require that the light emerge through the substrate, which can lead to transmission losses. One way to enhance output and make emission more directional is to etch a hole in the substrate, to produce what is called a Burrus diode, after its inventor Charles A. Burrus of AT&T Bell Laboratories. A fiber can be inserted into the hole to collect light.

A fundamentally different configuration is the edge-emitting diode, shown in Figure 6.5. Electrical contacts cover the top and bottom of an edge emitter, so light cannot emerge there. The LED confines light in a thin, narrow stripe in the plane of the p/n junction. This is done by surrounding that stripe with regions of lower refractive index, creating a waveguide that functions like an optical fiber, and channeling light out both ends where it can be coupled into a fiber. One disadvantage is that this increases the amount of heat the LED must dissipate.

● *An edge-emitting diode emits light from its ends.*

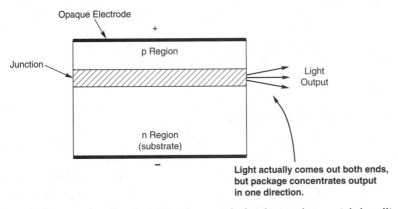

FIGURE 6.5.

An edge-emitting LED.

In general, the more complex the LED structure, the brighter and more tightly collimated is the emitted light. Concentration of the emitting area and the region through which current passes also decreases rise time and, thus, enhances possible modulation bandwidth. Of course, as with other devices, the greater complexity comes at higher cost.

Semiconductor Laser Sources

Semiconductor lasers are superficially like LEDs, but they produce light in a different way that results in higher output powers and more directional beams. Understanding their operation requires a quick explanation of laser physics. I will concentrate initially on the semiconductor diode lasers that are most widely used in fiber-optic communications, then describe other lasers sometimes used in fiber-optic systems later in this chapter.

Stimulated Emission

Laser emission is stimulated.

Light is emitted when something (e.g., an electron in a semiconductor) drops from a high energy level to a lower one, releasing the extra energy. Normally, it emits light without outside influence, in what is called spontaneous emission, but it does not do so the instant it is first able to. It takes its time to get around to spontaneous emission. Suppose, however, that the electron is sitting in the upper energy level waiting to emit its extra energy, and another photon comes along, with just the amount of energy the electron needs to emit. That external photon can stimulate the electron in the upper energy level to drop to the lower one and emit its energy as light of the same wavelength. The result is a second identical photon. The process is called Light Amplification by the Stimulated Emission of Radiation, and the acronym spells *Laser*.

Population Inversion

Special conditions are needed for laser emission. One requirement is that there be more electrons (or atoms or molecules) in the upper energy level than in the lower one. Specialists call this a population inversion because normally more electrons are in lower levels. This is necessary because whatever is in the lower energy level can absorb the emitted light. If more things are in the lower level than in the upper level, they would absorb the light faster than it could be emitted. However, if only that condition is met, the stimulated emission will go off in every direction, more like a light bulb than what we think of as a laser.

Laser Beam Formation

A laser beam is formed by a resonator that reflects light back and forth through the laser medium.

The laser beam is formed by a resonator, which confines light and makes it pass again and again through the excited medium. As shown in Figure 6.6 for a semiconductor laser, this resonator can be a pair of mirrors, one at each end of the region occupied by excited atoms (the recombination layer in a diode laser). Light emitted straight toward one mirror will be reflected back and forth, stimulating emission from electrons ready to recombine as the light passes through the junction plane. Light emitted in other directions will leak away. Thus, only the light traveling back and forth along the narrow stripe of the active region will be amplified and build up into a beam.

A closer look shows that the process is a little more complex. The mirrors are really the ends of the semiconductor crystal, called facets. The rear facet usually lets some light escape (which is often collected by a detector that monitors laser output power), but it may be coated to reflect most light back into the laser. In practice, the front facet is left uncoated so that most light escapes but some is reflected back into the semiconductor. Semiconductors are such strong stimulated emitters that this is all the feedback that is needed to produce laser action.

Some LED materials are not suitable for use in semiconductor lasers. However, all semiconductor laser materials can be operated as LEDs.

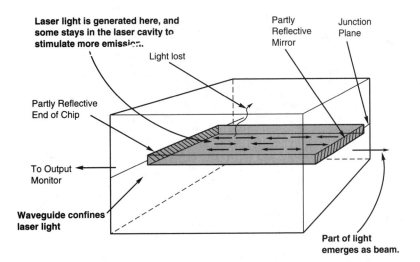

Laser light is generated here, and some stays in the laser cavity to stimulate more emission.

Light lost

Partly Reflective Mirror

Junction Plane

Partly Reflective End of Chip

To Output Monitor

Waveguide confines laser light

Part of light emerges as beam.

FIGURE 6.6.

Basic operation of a semiconductor laser.

Functional Differences

There are two important functional differences between LEDs and diode lasers. One is that LEDs lack reflective facets and, in fact, may be designed to minimize reflection back into the semiconductor. The other is that lasers must operate at higher drive currents to get the high density of ready-to-recombine electrons needed at the p/n junction.

The output of a semiconductor laser depends on the drive current passing through it, as long as the bias voltage is above the minimum required (the band-gap energy). At low currents, the laser emits feeble spontaneous emission, operating as an inefficient LED. However, as drive current passes a threshold value, the device shifts over to laser emission, and output rises steeply, as shown in Figure 6.7. For most diode lasers, LED emission is so weak it can be ignored.

These internal differences lead to some important functional differences. Lasers convert electrical input power to light more efficiently than LEDs and also have higher drive currents, so lasers are much more powerful than LEDs. The concentration of stimulated emission leads to a beam narrower than from an LED (although semiconductor laser beams are broad by laser standards). The higher drive currents and optical power levels make laser lifetimes shorter than LEDs. The amplification inherent in laser action tends to concentrate emission in a much narrower spectral width than LED output, normally about a couple of nanometers. In essence, the center of the emission curve is amplified much more than the fringes, making the laser curve much more steeply peaked, as shown in Figure 6.1.

Lasers are much more powerful than LEDs and emit a narrower range of wavelengths.

FIGURE 6.7.

Laser emission.

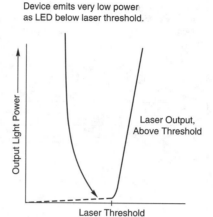

Device emits very low power
as LED below laser threshold.

Output Light Power →

Laser Output,
Above Threshold

Laser Threshold

Drive Current →

Structural Differences

Semiconductor lasers have more complex structures than LEDs, reflecting the more de-manding conditions for successful laser operation. The first semiconductor lasers were made in 1962 using simple structures, but they required liquid-nitrogen cooling and were se-verely limited in lifetime and duty cycle. Developers soon found ways to concentrate drive current and light generation so the laser would work better. The first step was the single-heterojunction laser in which a "heterojunction" between semiconductor materials with slightly different refractive indexes helped confine light on one side of the junction. The next step was the double-heterojunction laser, where a pair of heterojunctions formed a sandwich that confined the emitted light to the central junction region (the active layer). Further refinements confined light horizontally as well as vertically by concentrating the drive current and optical power in a narrow stripe running the length of the chip.

Lasers emit light
from a small region
similar in size to
the cores of
multimode fibers.

Most refined designs confine laser emission to a small region. State-of-the-art lasers have active stripes just a few micrometers across and a fraction of a micrometer high. Like an optical fiber, the stripe functions as a waveguide. Because of its narrow dimensions, it is a single-mode waveguide. Light is emitted from a region at the end of the chip the same size as the cross section of the active area. In state-of-the-art lasers, that region is nicely matched to the cores of single-mode fibers. The beam spreading or divergence is comparable to the acceptance angle of fibers.

Wavelengths

The output wavelengths of diode lasers depend on composition of the junction layer, like those of LEDs. The primary compositions used in diode laser light sources for fiber optics are variations on the standard III-V semiconductor compounds that can be fabricated on substrates of gallium arsenide or indium phosphide:

- $Ga_{(1-x)}Al_xAs$ on GaAs for 780 to 850 nm
- $In_{0.73}Ga_{0.27}As_{0.58}P_{0.42}$ on InP for 1310 nm
- $In_{0.58}Ga_{0.42}As_{0.9}P_{0.1}$ on InP for 1550 nm

Other InGaAsP mixtures are used for other wavelengths between about 1100 and 1600 nm. InGaAs on GaAs can generate 980 nm, a wavelength used as a power source for erbium-doped fiber amplifiers described briefly later in this chapter. The major constraints are the need for the proper band-gap in the active layer and for interatomic spacings reasonably close to those of readily available substrates (a restriction that has been relaxed recently with the development of strained-layer structures).

OUTPUT SPECTRUM

Diode lasers have a much narrower spectral width than LEDs, allowing the use of standard diode lasers to carry high-speed signals through fiber with low dispersion at the laser wavelength. However, the linewidth of 1 to 3 nm is large enough to cause dispersion problems if the nominal zero-dispersion wavelength of single-mode fiber does not match that of the laser. The main concern is with the use of 1550 nm lasers and step-index single-mode fibers with nominal zero dispersion at 1300 nm. Carrying high-speed signals requires lasers that emit a much narrower spectral width.

The output spectrum of a conventional narrow-stripe semiconductor laser is shown in Figure 6.8. Although much of the power is concentrated at one wavelength, laser oscillation also produces other wavelengths. The only wavelengths that are amplified in the laser are those for which the round-trip distance between the mirrors is an integral number of wavelengths λ:

$$2D=N\lambda$$

where 2D is the round-trip distance and N is an integer. In addition, the wavelength λ must be within the laser's gain curve, the range of wavelengths where the laser light can be amplified by stimulated emission, which is much broader than the individual peaks shown in Figure 6.8. The result is a series of narrow-wavelength spikes, called longitudinal (for along the length of the laser) modes. Unfortunately, the range of wavelengths covered by those multiple longitudinal modes is large enough for significant dispersion to accumulate at 1550 nm in step-index single-mode fibers, seriously limiting transmitted bandwidth.

SINGLE-FREQUENCY LASERS

To restrict oscillation to a single longitudinal mode, researchers have developed laser resonators more elaborate than simple facets on the ends of a semiconductor chip. Three leading approaches are shown in Figure 6.9. One is the distributed feedback laser, in which a series of corrugated ridges on the semiconductor substrate (replacing the mirrored end facts)

⬤
Single-frequency lasers are needed in fibers with zero-dispersion wavelength that does not match laser output.

⬤
Distributed feedback and external cavity lasers emit only a single frequency.

reflect only certain wavelengths of light back into the laser, and light at only one resonant wavelength is amplified. The distributed Bragg reflection laser works in much the same way, but the grating is etched outside the part of the laser that is pumped by electric current. In the external cavity laser, one or both facets are coated to suppress reflection, and one or two external mirrors are added around the chip to form a long resonator that limits oscillation to a single wavelength.

FIGURE 6.8.

Wavelengths in multiple longitudinal modes.

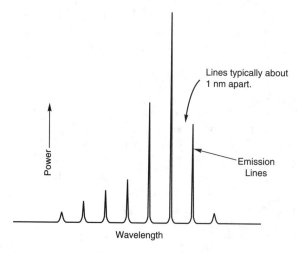

Lines typically about 1 nm apart.

Emission Lines

Power

Wavelength

External modulation of lasers is needed to achieve the best possible high-speed modulation.

Even further refinements are needed to achieve the utmost in fiber-optic transmission. The problem comes from directly modulating the laser by changing the drive current passing through it. Although that makes diode lasers much easier to modulate than other types, it also subtly shifts their wavelength. As the density of electrons in the semiconductor changes, so does the refractive index of the material, effectively changing the optical length of the laser cavity (D = nL, where n is the refractive index and L is the physical separation of the cavity mirrors). From the earlier equation for the wavelength at which the laser resonates, you can see that this means the wavelength λ changes by an amount $\Delta\lambda$:

$$\Delta\lambda = 2(\Delta n \times L)/N$$

where Δn is the change in refractive index and N is an integer, the number of wavelengths in a round-trip of the cavity. Although that change is small, it occurs during the laser pulse, causing each pulse to contain a range of wavelengths. The different wavelengths travel at slightly different velocities in a high-dispersion fiber, limiting performance.

FIGURE 6.9.

Three single-frequency lasers.

Grating limits emission
to one frequency.

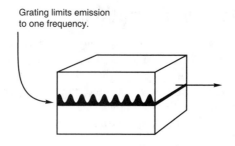

a. Distributed Feedback Laser

Drive current only
through this region

Active
Layer

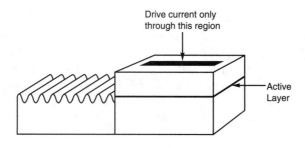

b. Distributed Bragg Reflection Laser

Facet coated to
prevent reflection

Diode Laser
(Size exaggerated)

Totally
Reflecting
Mirror

Output Beam

Output Facet

Length of Laser Cavity

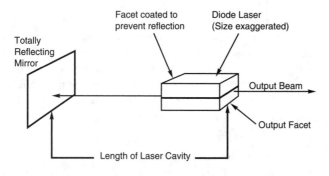

c. External Cavity Laser

Lifetimes

The output power of semiconductor lasers declines with age. Although lifetimes have improved greatly, aging and increasing temperature can cause laser output at a steady drive current to drop slowly. Eventually, power drops to a point where the laser no longer meets system needs and is pronounced "dead," although it may still emit some light. To deal with this problem, sophisticated laser transmitters include circuits to stabilize light output and temperature, as you will see in Chapter 7.

⬤
Output power of
semiconductor
lasers drops
with age.

Semiconductor Laser Amplifiers

You saw earlier that lasers work by amplifying stimulated emission. It is possible to use a semiconductor laser to amplify optical signals that originate outside it. The ends of the laser are coated to suppress the reflection of light back into the semiconductor chip. Light from an external laser or fiber is then directed into the junction layer, where stimulated emission amplifies it. The amplified signal then emerges from the other face of the diode amplifier. I will talk more about optical amplifiers in Chapter 9.

Fiber Amplifiers and Lasers

Semiconductor lasers are only one of several types of lasers. All work by amplifying stimulated emission, but they do so in different ways. One important family consists of glasses or crystals doped with small quantities of elements that can be excited by light from an external source, then stimulated to emit light at a longer wavelength. (Laser specialists often call these "solid-state" lasers, but in the laser world solid-state is not synonymous with semiconductor.) Traditionally, the laser material was made in a rod shape, but it can also be made in the form of a fiber, with the core doped with the light-amplifying impurity. This doped fiber can serve either as a laser or as an amplifier (described in more detail in Chapter 9).

Figure 6.10 shows the basic operation of one type of fiber amplifier, the erbium-doped fiber amplifier (often called an EDFA) at 1550 nm. Light from an external diode laser emitting at 980 or 1480 nm excites erbium atoms in the fiber. Those specific wavelengths are needed to excite the erbium; other wavelengths do not transfer their energy to the erbium atoms. When a weak signal at 1550 nm enters the fiber, those light waves can stimulate the erbium atoms to release their stored energy as additional 1550 nm light waves. The process continues as the signal makes a single pass through the fiber, building a stronger and stronger signal. Typical gains are 20 to 40 decibels, a factor of 100 to 10,000. Output powers can exceed 100 mW. These and other properties make the erbium-doped fiber amplifier nearly ideal as an optical amplifier.

Fiber amplifiers work better than semiconductor laser amplifiers and have many characteristics that make them very attractive for use in fiber-optic systems. As single-mode fibers, they are easy to connect to other single-mode fibers with minimal loss. They have very low signal distortion; the pulses that come out are quite similar to those that go in. Noise and crosstalk are also low, and fiber amplifiers are not sensitive to polarization of the input light. They respond very rapidly to input changes. Signals can pass through hundreds of them in series—separated by tens of kilometers of fiber—and still be recognizable at the end.

Erbium-fiber amplifiers are insensitive to signal speed and format, so they can be used with many different transmitters. They also have gain over a broad range of wavelengths, from 1530 to 1570 nm, so they can amplify signals from two or more sources at different wavelengths in that band—without causing overlap or crosstalk. One limitation is that the gain is not uniform over the entire range, so the relative strengths of signals at different wavelengths can change after passing through a series of erbium-fiber amplifiers.

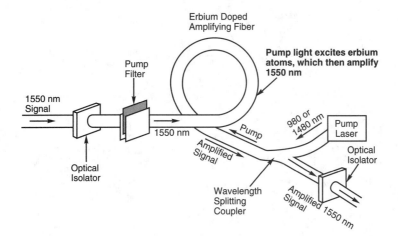

FIGURE 6.10.
An erbium-doped fiber amplifier.

No technology is ever completely ideal for any purpose, and erbium-doped fibers are no exception. If the pump laser burns out, an erbium-doped fiber becomes a strong absorber of the light it is supposed to amplify. However, there have been far more complaints that fiber amplifiers don't work at the 1300 nm wavelength of most current fiber systems. I'll talk about alternatives in Chapter 9.

Fiber Lasers

Put mirrors on both ends of an erbium-doped fiber and it can become a laser. The mirrors provide the resonant cavity needed for the laser to generate its own light or "oscillate." As in a diode laser, a few of the excited atoms release their excess energy as light, and this light can stimulate the emission of more photons, building up a beam. The fiber laser may be arranged with mirrors at each end, or in a loop, with couplers bringing light from the pump laser and splitting off the output beam, as shown in Figure 6.11.

An erbium-fiber laser cannot be directly modulated with an electrical signal like a diode laser. Its power comes from an external light source, usually a diode laser, which emits a steady beam. (Turning the pump laser off and on would turn the erbium-fiber laser off and on, but the results are not fast or reliable enough to use for signal modulation.) Instead, erbium-fiber lasers must be used with external modulators.

There are two ways erbium-fiber lasers can operate with external modulators. The fiber laser can be made to generate a steady or continuous beam at a narrow range of wavelengths. This stable single-wavelength output does not suffer the chirp that comes when a diode laser is directly modulated, so modulating it produces signals that suffer little dispersion in a fiber with high dispersion at 1550 nm. Erbium fiber lasers can also generate

Erbium-doped fibers can be made into lasers that generate a continuous beam or a train of ultrashort pulses at 1550 nm.

higher power than standard diode lasers, so a single erbium fiber laser could send signals to more terminals.

FIGURE 6.11.

Erbium-fiber ring laser.

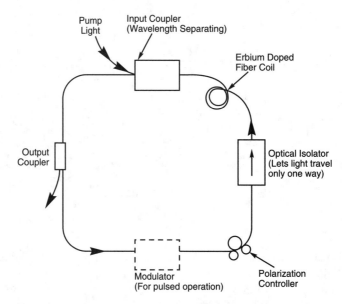

Alternatively, an erbium-fiber laser can be operated in a mode that generates a series of pulses shorter than one picosecond (10^{-12} second), separated by the time light takes to make a round-trip of the fiber laser. An optional modulator (shown dashed in Figure 6.11) controls generation of the pulses, by a technique called modelocking. The pulses are spaced uniformly in time, like a series of 1s in digital code, so an external modulator is needed to convert them into a signal. However, the shortness of the pulses makes them useful for high-speed transmission.

Other Solid-State Laser Sources

Diode-pumped neodymium lasers can generate more than a watt near 1300 nm.

Other solid-state lasers can also serve as signal source in fiber-optic systems. The most important of these are crystalline neodymium lasers, in which the rare earth neodymium is doped into crystals called YAG (for yttrium aluminum garnet) or YLF (for yttrium lithium fluoride). These lasers can be excited by semiconductor lasers of gallium arsenide emitting near 800 nm. The primary neodymium line is near 1060 nm, but there are secondary lines at 1313 and 1321 nm in YLF and 1319 nm in YAG. Those fall right in the 1300 nm fiber window.

Like erbium-fiber lasers, solid-state neodymium lasers cannot be modulated directly; they require external modulators. Their big attraction is their ability to generate high power—more than a watt near 1300 nm. That's much more than you want to send signals through a single length of fiber, but it can be split among many fibers to carry the same signals to many terminals, for network communications or cable television signal distribution.

What Have You Learned?

1. The most common light sources for short silica fiber-optic systems are GaAlAs LEDs operating at 820 or 850 nm. InGaAsP lasers and LEDs emit at 1300 and 1550 nm.

2. Red LEDs are the usual signal sources for plastic fibers.

3. LEDs and semiconductor lasers are both semiconductor diodes that emit light when forward biasing causes current carriers to recombine at the junction between p- and n-doped materials.

4. The wavelength emitted by an LED or diode laser depends on the material from which the diode is made.

5. Laser light is produced by the amplification of stimulated emission, which can occur at a semiconductor diode junction when there is a population inversion. A resonant cavity creates a laser beam.

6. Semiconductor lasers are more complex in structure than LEDs and can emit higher powers. Unlike LEDs, they do not operate below a certain threshold current. Above that threshold, they are much more efficient.

7. LEDs are longer-lived and less expensive than semiconductor lasers and do not require as careful control of operating conditions.

8. Special types of lasers are needed to provide the single-frequency light needed to transmit signals through high-dispersion fibers with large bandwidth.

9. Single-frequency lasers can be operated continuously with an external modulator to reduce dispersion.

10. Fibers doped with erbium can serve as optical amplifiers for 1550 nm signals.

11. The addition of resonator mirrors allows erbium-doped fibers to become lasers that generate their own signals.

12. Solid-state neodymium lasers can generate high continuous powers at 1300 nm but require external modulators.

What's Next?

Now that I have discussed light sources, Chapter 7 will examine how they are incorporated into fiber-optic transmitters.

Quiz for Chapter 6

1. Operating wavelengths of GaAlAs LEDs and lasers are

   a. 820 and 850 nm.

   b. 665 nm

   c. 1300 nm.

   d. 1550 nm.

   e. none of the above.

2. Light emission from an LED is modulated by

   a. voltage applied across the diode.

   b. current passing through the diode.

   c. illumination of the diode.

   d. all of the above.

3. Which of the following statements about the difference between semiconductor lasers and LEDs are true?

   a. Lasers emit higher power at the same drive current.

   b. Lasers emit light only if drive current is above a threshold value.

   c. Output from LEDs spreads out over a broader angle.

   d. LEDs do not have reflective end facets.

   e. All of the above.

4. Laser light is produced by

   a. stimulated emission.

   b. spontaneous emission.

   c. black magic.

   d. electricity.

5. The spectral width of a semiconductor laser is about

   a. 2 nm.

   b. 30 nm.

   c. 40 nm.

   d. 850 nm.

   e. 1300 nm.

6. A distributed-feedback laser is

   a. a laser that emits multiple longitudinal modes from a narrow stripe.

   b. a laser with a corrugated substrate that oscillates on a single longitudinal mode.

   c. a laser made of two segments that are optically coupled but electrically separated.

   d. a laser that requires liquid-nitrogen cooling to operate.

7. Which of the following is an important advantage of external modulation of lasers?

   a. Simpler operation.

   b. Does not require electrical power.

   c. More precise control over output.

   d. Avoids wavelength chirp that could cause dispersion.

8. Which of the following is not an advantage of erbium-doped fiber amplifiers?

     a. High gain.

     b. Insensitive to signal speed.

     c. Operation near 1300 nm.

     d. Broad amplification bandwidth at 1550 nm.

9. What is the power source for erbium-doped fiber amplifiers?

     a. Electric current passing through the fiber.

     b. They require no power.

     c. Diode lasers emitting at 980 or 1480 nm.

     d. Power is drawn from the optical signal.

10. How can erbium-doped fibers be made into lasers?

     a. They cannot be made into lasers.

     b. By adding mirrors to form a resonant cavity.

     c. By adding pump lasers.

     d. By adding external modulators.

Transmitters

About This Chapter

Optical transmitters generate the signals sent through fiber-optic cables. They come in many types—from cheap LEDs directly driven by signal sources to sophisticated transmitters based on costly semiconductor lasers. Some operate at telephone-like speeds and bandwidths over several meters; others send hundreds of thousands of megabits per second through tens of kilometers of fiber. All are based on the light sources described in Chapter 6.

This chapter examines transmitters and how they work. It begins by describing the major functional considerations for transmitters, then outlines how they generate signals. Because my emphasis is on optics, I don't go into detail about the electronics. I will cover signal amplifiers and regenerators later, in Chapter 9.

A Few Words About Terminology

Strictly speaking, transmitters are the devices that generate the signal sent through optical fibers. However, the terminology and packaging of commercial equipment can be confusing and deserves more explanation.

Fiber-optic transmitters are often packaged with receivers and cables and sold as systems. For short-distance digital transmission between two points, these systems are labeled data links. (Some are called fiber-optic modems, although not all of them work over fibers at telephone data rates.) A transmitter may be packaged with a receiver as a "transceiver" to provide two-way service to a single terminal or node. Local area networks interconnect multiple terminals spread over relatively small distances (typically no more than a few kilometers); only a small fraction use fiber optics. MANs (metropolitan area networks) and WANs (wide area networks) cover wider areas than LANs, from a campus-sized area to an entire city; more of them use fibers.

Industry terminology is vague, and more than one short analog system has been called a data link. Details of how transmitter, receiver, cable, and other components come together to make fiber-optic systems, and the specialized terminology of different applications, will be described in later chapters on system design and applications.

Fiber-optic transmitters may be packaged with receivers as systems. Short digital systems are called data links.

Performance of a fiber-optic transmitter is affected by signal type, speed, operating wavelength, and light source.

Transmitter Performance

Several factors enter into the performance of a fiber-optic transmitter, including the type of signal being sent, the speed, the operating wavelength, the type of light source, and the cost. Each of these deserves a brief explanation.

Analog versus Digital Transmission

Fiber-optic systems can transmit analog or digital signals.

Although much of the communications world has shifted from analog to digital transmission, not everyone has. Some applications require continuous analog waveforms; others operate best with discrete digital pulses. In theory, a simple LED source could be modulated by either an analog or digital signal, but in practice, transmitters are designed for one or the other type of modulation.

As you saw in Chapter 3, digital signals can withstand distortion better than analog signals. Figure 7.1 shows how the inherently analog process of signal transmission can distort signals by not precisely reproducing the input waveform. Distortion presents a problem for analog signals because the output should be a linear reproduction of the input. That is, if the input signal is $F(t)$, the output should be $cF(t)$, where c is a constant. Digital systems can tolerate distortion because they need to detect only the presence or absence of a pulse—not its shape.

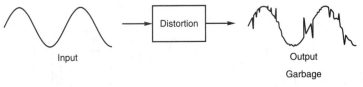

a. Analog

FIGURE 7.1.

Effects of distortion on analog and digital signals.

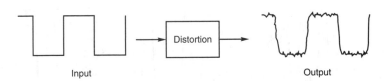

b. Digital

Digital transmission, on the other hand, demands faster response than analog to follow the rapid rise of signals. Breaking a digital signal down into its component frequencies shows that the sharp edge of a digital pulse is made up of high frequencies. However, the system need detect only the difference between off and on, so the sharp edges can be somewhat blurred without introducing errors.

Thus, analog systems must accurately reproduce inputs, but speed is not as crucial as in digital systems, which must be fast but need not be accurate. These differences mean that analog and digital transmitters should use different designs. They may be capable of using the same light source but with different electronics.

Bandwidth and Data Rate

The operating speed of a fiber-optic transmitter is measured in two ways: bandwidth for analog signals and data rate for digital signals. Both refer to the amplitude modulation of light from the source, which is shown in Figure 7.2. (Light waves are actually much smaller than shown. If each digital pulse was 1 ns [10^{-9} s] long, it would contain 300,000 waves of 1 mm light.) Analog bandwidth is normally defined as the modulation frequency where the modulated signal amplitude drops 3 dB below the low-frequency modulated signal (equivalent to a reduction of 50% in power). The digital data rate is the maximum number of bits per second that can be transmitted with an error rate below a specified level (typically one error in 10^9, or a bit error rate of 10^{-9}).

●
Bandwidth usually defines the capacity of an analog system. Digital capacity is measured as data rate.

FIGURE 7.2.

Light modulation by digital and analog signals.

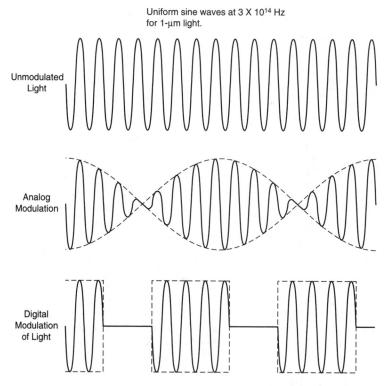

Uniform sine waves at 3 X 10^14 Hz for 1-µm light.

Unmodulated Light

Analog Modulation

Digital Modulation of Light

Neither of these quantities actually measures the limiting characteristic of light sources: their response time. The rise time is how long it takes light output to rise from 10% to 90% of the steady-state level; fall time is the inverse. If rise and fall times are equal (they aren't always), this can be used to approximate the bandwidth:

$$BW = 0.35/RISE\ TIME$$

where BW is in megahertz if rise time is in microseconds. The precise relationship between bandwidth and rise time differs among light sources and transmitters. Rise time is an important variable in light source selection. Many semiconductor lasers have rise times of a fraction of a nanosecond. LED rise times range from a few nanoseconds to a few hundred nanoseconds, depending on design.

Operating Wavelength

Source wavelength affects performance of fiber-optic systems because fiber transmission depends on wavelength. One effect is signal attenuation, which you saw in Chapter 4 is a function of wavelength. This limits transmission distance directly, by reducing power available at the receiver.

Pulse spreading or dispersion is a more subtle effect that limits bandwidth or data rate by spreading successive pulses so much that they interfere with each other. As you saw in Chapter 4, dispersion in a single-mode fiber depends on the wavelength and on the range of wavelengths in the pulse. Dispersion is measured in picoseconds of spreading per nanometer of source bandwidth and kilometer of fiber length—which means that pulse spreading increases with spectral width and distance traveled. Dispersion can limit transmission distance by blurring successive pulses so they cannot be distinguished, even though the receiver does detect the light.

Output Power

Output power is also critical to transmitter performance. The critical power level is what is delivered to the fiber, which is inevitably lower than what the light source delivers, because of transfer losses. Further power is lost in transmission to the receiver, where the power must be sufficient for detection but not so high that it overwhelms the receiver. I will cover basic receiver characteristics in the next chapter.

Transmitter Design

The light source is important, but it's not the only part of the transmitter. A housing is required to mount and protect the light source and to interface with the electronic signal source and the transmitting optical fiber. Internal components may be needed to optimize coupling to the fiber. Drive circuitry is usually needed, and temperature control and output monitoring can be crucial for sophisticated lasers.

● A transmitter includes housing, drive circuitry, and monitoring equipment, as well as a light source.

The practical boundaries between transmitters and light sources can be vague. Simple LED sources can be mounted in a case with optical and electronic connections and little or no drive circuitry. On the other hand, a high-performance laser may be packaged as a transmitter in a case that also houses an output monitor and thermoelectric cooler, and perhaps an external modulator. Then that whole package may be incorporated into a larger transmission system that performs electronic functions such as multiplexing. Thus, you can say that some transmitters can contain transmitters.

Elements of Transmitters

The basic elements that may be found in transmitters, as shown in Figure 7.3, are as follows:

● Several elements make up fiber-optic transmitters.

- Housing
- Electronic interfaces
- Optical interfaces
- Drive circuitry

- Temperature sensing and control
- Optical output sensing
- Electronic signal preprocessing (e.g., data buffers)
- External modulator (if the light source is not modulated directly)
- Attenuator (for short systems)

Note that the figure does not show an actual physical arrangement, only the relationships of the elements. Many transmitters do not contain all of these elements.

FIGURE 7.3.

Elements used in fiber-optic transmitters.

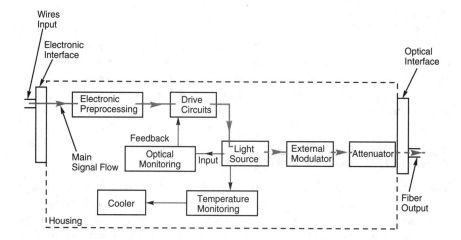

HOUSING

The housing for a fiber-optic transmitter, in its simplest form, is just a box designed to be mounted conveniently. Screws or other mounting equipment attach it mechanically to printed circuit boards or other electrical components. Some are quite compact, only an inch or two long, including integral optical connectors. Others are packaged to mount on standard equipment racks.

ELECTRONIC INTERFACES

Electronic interfaces may be wires, standard electronic connectors, or pins emerging from packages. Some simple transmitters can be driven directly by electronic input signals. More complex transmitters may require power and may accept multiple electronic inputs (and provide one or more electronic outputs as well).

OPTICAL INTERFACES

The optical interface between light source and fiber can take various forms, as shown in Figure 7.4. One of the most common is integration of a fiber-optic connector in the housing. Light is delivered to that connector by internal optics, including a collimating lens and sometimes a short fiber segment. Another approach is a short fiber pigtail that collects light from the emitting area and delivers it outside the case, where it can be spliced to an external fiber. The choice between these types depends on factors including cost, whether connections are to be temporary or permanent, type of fiber, importance of minimizing interconnection losses, and operating environment.

●
Common optical interfaces are connectors and fiber pigtails.

FIGURE 7.4.

Transmitter optical interfaces.

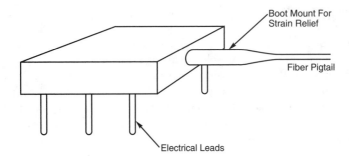

a. Transmitter with Fiber Pigtail

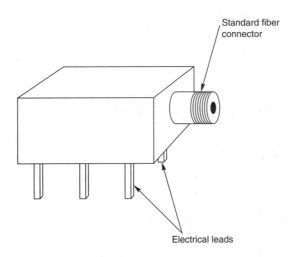

b. Transmitter with Integral Connector

DRIVE CIRCUITRY

The drive circuitry depends on application requirements, data format, and the light source. LEDs can be directly driven by a suitable current source (although most signals are in the form of voltage and must be converted to current). However, semiconductor lasers must be biased to a current level near laser threshold. Some LEDs work better with drive circuitry to tailor electrical input. For example, the proper drive waveform can effectively reduce the rise time of an inexpensive LED and allow its use at higher-than-specified bandwidths.

ELECTRONIC PREPROCESSING

The drive circuitry of some transmitters electronically preprocesses the input electrical signals to put them into a form suitable for driving the light sources. A simple example is the conversion of signals from the voltage variations that drive electronic circuitry to the current variations that modulate lasers and LEDs. Other preprocessing may change signal formats to types better suited for fiber transmission, or provide a buffer that holds input data until it is transmitted (necessary in local area networks where data is not sent as fast as it arrives).

OUTPUT SENSING AND STABILIZATION

Laser transmitters typically include output-stabilization circuits. A photodetector monitors light emitted from the rear facet of the laser and drives a feedback circuit that adjusts drive current so total output power remains stable, avoiding age-induced drops in laser output.

TEMPERATURE SENSING AND CONTROL

Operating characteristics of a semiconductor laser—notably threshold current, output power, and wavelength—change with temperature. The threshold current increases roughly exponentially with the relative change in temperature, $\Delta T = (T_1 - T_2)/T_1$, where T_1 is initial temperature and T_2 is final temperature. Because optical output is proportional to how much the drive current exceeds the threshold current, output power decreases as temperature rises. Also, more electrical power must be dissipated within the laser, so laser lifetime decreases with operating temperature. The change in wavelength is more subtle. It is caused by temperature-induced changes in the semiconductor's refractive index and thus in the effective length of the laser cavity.

Thermoelectric coolers in transmitters can assure operation at a stable temperature. Temperature control of the laser can reduce temperature-induced degradation of output power. Stable output power is important for proper receiver operation. As detected power decreases, bit-error rate increases in digital systems, and signal-to-noise ratio decreases in analog systems, reflecting degraded performance. The more subtle wavelength changes are not important in most present systems, but they could pose problems in future wavelength-division multiplexed systems that require transmission of specific wavelengths. Some newer lasers are less sensitive to temperature increases, so they do not require active cooling.

Attenuators

Transmitters are made to produce standard power levels, but in some cases those levels may be higher than desired. This is most likely in networks that have legs of much different lengths or that contain many more fiber junctions than others. Because receivers can handle only limited input powers, attenuators are sometimes used to reduce transmitter output to safe levels for the receiver. Note, however, that they are comparatively rare; they may also be used at the receiver.

External Modulators

You saw in Chapter 6 that direct modulation of a diode laser can cause "chirp," which spreads out the wavelength range emitted by a laser. If the transmission wavelength is one where the fiber has high dispersion, this can cause excessive pulse dispersion, limiting bandwidth and data rate. One way to circumvent this limitation is by building a transmitter that drives the laser to produce a steady beam that is modulated in intensity externally.

Types of Transmitters

The block diagrams of fiber-optic transmitters may look similar, but they can differ greatly in detail, depending on performance requirements, particularly data rate, power level and (for analog systems) noise level. First you will look at major differences among transmitters, then look more closely at some factors that cause these differences, such as modulation format.

Functional Differences

Most major functional differences are in the modulation scheme and speed. Analog modulation is typically for audio, video, or radar signals. Audio signals require bandwidths measured in tens of kilohertz at most, whereas each standard (NTSC) video signal requires 6 MHz. Much higher bandwidths are needed for multichannel cable television systems or for transmitting radar signals.

●
Transmitters differ in modulation scheme and speed.

Digital transmitter circuits must have fast response to produce fast rise-time digital pulses. Again, the speed depends on the application. Some simple computer data links may require no more than tens of thousands of bits per second, but long-haul telephone systems now operate at speeds to 2.5 Gbit/s, and systems carrying 5, 10, and 20 Gbit/s are in development.

The choice among light sources depends primarily upon speed, type of signal, and operating distance. The higher the signal speed and the longer the transmission distance, the more likely lasers are to be used. Longer-wavelength sources are generally preferred for long-distance transmission. Care must be taken to pick light sources with linear response for analog transmission, but linearity is not critical for digital systems.

Important differences lie in the transmitter electronics. Suitable circuitry can increase the effective speed of transmitters, for example. Logic circuits can change digitally encoded signals from one format to another. Drive circuits can compensate for nonlinearities in light source analog response.

The design of transmitter and receiver circuitry is a specialized realm of electronics, which I will not explore deeply because this book is concentrating on optics. Sophisticated transmitters and receivers are typically manufactured and sold to users, who see them as complete, functional black boxes, or as fiber-optic modems that mate on each end with electronic devices and are linked by a fiber-optic cable.

Intensity Modulation

Most fiber-optic transmitters send optical signals in a form called intensity modulation. Simply put, the signal is proportional to the intensity of the light sent down the fiber, as in Figure 7.1. Intensity modulation works for both analog and digital transmission. It is implemented simply by converting the input electrical signal to a drive current applied to a diode laser or external modulator.

Most fiber-optic transmitters use intensity modulation.

Multiplexing

Multiplexing is the combination of multiple signals into a single signal for transmission. Various techniques can be used, but all serve the same fundamental purpose. They are implemented in different ways for fiber-optic transmission. The multiplexing function may be independent of the transmitter, incorporated in it, or packaged with the transmitter although logically separate. There are three important types.

Signals can be combined by time-division, frequency-division, or wavelength-division multiplexing.

- **Time-division multiplexing** combines two or more digital signals, essentially by interleaving the bits from the separate data streams. This function is logically independent of the transmitter but may be incorporated into the transmitter package. For example, 24 voice phone lines could plug into a fiber-optic transmitter that generates a multiplexed digital 1.5 Mbit/s signal that carries all 24 voice circuits. The multiplexer and transmitter are in the same box, but the operations could be performed separately.

- **Frequency-division multiplexing** combines two or more analog signals by placing them at different frequencies in a broad-bandwidth analog signal. A cable television system does this when it transmits many video signals to your home through the same cable. The transmission frequencies may not be the same as the frequencies of broadcast signals (which is why it matters if you program your television to select UHF channels from cable or broadcast television).

● **Wavelength-division multiplexing** is the use of two or more wavelengths to carry signals through the same optical fiber. The signals are generated by separate light sources in the same transmitter, which are combined for transmission through a single fiber. Then they are separated optically at the other end before being detected. Thus a single fiber could carry one signal at 1300 nm and another at 1550 nm. In fact, depending on the optics, the same fiber could carry other signals spaced more closely in wavelength. Wavelength-division multiplexing is not widely used today, but it could multiply the capacity of existing fiber-optic cables.

Coherent Transmission

Other transmission schemes are in development. One interesting approach is called coherent or heterodyne transmission. The form shown in Figure 7.5 operates like a heterodyne radio system and requires two lasers, one at the transmitter and a second (a local oscillator) at the receiver. The two lasers transmit at slightly different frequencies, v_1 and v_2. At the receiver, the two beams are mixed together to obtain the difference frequency, $v_1 - v_2$, in the microwave region. That microwave signal—at about 1 GHz in many experiments—is amplified and demodulated to give the desired signal.

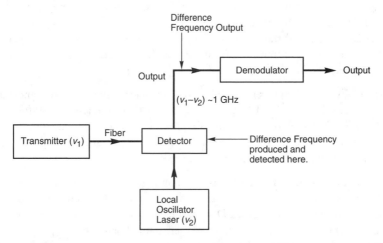

FIGURE 7.5.

Coherent transmission scheme.

Why go through all that trouble? Because coherent receivers can pick out weaker signals from a noisy background than conventional direct-detection receivers, just as in radio sets. Impressive performance has been demonstrated in laboratory systems. However, much work remains to be done. Critical issues include stabilizing the frequencies of the semiconductor lasers and limiting emission to a narrow frequency range. In practice, coherent transmission may lose out to a newer idea, solitons.

Soliton Transmission

Technically speaking, a soliton is a special solution of a complex equation for wave propagation, the sort of thing that normally belongs in the realm of esoteric theory. However, solitons have an important meaning in fiber-optic systems—they are pulses that can retain their original shapes after traveling long distances. The result sounds a little like magic, but it isn't.

Solitons work because of two competing effects in optical fibers. One is the pulse dispersion I discussed earlier, which causes light pulses containing a range of wavelengths to spread out as they travel along the fiber. The other is called self-phase modulation, which spreads the pulse out over a range of wavelengths. In essence, the two effects cancel each other out, so the pulse returns to its original form after traveling through the fiber. Attenuation does weaken the pulse, but optical amplifiers can restore its strength. The self-restoring nature of solitons allows transmission over tremendous distances at high speeds, as you will see in Chapter 15.

Sample Transmitters

Although I won't go into detail on transmitter design, it is useful to have a general idea of what is inside the black boxes you're likely to encounter. The best way to get a feeling for the internal workings of transmitters is to look at some simple circuits. As you would expect, the complexity of transmitters increases rapidly with their performance level, but the basic concepts remain the same.

An LED emits light when it is forward biased at a voltage higher than the band-gap voltage, about 1.5 V for GaAlAs LEDs emitting at 800-900 nm and around 1 V for 1300 nm InGaAsP LEDs. As with other forward-biased semiconductor diodes, you can consider the voltage drop across the diode constant regardless of current. To modulate the light output, you modulate the drive current, not the voltage.

A very simple drive circuit for an LED is shown in Figure 7.6. A modulated voltage is fed into the base of a transistor, causing variations in the current passing through the LED and current-limiting resistor. The LED can be on either side of the transistor, but the transistor or some other circuit element must be there to modulate the current. The resistor is needed to limit LED current. The auxiliary drive circuitry (not shown) becomes increasingly complex in more sophisticated transmitters as it is called upon to perform more functions (e.g., converting digital signal encoding format). However, the basic idea remains the same.

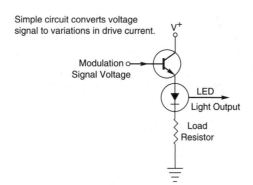

Simple circuit converts voltage
signal to variations in drive current.

Modulation
Signal Voltage

LED
Light Output

Load
Resistor

FIGURE 7.6.

Simplified drive circuit for LED.

Simple commercial transmitters come in simple packages, such as the one shown in Figure 7.7, along with a block diagram of the circuit. Note that the transmitter includes a buffer to hold input data until it can be transmitted. The whole package is about 2 inches (5 cm) long, including an integral ST-type connector.

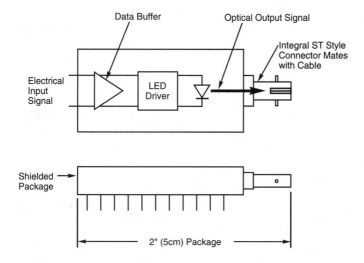

FIGURE 7.7.

A packaged LED-based transmitter with block diagram.

Laser Transmitters

The electrical characteristics of a semiconductor laser are similar to those of an LED because both are semiconductor diodes. However, the laser does not emit much light until drive current passes a threshold value well above zero. Also, because lasers need much higher drive currents than LEDs, their current-limiting resistors are usually smaller.

Laser drive circuits must deliver higher currents than those for LEDs.

Laser diodes are often packaged in modules like the 14-pin dual-in-line package shown in Figure 7.8. The package includes only components that must be physically mounted in the same case as the laser diode. One is a thermoelectric cooler, needed to maintain stable operating temperature. Another is a thermistor, which monitors temperature inside the case. Also included is a photodiode, which detects light from the rear facet of the laser to monitor power levels. Pins on the package carry signals to the rest of the transmitter circuitry.

FIGURE 7.8.

Laser diode packaged for mounting on a transmitter board.

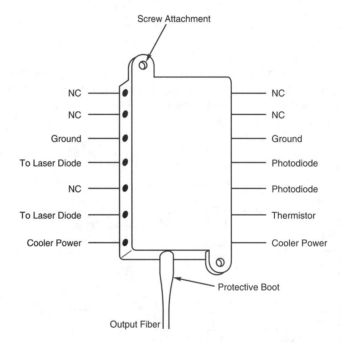

The circuit in Figure 7.9 is an example of the type used in laser transmitters. The drive signal is applied as a voltage to the base of a transistor, so it modulates the drive current through the laser diode. The detector packaged with the laser monitors output, providing feedback to an amplifier (at right), which adjusts the bias applied to the laser diode, thus controlling average power level. The circuitry can get considerably more complex. Simple circuits can be used to drive lasers that emit continuous beams which are modulated externally.

Typically, the drive circuit pre-biases the laser with a current close to, but still below, the threshold. The modulated drive signal should raise it above the threshold. This approach enhances speed by avoiding the delay needed to raise drive current above the threshold. At high speeds, it may be useful to bias the laser so that it emits a little light in the off state but a much higher power when turned on. One concern with pre-biasing is that the small amount of light emitted in the off state could confuse the receiver if transmission losses are too low. However, careful design can minimize the cost in receiver sensitivity.

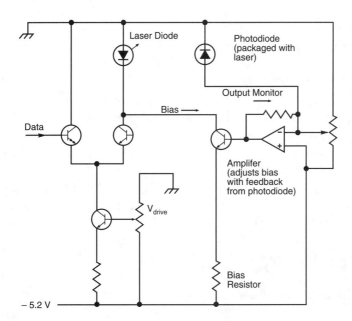

Transceivers, Repeaters, and Regenerators

Transmitters may be packed with other equipment for some applications. A transceiver is a package that includes both transmitter and receiver, which send and receive signals to and from a single attached source, like a computer terminal. Normally, the signals are sent over one fiber and received over a second fiber. Circuitry for the transmitter and receiver are separate, as are the electrical signal input and output. In fact, in most cases the transceiver is simply a way to package a separate transmitter and receiver when two-directional transmission is required.

In Chapter 9, I will talk about how transmitters can be packaged in series with receivers to serve as repeaters and regenerators.

●
Transmitter and receiver can be packaged in parallel as a transceiver.

What Have You Learned?

1. Both digital and analog transmitters rely on direct amplitude modulation of light intensity. Changes in drive current change the light output of LEDs and semiconductor lasers.

2. The most common light sources for short fiber-optic systems are GaAlAs LEDs operating at 820 or 850 nm. 1300 nm InGaAsP LEDs are used in some networks; laser sources are used at high speeds and long distances.

3. Performance of a fiber-optic transmitter is affected by signal type, speed, operating wavelength, and light source.

4. Bandwidth usually defines capacity of an analog system. Digital capacity is measured as data rate. Both depend on rise time of the transmitter.

5. A transmitter includes housing, drive circuitry, and monitoring equipment, as well as a light source.

6. Common optical interfaces are connectors and fiber pigtails.

7. Most fiber-optic transmitters use intensity modulation. Frequency modulation has been demonstrated in the laboratory.

8. Signals can be combined by time-division, frequency-division, or wavelength-division multiplexing.

9. Modulation of a voltage fed into a transistor causes variations in current passing through an LED or laser diode—and in its light output. Lasers require higher drive currents than LEDs.

What's Next?

Now that I have discussed fiber-optic transmitters, Chapter 8 will examine the other end of the system, the receiver, and Chapter 9 will describe the roles of repeaters, regenerators, and optical amplifiers.

Quiz for Chapter 7

1. Digital transmission capacity is measured as
    a. bandwidth in megahertz.
    b. rise time in microseconds.
    c. frequency of 3 dB point.
    d. number of bits transmitted per second.

2. Analog transmission capacity is measured as
    a. bandwidth in megahertz.
    b. rise time in microseconds.
    c. frequency of 3 dB point.
    d. number of bits transmitted per second.

3. If the rise time of a transmitter is 1 nanosecond, what is its theoretical bandwidth?
    a. 1 Gbit/s.
    b. 100 MHz.
    c. 350 MHz.
    d. 350 kHz.
    e. None of the above.

4. Which of the following is not found in fiber-optic transmitters?
    a. Electronic drive circuitry.
    b. Helium-neon gas laser.
    c. Photodiode to sense laser output.
    d. Temperature sensor.

5. Which of the following is not a standard optical interface with transmitters?

   a. An integral fiber-optic connector.

   b. An output window.

   c. A fiber-optic pigtail.

6. What provides feedback to help stabilize laser intensity in a transmitter?

   a. A signal relayed from the receiver.

   b. Changes in input impedance.

   c. Output from the rear face of the laser monitored by a photodiode.

   d. Light scattered from the optical interface with the output fiber.

   e. No feedback is used.

7. What is the usual modulation method for fiber-optic transmitters?

   a. Intensity modulation.

   b. Frequency modulation.

   c. Wavelength modulation.

   d. Voltage modulation.

8. What does time-division multiplexing do?

   a. Transmits different signals at different wavelengths.

   b. Shifts the frequencies of several analog signals to combine them into a single output.

   c. Encrypts signals for secure transmission.

   d. Interleaves several digital signals into a single stream of bits.

9. The voltage drop across a 1300 nm LED is

   a. 1.5 V.

   b. 1 V.

   c. 0 V.

   d. changing constantly as the LED is modulated.

10. What circuit element receives the drive voltage in a fiber-optic transmitter?

   a. A filter capacitor.

   b. A load-limiting resistor.

   c. A temperature sensor.

   d. A transistor.

Receivers

About This Chapter

The receiver is as essential an element of any fiber-optic system as the optical fiber or the light source. The receiver's job is to convert the optical signal transmitted through the fiber into electronic form, which can serve as input for other devices or communication systems, including repeaters.

This chapter discusses the basic types of receivers, important performance considerations, and how they work. It focuses particularly on the types of detectors and to a lesser extent on the electronics.

Basic Elements of Receivers

Fiber-optic receivers come in many varieties, from simple types that are little more than packaged photodetectors to sophisticated systems that process a weak signal to allow accurate, high-speed transmission. The basic elements of digital and analog fiber-optic receivers are shown in the block diagram of Figure 8.1.

The basic functional elements of a receiver are as follows:

1. The detector (to convert the received optical signal into electrical form).
2. Amplification stages (to amplify the signal and convert it into a form ready for processing).
3. Demodulation or decision circuits (to reproduce the original electronic signal).

In practice, the functional distinctions can be hazy, because some detectors (e.g., avalanche photodiodes) have internal amplification. Some receivers do not have separate demodulation or decision circuits because the electrical signal from the amplification stages is good enough for use by other electronic equipment. Some receivers may be preceded by an optical preamplifier at the end of the fiber, an optical amplifier that boosts the weak signal to improve receiver performance.

The entire receiver is packaged in a housing designed to meet user requirements, which provides interfaces with the optical fiber and with whatever is to receive the signal.

FIGURE 8.1.

Basic elements of analog and digital receivers.

a. Analog Receiver

**Note similarity to analog receiver –
this is an analog preprocessing stage.**

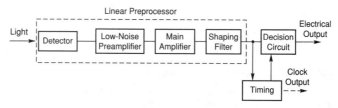

b. Digital Receiver

•
Receiver functions include detection, amplification, and demodulation.

•
Analog and digital receivers use similar detectors and amplifiers; the differences are in the demodulation or decision circuits.

The similarities between analog and digital receivers are striking. The initial stages are the same in the two types of receivers; the differences come where the signal is converted into final form for output to other equipment. Why is there an analog receiver at the front end of a digital receiver? Because the real world is analog. The signal reaching the detector may start in digital form, but by the time it reaches the receiver it varies continuously in level like an analog signal. It must be converted to electrical form and amplified as an analog signal before electronic decision circuits can convert it back to digital form.

Then why bother with digital transmission and receivers? Because analog transmission requires precise replication of the original waveform. Any changes are distortion. Once distortion is added, the electronics don't know what is distortion and what is signal. Digital transmission, on the other hand, does not require precise replication of the waveform. Instead, it requires only the ability to decide whether the signal is off or on, which can be done even in the presence of distortion (although severe distortion can lead to bit-interpreting errors).

Detector Basics

The detectors used in fiber-optic communications are semiconductor photodiodes or photodetectors, which get their name from their ability to detect light. The simplest semiconductor detectors are solar cells, where incident light energy raises valence band electrons to the conduction band, generating an electric voltage. Unfortunately, such photovoltaic detectors are slow and insensitive.

Photodiodes are much faster and more sensitive if electrically reverse-biased, as shown in Figure 8.2. (This is the opposite of LEDs and lasers, which are forward-biased to emit light.) The reverse bias draws current-carrying electrons and holes out of the junction region, creating a depleted region, which stops current from passing through the diode. Light of a suitable wavelength can create electron-hole pairs in this region by raising an electron from the valence band to the conduction band, leaving a hole behind. The bias voltage causes these current carriers to drift quickly away from the junction region, so a current flows proportional to the light illuminating the detector. Several types of detectors can be used in fiber-optic systems, as described below.

● Semiconductor photodiodes are reverse-biased to detect light; they produce a current proportional to the illumination level.

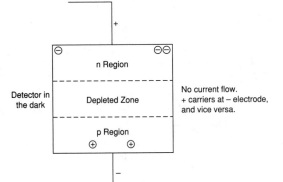

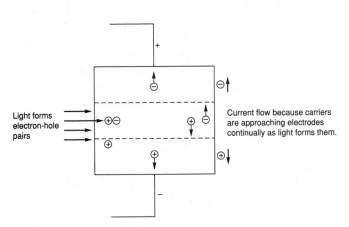

FIGURE 8.2.

Photodetector operation.

Photodetectors can be made of silicon, gallium arsenide, germanium, indium phosphide, or other semiconductors. Their wavelength response depends on their composition. To produce a photocurrent, photons must have enough energy to raise an electron across the band gap from the valence band to the conduction band. This gives most photodetectors

● Wavelength sensitivity of photodetectors depends on the materials from which they are made.

a fairly sharp cut-off at long wavelengths; the sensitivity of silicon, for example, drops sharply between 1000 and 1100 nm. Other effects, such as absorption in other parts of the device, cause response to drop more gradually at shorter wavelengths. The approximate operating ranges for the most important detector materials are shown in Table 8.1.

Table 8.1. Detector operating ranges.

Material	Wavelength (nm)
Silicon	400-1000
Germanium	600-1600
GaAs	800-1000
InGaAs	1000-1700
InGaAsP	1100-1600 (doping dependent)

Some detectors can be integrated with electronic circuits. Silicon and germanium are usable both as detectors and in electronic circuits. Other detector compounds, like InGaAs, are made on substrates of gallium arsenide, another electronic material. Detectors can also be combined with electronic components in hybrid circuits. I'll look at the details later.

Performance Considerations

Many interrelated factors affect receiver performance.

The factors influencing receiver performance are complex and often interrelated. At first glance, they might seem to be sensitivity, speed, and cost. In practice, these factors often depend on other factors, including operating wavelength, choice of fiber and transmitter, dark current, noise-equivalent power, and nature of transmission coding. They depend upon both the response of the detector itself and the processing performed by the electronics. Let's look at the most important receiver parameters.

Sensitivity

Sensitivity measures how well a receiver responds to a signal as a function of its intensity.

Although sensitivity sounds like a simple concept, it is actually a conceptual umbrella covering how detectors and receivers respond to signal intensity. To help you understand the nature of the problem, I will stop briefly and look at how a receiver handles a weak signal.

SIGNAL QUALITY

The role of a receiver is to accurately reproduce the signal it receives through an optical fiber. Two fundamental characteristics affect how well this can be done: signal strength and the noise level, which tends to obscure or degrade the signal. For analog systems, the signal-to-noise (S/N) ratio (the signal power divided by noise power, normally expressed in decibels) measures quality. The higher the signal-to-noise ratio, the better the received signal. High noise, like scratches on an old-fashioned phonograph record, can overwhelm the signal. The practical definition of a good S/N ratio depends on the application. In many fiber-optic systems, 40-50 dB is considered good to excellent, but S/N ratios in the 30 dB range are acceptable for many applications.

In digital systems, where the received information is either 1 or 0, quality is measured as the probability of incorrect transmission, the bit-error rate. That tells how accurately the receiver can tell 0s from 1s, which in turn depends on factors such as received power, sensitivity, noise, and transmission speed. For a given receiver, the dependence on received power is striking, as shown in Figure 8.3. In certain ranges, a 5 dB decrease in received power can make the error rate soar from 10^{-12} to 10^{-3}. The bit-error rate also depends on data rate; the slower the data rate, the lower the error rate. The typical goal for telecommunications is a bit-error rate of 10^{-9}, or one error per billion bits, but even lower rates (10^{-12}) are needed for computer data transmission.

Signal-to-noise ratio and bit-error rate are introduced here because they become important at the receiver. However, they really measure overall system performance. Receiver performance enters the picture because of the importance of sensitivity, but overall performance also depends on transmitter output and losses in transmission. The effect on overall system design will be described later.

> Signal strength and noise affect how well a receiver can reproduce a signal.

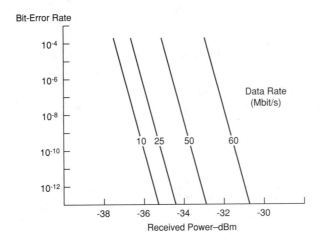

FIGURE 8.3.

Bit-error rate as a function of power. (Courtesy of AMP Inc.)

WAVELENGTH

Receiver sensitivity generally refers only to how well a receiver responds to a signal of a given amplitude—not to its time response or bandwidth. The sensitivity depends on the detector itself and the circuits in the receiver that amplify and process the electrical signal from the detector. The detector parameters themselves depend on wavelength of the light and operating conditions.

RESPONSIVITY AND QUANTUM EFFICIENCY

One measure of photodetector sensitivity is responsivity, the ratio of electrical output from the detector to the input optical power. Most fiber-optic detectors generate signals as current, so this is normally measured as amperes per watt (A/W). (Because input optical powers are in the microwatt range, responsivity might be more properly given as microamperes per microwatt [$\mu A/\mu W$] and sometimes is, but the two measurements are equivalent.) If electrical output signals are voltages, response can be measured in volts per watt (V/W).

A closely related quantity for detectors is quantum efficiency. This measures the fraction of incoming photons that generate electrons in the output signal:

$$QUANTUM\ EFFICIENCY = ELECTRONS/PHOTONS$$

This and responsivity both depend on wavelength, as can be seen in the plots of quantum efficiency in Figure 8.4. Like responsivity, quantum efficiency depends on the detector material and structure. Shape of a quantum efficiency curve differs from that of a responsivity curve, because photon energy changes with wavelength. A 400 nm photon carries twice as much energy as an 800 nm photon, so only half as many 400 nm photons are needed to generate a given power.

FIGURE 8.4.

Typical wavelength response of photo-diodes. (Courtesy of Tran V. Muoi)

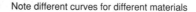

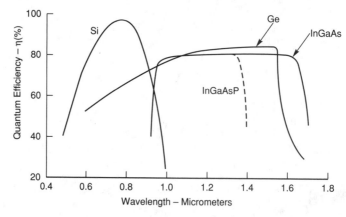

Quantum efficiency cannot be larger than 1.0 by that formal definition. However, values larger than 1.0 are possible if quantum efficiency is defined as the ratio of output electrons to incoming photons. That can be greater than 1.0 in detectors (or receivers) with internal amplification, because amplification may produce multiple electrons in the output current of a single electron generated by light at the p/n junction.

DIFFERENT MATERIALS

The dependence of detector response on wavelength means that different detectors must be used at different wavelengths. Silicon is fine for 600-900 nm, but it is insensitive at wavelengths longer than 1000 nm. Germanium, InGaAsP, and InGaAs photodiodes can be used at longer wavelengths, as shown in Figure 8.4, because their smaller band-gap energies let them respond to less-energetic photons. Although the shape of the responsivity curve depends mostly on the material, the value of the responsivity also depends on device structure. Typical values are about 0.6 A/W for simple silicon photodiodes at 800 nm, and about 0.8 A/W for InGaAs at 1300 nm.

Silicon detectors are insensitive to wavelengths longer than 1000 nm, so germanium, InGaAs, or other materials must be used.

AMPLIFICATION

A final factor that enters into sensitivity is amplification. The figures I cited are given in terms of the optical power reaching the detector. Placing an optical amplifier in front of the detector, so it can act as a preamplifier, will increase the optical power reaching the detector, enabling it to detect fainter signals.

Adding internal amplification stages, as in a phototransistor or avalanche photodiode, can multiply responsivity. Electronic amplification can multiply receiver responsivity. However, amplification is indiscriminate; it multiplies noise as well as signal and can add some of its own noise.

Dark Current and Noise-Equivalent Power

The electrical signal emerging from a detector includes noise as well as signal. Some noise comes with the input optical signal (optical noise), and some is generated within the detector. Other noise is added by the amplifier. Electromagnetic interference causes noise when stray electromagnetic fields induce currents in conductors in the receiver. Noise mechanisms are complex enough to be a field in themselves, and the details are beyond the scope of this book. However, it is important to understand two key concepts: dark current and noise-equivalent power.

Detector output includes noise as well as signal.

Any detector will produce some current when it is operated in the normal manner but not exposed to light (i.e., kept in the dark). This dark current measures inherent electrical noise within the detector, which will also be present when the detector is exposed to light. It sets a floor on the minimum detectable signal, because for a signal to be detected it must produce measurably more current than the dark current. Dark current depends on operating temperature, bias voltage, and the type of detector.

Dark current is the noise current a detector produces when it is not illuminated.

● Noise-equivalent power is optical power needed to generate output current equal to root-mean-square noise.

Noise-equivalent power (NEP) is the input power needed to generate an electrical current equal to the root-mean-square noise of the detector (or receiver). This is a more direct measurement of the minimum detectable signal because it compares noise level directly to optical power. NEP depends on frequency of the modulated signal, the bandwidth over which noise was measured, area of the detector, and operating temperature. Its units are the peculiar ones of watts divided by the square root of frequency (in hertz) or $W/Hz^{1/2}$. Specified values are normally measured with a 1 kHz modulation frequency and a 1 Hz bandwidth.

Operating Wavelength and Materials

● Each photodiode material has a unique response as a function of wavelength.

As you saw earlier, the sensitivity of a detector—and hence of the entire receiver—depends strongly on wavelength, because of differences among materials. Individual photodiodes are often designed to work best in part of the material's range. For fiber optics, silicon may be optimized for either the red region, where plastic fibers transmit best, or the 800-900 nm region of GaAlAs emitters. Germanium, InGaAs, and InGaAsP are typically designed for use at 1300 to 1550 nm, although they may be optimized for one wavelength. The spectral response of InGaAs and InGaAsP depends on the fractions of indium, gallium, arsenide, and phosphorous they contain, which affects their band gaps and other properties, as in diode lasers.

Photodetector sensitivity also depends on temperature. The effect is not pronounced at the wavelengths used in present fiber-optic systems, but detector sensitivity at wavelengths much longer than 2 or 3 μm is greatly improved by cooling.

The choice of material affects more than the wavelengths that can be detected. Electrical characteristics, such as speed and dark current, also differ among materials. Germanium detectors tend to be noisier and slower than those made of silicon or the III-V materials such as GaAs and InGaAs, and InGaAs is often preferred for use at 1300 and 1550 nm.

Speed and Bandwidth

● Detectors do not respond instantly to changes in input.

Detectors take finite times to respond to changes in input. That is, there is a delay between input of an optical signal to a detector and its production of an electrical current. The delay depends on the material and the device design.

A second internal speed limit affects fiber-optic systems more directly: the time it takes the electrical output signal to rise from low to high levels and the corresponding fall time. The rise time is normally defined as the time the output signal takes to rise from 10% to 90% of the final level after the input is turned on abruptly. Analogously, fall time is how long the output takes to drop from 90% to 10% after the input is turned off. Device geometry, material composition, electrical bias, and other factors all combine to determine rise and fall times, which may not be equal. Generally the longer of these two quantities is considered the device response time. (The fall time of some detectors is markedly slower than their rise time.)

Although internal delay does not directly affect bandwidth or bit rate of a fiber-optic receiver, the rise and fall times do. Propagation delays shift the signal in time, but rise and fall times spread it out. A 10 ns delay, for example, means that a 10 ns pulse arrives at the output 10 ns late, but still only 10 ns long. However, a 10 ns response time doubles the pulse length to 20 ns, in effect halving the bit rate. Frequency response and limiting bandwidth of a detector are inversely proportional to response time, as is maximum bit rate.

For relatively slow devices, response time is proportional to the RC time constant—the photodiode capacitance multiplied by the sum of the load resistance and the diode series resistance. Speed can be increased by reducing equivalent capacitance. As speeds increase, two other factors can limit response time: diffusion of current carriers in the photodiode and time needed for carriers to cross the depletion region.

There are wide differences among detector response times. The slowest are photodarlingtons, with response times measured in terms of microseconds. With avalanche photodiodes and fast *pin* photodiodes, response time can be well under a nanosecond. (The fastest commercial detectors have response times in the 10-picosecond range.) I'll get into these characteristics when I examine the different types of detectors.

Signal Coding, Analog and Digital Modulation

Signal format also affects performance of a fiber-optic receiver. The simplest type is straightforward analog intensity modulation. In this case, it is the job of the receiver to reconstruct the transmitted waveform with as little distortion as possible. Accuracy of the reproduced waveform depends on intensity of the received signal, linearity and speed of the receiver, and noise levels in the input signal and the receiver.

The detection of digital signals is strongly influenced by the coding scheme. Bit rates in digital fiber-optic systems are so much slower than the frequency of the light waves that digital modulation schemes can be considered as being superimposed on a steady optical carrier. The modulation is strictly binary; the light is either off or on. (However, in some high-speed systems, off may actually be a very low-level light signal when a laser light source is pre-biased.)

The two most common ways of optically encoding this kind of modulation are shown in Figure 8.5. One is return-to-zero (RZ) coding, where the signal level returns to a nominal zero level between bits. The other is no-return-to-zero (NRZ) coding, in which the signal does not return to zero but remains at one if two successive 1 bits are transmitted. The two differ in their effective speed, because RZ signals have twice as many pulses. Each modulation scheme has its own advantages.

● Rise and fall times limit receiver bandwidth and bit rate.

● Coding format is important for digital signals.

FIGURE 8.5.

Signal levels for digital coding.

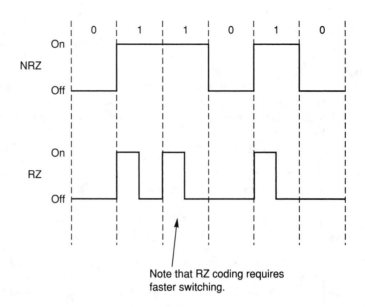

Note that RZ coding requires faster switching.

Device Geometry

As mentioned earlier, detector geometry can influence detector speed and sensitivity. You should also be aware of another geometrical factor: detector active area.

> All light emerging from a fiber should fall onto a detector's light-sensitive area.

For a detector to operate efficiently, all light emerging from the fiber should fall onto its light-sensitive (active) area. This means that the detector's active area should be larger than the fiber core (how much larger depends on how light is transferred to the detector). If the output fiber has core diameter d and half-acceptance angle θ, and is a distance S from the detector; then it will project onto the detector a spot with diameter D:

$$D = d + 2S \tan\theta$$

If that spot size is larger than the active area of the detector, some light is lost and sensitivity is reduced.

Most detectors have active areas larger than the cores of multimode fibers, so normally little light misses them. However, losses can occur if the fiber is far from the detector, if the fiber core is large, or if the fiber end and detector are misaligned.

On the other hand, detector speed is limited by its area—the larger the device, the longer it takes to respond. For the simple *pin* detectors described later, bandwidth decreases with increasing device width, according to

$$\text{BANDWIDTH} = \text{CONSTANT/DEVICE WIDTH}$$

The fastest detectors have active areas only a little wider than the core of a single-mode fiber.

Dynamic Range

Another concern in receivers is dynamic range—the range of input power over which they produce the desired output. At first glance, it might seem that the higher the input power, the better the response would be. However, that isn't the case, because any receiver responds linearly to input over only a limited range. Once input power exceeds the upper limit of that range, signals are distorted, leading to high noise in analog systems and errors in digital transmission.

The basic problem can be seen in Figure 8.6, which shows receiver output as a function of input light power. In the lower part of the curve, the response is linear. An increase in input power by an amount Δp produces an increase in output current Δi, where $\Delta i/\Delta p$ is the slope of that part of the curve. Typically, the response is nearly linear at low levels, although noise overwhelms the signal at the lowest powers. However, if the input power is too high, the receiver response saturates and output power falls short of the expected level. The result is distortion, much as when an audio speaker is driven with more power than it can handle. Exceeding the dynamic range of a digital receiver can likewise increase the bit-error rate. In either case, inserting attenuators can reduce average signal intensity to a level within the receiver's dynamic range.

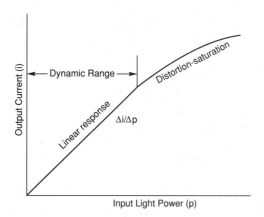

FIGURE 8.6.

Receiver output as a function of input light power.

The dynamic ranges of detectors vary widely, with those intended for short-distance communication often having the widest ranges because they may be used with different lengths of fiber. Some receivers used in long-distance systems have much more limited dynamic ranges and may be limited to input powers below 100 μW. However, some receivers include automatic gain-control circuits to prevent high input signals from overloading the output. Important limits on dynamic range come from the receiver circuitry; as load resistance rises, nonlinear response begins to occur at lower input powers.

Detector Types

The detectors used in fiber-optic receivers are more complex than the simple photodiodes described earlier in this chapter. More sensitive detectors, with higher current outputs, are normally used. Some provide enough output current to serve as receivers by themselves (when packaged), whereas others require external electronics to amplify and process electrical signals. The same basic structures can be used in any semiconductor, but in practice the lowest-cost types (phototransistors and photodarlingtons) are used only in silicon.

pn and *pin* Photodiodes

Photodiode sensitivity is improved by sandwiching an undoped intrinsic region between the p and n regions.

You saw earlier that fiber-optic photodiodes are reverse-biased so the electrical signal is a current passing through the diode. Such a detector is said to be operating in the photoconductive mode because it produces signals by changing its effective resistance. However, it is not strictly a resistive device, because it includes a semiconductor junction, forming a *pn* photodiode. (True "photoconductive" detectors exist, in which light produces current carriers that increase the conductivity of a bulk semiconductor that lacks a junction layer, but they are not used in fiber-optic systems.)

Reverse biasing draws current carriers out of the central depleted region, blocking current flow unless light frees electrons and holes to carry current. The amount of current increases with the amount of light absorbed, and the light absorption increases with the thickness of the depleted region. Depletion need not rely entirely on the bias voltage. The same effect can be obtained if a lightly doped or undoped intrinsic semiconductor region is between the p- and n-doped regions shown in Figure 8.7. In a sense, such *pin* (p-intrinsic-n) photodiodes come pre-depleted because the intrinsic region lacks the impurities needed to generate current carriers in the dark. This design has other practical advantages. By concentrating absorption in the intrinsic region, it avoids the noise and slow response that occur when the p region of ordinary *pn* photodiodes absorbs some light. The bias voltage is concentrated across the intrinsic semiconducting region because it has higher resistivity than the rest of the device, helping raise speed and reduce noise.

pin detectors can have response times well under 1 ns and dynamic ranges of 50 dB.

The speed of *pin* photodiodes is limited by variations in the time it takes electrons to pass through the device. This time spread can be reduced in two ways—by increasing the bias voltage and/or by decreasing the thickness (and width) of the intrinsic layer. Reducing intrinsic layer thickness must be traded off against detector sensitivity because this reduces the fraction of the incident light absorbed. Typical biases are 5-20 V, although some devices have specified maximum bias above 100 V. Typical response times range from a few nanoseconds to about 10 picoseconds. Sensitivity of silicon *pin* detectors is about 0.7 A/W at 800 nm; InGaAs is somewhat more sensitive at longer wavelengths. An important attraction of *pin* photodiodes is a large dynamic range; their output-current characteristics can be linear over six decades (50 dB).

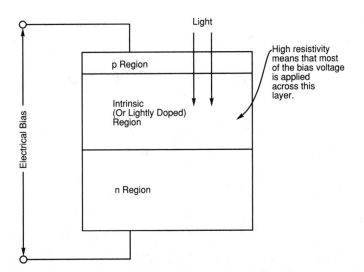

FIGURE 8.7.

A simple pin *photodiode.*

The speed and sensitivity of *pin* photodiodes are more than adequate for most fiber-optic applications, and they are widely used even in high-performance systems. Often they are attached to electronic preamplifiers to boost sensitivity.

pin photodiodes are widely used because of their high speed and good sensitivity.

The designs of actual *pin* photodiodes are more complex than this simple example, particularly at high speeds, like the multigigahertz detector in Figure 8.8. Signals at 1200 to 1600 nm pass through the antireflection coating and upper layer of InP (which are transparent at those wavelengths) and are absorbed in the intrinsic InGaAs layer. Other designs direct light through the InP substrate, which is also transparent to the signal wavelengths.

Phototransistors

The simplest detector with internal amplification is the phototransistor. Those familiar with transistors can view it as a transistor in which light generates the base current. An alternative is to see a phototransistor as a *pn* photodiode within a transistor. In practice, light normally reaches the base through or around the emitter, which has a wide band gap so it is transparent to the wavelengths detected. Most commercial phototransistors are made of silicon; their most common uses are in inexpensive sensors, but the same devices can be adapted for low-cost, low-speed fiber-optic systems.

Phototransistors both sense light and amplify light-generated current; they are used only in low-cost, low-speed systems.

The photocurrent generated in the base-emitter junction is amplified, like base current in a conventional transistor, giving much higher responsivity than a simple photodiode. However, this increase comes at a steep price in response time and linearity and at some cost in noise. You can see this by looking at Table 8.2, which compares silicon photodetectors. In practice, phototransistors are limited to systems operating below the megahertz range.

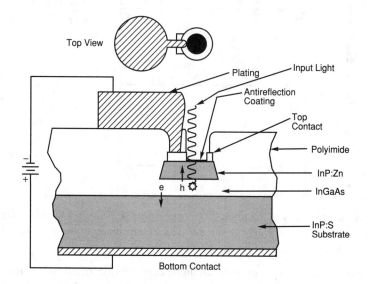

Table 8.2. Typical detector characteristics.

Device	Responsivity	Rise Time	Dark Current
Phototransistor (Si)	18 A/W	2.5 μs	25 nA
Photodarlington (Si)	500 A/W	40 μs	100 nA
pin photodiode (Si)	0.5 A/W	0.1-5 ns	10 nA
pin photodiode (InGaAs)	0.8 A/W	0.01-5 ns	0.1-3 nA
Avalanche photodiode (Ge)	0.6 A/W	0.3-1 ns	400 nA
Avalanche photodiode (InGaAs)	0.75 A/W	0.3 ns	30 nA
Si *pin*-FET (detector-amp)	15,000 V/W	10 ns	NA (output is V)
InGaAs *pin*-FET (detector-amp)	5,000 V/W	1-10 ns	NA (output is V)

Recent research has shown that much faster operation is possible in phototransistors containing heterojunctions of two materials, such as GaAs and GaAlAs, similar to the heterojunctions in lasers described in Chapter 6. That structure reduces the transit times and device capacitance that slow response of silicon phototransistors. (Heterojunctions cannot be made in silicon or germanium, only in III-V semiconductors, such as gallium arsenide and indium phosphide.) Laboratory devices have shown bandwidths well over 100 MHz, but the technology has yet to find practical use.

Photodarlingtons

The photodarlington is a simple integrated darlington amplifier in which the emitter output of a phototransistor is fed to the base of a second transistor for amplification. Adding the second transistor increases responsivity but lowers speed and increases noise. Thus, the photodarlington's uses are even more narrowly constrained than the phototransistor's, but it offers higher responsivity for low-cost, slow-speed applications.

Avalanche Photodiodes

A different approach to providing internal amplification within a detector is the avalanche photodiode (APD). It relies on avalanche multiplication, in which a strong electric field accelerates current carriers so much that they can knock valence electrons out of the semiconductor lattice. The result at high enough bias voltage is a veritable avalanche of carriers—thus the name.

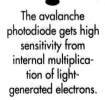

The avalanche photodiode gets high sensitivity from internal multiplication of light-generated electrons.

Multiplication factors—the degree to which an initial electron is multiplied—typically range from 30 to about 100. The multiplication factor M is

$$M = 1/[1-(V/V_B)^n]$$

where V is the operating voltage, V_B is the voltage at which the diode would break down electrically, and n is a number between 3 and 6 dependent on device characteristics. Care must be taken in operation because exceeding the breakdown voltage could fatally damage the device. A representative plot of this characteristic as bias voltage approaches breakdown is shown in Figure 8.9. (Note that to reach multiplication factors of 100, bias must approach within a few percent of breakdown.) Breakdown voltages normally are well over 100 V and in some devices can range up to a few hundred volts.

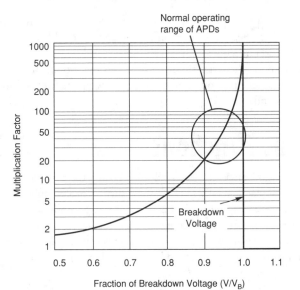

FIGURE 8.9.

Increase of multiplication factor in APD.

APDs are fast, but the uneven nature of multiplication introduces noise. Avalanche gain is an average; not all photons are multiplied by the same factor. Signal power increases roughly with the square of the multiplication factor M; for moderate values of M, it increases faster than noise. However, as M increases to high levels, noise increases faster than the square of M (approximately as $M^{2.1}$). As a result, avalanche photodiodes have an optimum multiplication value, typically between 30 and 100.

The biases used for avalanche photodiodes are much higher than the few volts normally used in semiconductor electronics. The need for special circuits to provide this drive voltage, and to compensate for the temperature sensitivity of APD characteristics, makes APD receivers more complex than *pin* types and has limited their applications.

pin-FET and Integrated Receivers

As mentioned earlier, external electronics can amplify the electrical signal from a photodetector. Some receivers integrate the functions of detector and amplifier (or preamplifier) in a single hybrid or integrated circuit that serves as a detector-(pre)amplifier. These devices are called integrated detector-amplifiers, detector-amplifiers, or *pin*-FETs (the last because the preamplifier uses field-effect transistor or FET circuitry).

Figure 8.10 shows the type of circuit used in a hybrid receiver with *pin* photodiode and low-noise FET preamplifier. This circuit amplifies the electrical signal before it encounters the noise associated with the load resistor, increasing S/N ratio and output power. The amplifier circuit also converts the current signal from the photodiode into a voltage signal, as used by most electronic devices. The voltage level depends on the circuitry. Often, the circuit includes automatic gain control so the voltage level will be compatible with later amplification stages. However, some *pin*-FET circuits do not limit gain; their responsivity can vary widely, but a typical value is around 10,000 V/W.

Integrated detector-preamplifiers have become popular for many moderate-speed fiber-optic applications because of their simplicity and reasonable cost. Unlike avalanche photodiodes, they do not require voltages above the 5 V levels normally needed by semiconductor electronics. Their rise times tend to be slower than the fastest discrete *pin* photodiodes or avalanche photodiodes, but they are adequate for transmitting hundreds of megabits per second.

Electronic Functions

Converting an optical signal into electrical form is only the first part of a receiver's job. The raw electrical signal generally requires some further processing before it can serve as input to a terminal device at the receiver end. Typically, photodiode signals are weak currents that require amplification and conversion to voltage. In addition, they may require such cleaning up as squaring off digital pulses, regenerating clock signals for digital transmission, or filtering out noise introduced in transmission. The major electronic functions are as follows:

1. Preamplification
2. Amplification
3. Equalization
4. Filtering
5. Discrimination
6. Timing

If you're familiar with audio or other electronics, you will recognize some of these functions. Not all are required in every receiver, and even some of those included may not be performed by separate, identifiable devices. A phototransistor, for example, both detects and amplifies. And many moderate-performance digital systems don't need special timing circuits. Nonetheless, each of these functions may appear on block diagrams. Their operation is described briefly below.

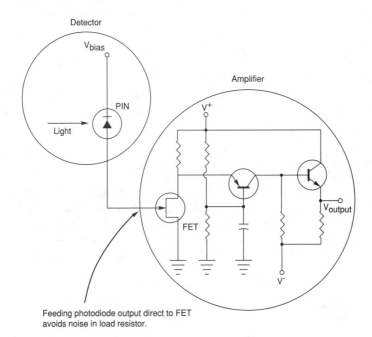

FIGURE 8.10.
Circuit for a pin-*FET receiver.*

Preamplification and Amplification

Typical optical signals reaching a fiber-optic receiver are 1-10 μW and sometimes lower. If a *pin* photodiode with 0.6 to 0.8 A/W responsivity detects such signals, its output current is in the microampere range and must be amplified for most uses. In addition, most electronics require input signals as voltage, not current. Thus, detector output must be amplified and converted.

●
Microampere-level
pin detector outputs
must be converted
to logic-level
voltages.

Receivers may include one or more amplification stages. Often the first is called pre-amplification because it is a special low-noise amplifier designed for weak input. (An optical amplifier placed in front of a detector is also called a preamplifier, as described in Chapter 9.) In some cases, as mentioned earlier, the preamplifier may be packaged with the detector. The preamplifier output often goes into an amplifier, much as the output of a tape-deck preamplifier goes to a stereo amplifier that can produce the power needed to drive speakers.

Equalization

Detection and amplification can distort the received signal. For example, high and low frequencies may not be amplified by the same factor. The equalization circuit evens out these differences, so the amplified signal is closer to the original. Much the same is done in analog high-fidelity equipment, where standard equalization circuits process signals from tape heads and phonograph cartridges so they more accurately represent the original music.

Filtering

Filtering blocks noise while transmitting the signal.

Filtering helps increase the S/N ratio by selectively attenuating noise. This can be important when noise is at particular frequencies (e.g., a high-frequency hiss on analog audio tapes). It is most likely to be used in fiber optics to remove undesired frequencies close to the desired signal, such as harmonics.

Discrimination

Discrimination circuits generate digital pulses from an analog input.

So far, the functions you've looked at are needed to regenerate the original waveform for both analog and digital receivers. However, a further stage is needed to turn a received analog signal back into a series of digital pulses—decoding and discrimination. Rectangular pulses that started with sharp turn-on and turn-off edges have been degraded into unboxy humps, as shown in Figure 8.11. Dispersion may have blurred the boundaries between pulses.

FIGURE 8.11.

Discrimination cleans up noise to regenerate digital pulses.

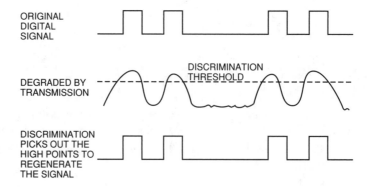

From a theoretical standpoint, this rounding of square pulses represents loss of high frequencies in the signal. Mathematically, a square-wave pulse is the sum of many sine waves of different frequency, the first with frequency equal to the square-wave frequency and others with frequencies that are integral multiples (harmonics) of the square-wave frequency. The highest frequencies (the highest-order harmonics) make up the sharp rising and falling edges of the pulse, so if they are lost, the pulses lose their square edges.

The remaining low-frequency components contain most of the information needed, but they are not clean enough to serve as input to other electronic devices. Regeneration of clean pulses requires circuitry that decides whether or not the input is in the on or off state by comparing it to an intermediate threshold level. The decision circuit generates an "on" pulse if the power is above the threshold; otherwise it produces an "off" signal. Care must be taken in selecting this threshold level to avoid misinterpreting input; too low a threshold, for example, could turn noise spikes in the off state into signal pulses.

● Pulse regeneration requires circuitry to decide if the input is on or off.

Timing

Another essential task in many receivers is resynchronization of the signal. Conventional digital signals are generated at a characteristic clock rate, such as once every microsecond. As a signal is transmitted, random errors in timing gradually build up and can eventually reach a level where they are comparable to the duration of a pulse. Then the receiver can confuse successive pulses, causing transmission errors.

● Timing of digital pulses often must be resynchronized.

Random timing errors are called jitter. They can pose serious practical problems. Suppose, for instance, that a digital signal passed through ten regenerators, each of which could reconstruct the series of pulses with accuracy of ±0.1 μs. If all those errors were negative, they would add to 1 μs; if that is the clock rate, the electronics at the receiving end might assume that the received bit was the bit supposed to come in the next time slot. Accumulated jitter can have unpredictable effects on external circuits. Retiming the pulses when they are regenerated can help combat jitter problems.

Packaging Considerations

As with transmitters, packaging is an important concern with receivers. The basic requirements are simple and easy-to-use mechanical, electronic, and optical interfaces. The main mechanical issues are mounts. The electronic interfaces must allow both for input of bias voltage and amplifier power (where needed) and for output of signals in the required format. The details can differ significantly.

● Mounting is the main mechanical issue in packaging.

Optical interface requirements are simpler than those for transmitters because transferring light from fibers to detectors is simpler. Detector active areas are larger than the emitting areas of most fibers. Thus, simply butting the fiber against the detector is adequate for most purposes. Most short-distance receivers using multimode fiber come with integral connectors that mate with fiber connectors. The connector case either delivers light directly to the detector or delivers the light to a fiber that carries it to the detector. Connector mounts are also provided on long-distance single-mode fiber receivers. Light transfer is simplified because the large core of a multimode fiber can collect virtually all the light delivered by the small core of a single-mode fiber.

In general, packaged receivers look very much like transmitters, and packaged detectors look like light sources. You really do have to read the labels to tell them apart. Detector modules are packaged inside receivers just as light-source modules are put inside transmitters. Internal design constraints become increasingly severe at high frequencies because of the problems inherent in high-frequency electronic transmission.

Sample Receiver Circuits

Details of receiver circuitry vary widely with the type of detector used and with the purpose of the receiver. For purposes of this book, I will cover only a few simple circuits for important devices and avoid detailed circuit diagrams.

Photoconductive Photodiodes

The typical *pin* or *pn* photodiode used in a fiber-optic receiver is used in a circuit with a reverse-bias voltage applied across the photodiode and a series load resistor, such as that shown in Figure 8.12. In this mode, the photodiode is photoconductive because the photocurrent flowing is proportional to the nominal resistance of the illuminated photodiode. This simple circuit converts the current signal from a photodiode to a voltage signal.

Photoconductive photodiodes have a load resistor in series with the bias voltage.

FIGURE 8.12.

Basic circuit for photoconductive pin *or* pn *photodiode.*

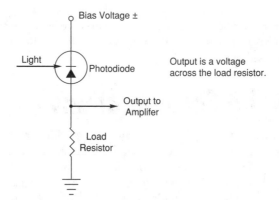

The division of the bias voltage between the photodiode and the fixed resistor depends on illumination level. The higher the illumination of the photodiode, the more current it will conduct and, thus, the larger the voltage drop across the load resistor. In the simple circuit shown, the signal voltage is the drop across the load resistor. Most circuits are more complex, with amplification stages beyond the load resistor, as in *pin*-FET and detector-preamplifier circuits. Figure 8.13 is an example that includes automatic gain control to increase dynamic range.

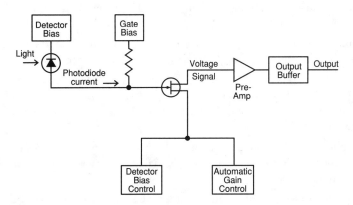

FIGURE 8.13.

Block diagram of pin-*FET receiver circuit.*

Avalanche Photodiode Circuits

The circuits used for avalanche photodiodes are conceptually similar to those used for photoconductive *pin* photodiodes. However, because of the high bias voltages required and the sensitivity of the photodiode to bias voltage, care must be taken to assure stable bias voltage. This adds to circuit complexity, as shown in the block diagram of Figure 8.14.

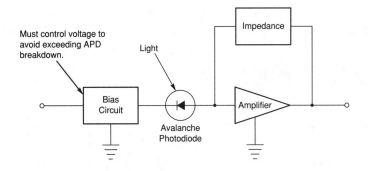

FIGURE 8.14.

Basic receiver circuit for avalanche photodiode.

What Have You Learned?

1. Basic elements of a receiver are the detector, amplification stages, and demodulation or decision circuits.

2. Many fiber-optic detectors are reverse-biased semiconductor photodiodes. Light striking the depleted region near the junction generates free electrons and holes, so a current flows through the diode; this current is the signal. Simple circuits can convert the current signal into a voltage needed for electronic processing.

3. Other fiber-optic detectors use avalanche photodiodes, with an internal amplification stage that requires high bias voltage.

4. Different detectors are needed for different wavelengths. Silicon is used at 800-950 nm. Germanium can be used at 1300 nm, and InGaAs is usable between 1200 and 1600 nm.

5. Phototransistors and photodarlingtons have internal amplification that gives them high responsivity, but they are too slow for most fiber-optic applications.

6. Detectors operate best over a limited dynamic range. At higher powers, they distort received signals, whereas at lower powers the signal can be lost in the noise. Automatic gain control can extend the dynamic range of *pin*-FET detectors.

7. *pin* photodiodes are often packaged with FET preamplifiers in *pin*-FETs that are common fiber-optic detectors. That packaging avoids load-resistor noise but slows response time.

8. The input stages of analog and digital receivers are similar because by the time the signals reach the receivers they are weak and rounded. The difference in those receivers is in the electronic processing after amplification.

9. Digital receivers include a discrimination stage to regenerate digital pulses from analog waveforms. They may also include retiming circuits.

What's Next?

In Chapter 9, I move on to optical amplifiers and to electro-optic repeaters and regenerators, which combine transmitters and receivers.

Quiz for Chapter 8

1. What is the main difference between an analog and a digital receiver?

    a. Special amplification circuitry.

    b. The presence of decision circuitry to distinguish between on and off signal levels.

    c. The two are completely different.

    d. Digital receivers are free from distortion.

2. Photodiodes used as fiber-optic detectors normally are

    a. reverse-biased.

    b. thermoelectrically cooled.

    c. forward-biased.

    d. unbiased to generate a voltage like a solar cell.

3. What bit-error rate is typically specified for digital telecommunication systems?

   a. 40 dB.

   b. 10^{-4}.

   c. 10^{-6}.

   d. 10^{-9}.

   e. 10^{-12}.

4. Silicon detectors are usable at wavelengths of

   a. 800-900 nm.

   b. 1300 nm.

   c. 1550 nm.

   d. all of the above.

5. The fastest photodetectors have response times of

   a. a microsecond.

   b. hundreds of nanoseconds.

   c. tens of nanoseconds.

   d. a few nanoseconds.

   e. around 10 picoseconds.

6. A *pin* photodiode is

   a. a point-contact diode detector.

   b. a detector with an undoped intrinsic region between p and n materials.

   c. a circuit element used in receiver amplification.

   d. a photovoltaic detector.

7. Match the characteristics listed below with the type of detector.

   a. *pin* photodiode

   b. avalanche photodiode

   c. *pin*-FET receiver

   d. photodarlington

   e. phototransistor

   A. Response time can be less that 1 ns; responsivity under 1 A/W.

   B. Rise time a few microseconds; responsivity about 20 A/W.

   C. Rise time about 10 ns; responsivity thousands of volts per watt.

   D. Response time 1 ns or less; responsivity tens of amperes per watt.

   E. Rise time tens of microseconds; responsivity hundreds of amperes per watt.

8. Which of the following does not include an internal amplification stage?

   a. Photodarlington.

   b. Phototransistor.

   c. Avalanche photodiode.

   d. *pin*-FET.

   e. *pin* photodiode.

9. A receiver's bandwidth can be limited by

   a. dynamic range.

   b. rise time.

   c. responsivity.

   d. quantum efficiency.

   e. bias voltage.

10. A discrimination or decision circuit

   a. filters out noise in analog receivers.

   b. tells off from on states.

   c. decides which pulses to amplify.

   d. controls input level to avoid exceeding a receiver's dynamic range.

Repeaters, Regenerators, and Optical Amplifiers

About This Chapter

In the last three chapters, I have talked about the components on the ends of fiber-optic systems: light sources, transmitters, and receivers. However, if you look carefully you will find some components in the middle.

Repeaters, regenerators, and optical amplifiers all serve to stretch transmission distances. Inserted into a fiber system at a point where the original signal is becoming weak, they generate a stronger signal, letting the system operate over much longer distances. Repeaters and regenerators convert the signal into electronic form before amplifying it; optical amplifiers work directly on light.

The Distance Problem

Signals fade away with distance when traveling through any type of cable. The further you go, the fainter they become, until they become too faint to detect reliably. As you saw in the last chapter, when a digital signal fades below a certain level, the bit-error rate rises rapidly. Likewise, an analog signal becomes noisy or distorted, like a distant radio station.

Communication systems add amplifiers or repeaters to cables to avoid this problem. The repeater detects the signal before it becomes too weak, then amplifies it to a higher level and sends it through another length of cable. (Repeaters don't do much good if they amplify a signal that is already distorted.) Figure 9.1 shows how repeaters work.

FIGURE 9.1.

Role of a repeater in a fiber-optic system.

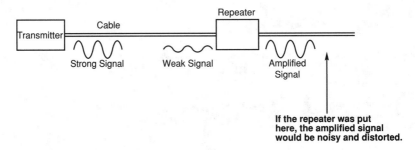

Signals require amplification if they travel far enough through either metal or fiber-optic cables. However, as you saw earlier, fiber optics attenuate high-speed signals much less than coaxial cables. That means that signals can travel much further between amplification stages in optical fibers than in coaxial cables. Exactly how far depends on the system design, but in general it is many times further. For example, coaxial cables need repeaters every few thousand feet (roughly a kilometer) to carry hundreds of megabits per second of digital signals, or dozens of cable television channels. Fiber can carry the same signals at least 20 to 30 miles (30 to 50 kilometers) before the signal grows weak enough to need amplification. It's no wonder cable and telephone companies prefer fibers for many applications.

> Signals require amplification because they fade with distance in both wires and fibers. Fibers can carry high-speed signals much further.

To be fair, it isn't always that simple because of other differences between metal cables and fibers. Optical signals do not divide the same way electrical signals do, so a fiber system in general can drive fewer terminals. Thus, if you're trying to split signals among many terminals that aren't very far away, fiber systems may need more amplification. However, the basic idea remains the same.

> Regenerators clean up signals as well as amplify them.

In its simplest form, a repeater simply amplifies the input signal. That means that it amplifies any noise that comes along with the signal, which can sometimes be significant. However, digital systems usually include "regenerators" rather than simple repeaters. As the name implies, regenerators use the input to try to regenerate or reconstruct the original signal.

First the regenerator amplifies the signal, noise and all. Then it looks at changes in the signal strength and decides that all signals above a certain level are 1s rather than 0s. This discrimination stage converts the noisy old input waveform into crisp new digital pulses, as shown in Figure 8.11. Then retiming circuits make sure that the pulses fall into the right time slots and don't drift away to be mistaken for a pulse in an earlier or later time slot.

Types of Amplification

So far I've been deliberately vague in talking about what goes on in the amplification stage. This is because two different types of amplification are possible—electronic and optical.

Most fiber-optic systems use electronic amplification. They first convert the input optical signal into electronic form, then amplify it (and clean it up) to drive another optical transmitter. You can think of an electronic (or electro-optic) repeater as a receiver and transmitter placed back to back, so the receiver output drives the transmitter, as shown in Figure 9.2a. In general, electro-optic repeaters also have electronic circuits that clean up or regenerate the signal, as described earlier. Their overall function is very similar to the electronic repeaters used for communications via wires or radio frequencies.

In contrast, an optical amplifier takes a weak input optical signal and amplifies it to generate a strong output optical signal, as shown in Figure 9.2b. The signal is never converted into electronic form at all. (If you see wires attached to an optical amplifier, they are to provide electrical power.) However, present optical amplifiers do not clean up or regenerate the input signal; they amplify it, noise and all.

●
Electro-optic repeaters must convert an optical signal into electronic form for amplification. Optical amplifiers directly amplify the optical signal.

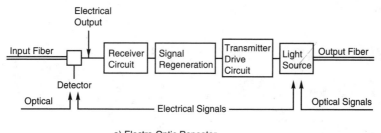

a) Electro-Optic Repeater

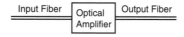

b) Optical Amplifier

FIGURE 9.2.
Electro-optic and optical amplifiers.

This conceptual simplicity is one attraction of optical amplifiers. The simplicity ultimately promises lower costs, because an optical amplifier requires fewer components. (In practice, they are much more expensive components, so optical amplifiers are costly today, but prices should come down with the traditional economies of scale and learning curve that reduce production costs.) Optical amplifiers also offer higher reliability, likewise because they contain fewer components. That is critical for undersea cables, because hauling dead repeaters up from the ocean floor is difficult and expensive.

●
Optical amplifiers are conceptually simpler than electro-optic regenerators.

●

Optical amplifiers
can accommodate
different signal
speeds.

However, for many applications the critical advantage of optical amplifiers is that they can amplify whatever signal a fiber-optic system is carrying, as long as it's within their operating range. Electro-optic repeaters must be designed to operate at a certain data rate, such as 45 Mbit/s. If you wanted to send 150 Mbit/s through the system, you would have to replace all the repeaters. That is not the case with optical amplifiers; they could carry 150 Mbit/s as easily as 45 Mbit/s, as long as both signals were in the wavelength range of the amplifier. Optical amplifiers also have very low signal noise levels.

Another potentially important advantage of optical amplifiers is their ability to simultaneously amplify multiple signals at different wavelengths in their operating band. For example, an erbium-doped fiber amplifier can amplify signals at 1530, 1540 and 1550 nm, without causing them to interfere with each other. That offers system designers another opportunity to upgrade transmission capabilities.

At this writing, few optical amplifiers are in use, but they are expected to spread rapidly. A major drawback is that current optical amplifiers are practical only at 1550 nm, but most current fiber systems carry signals at 1300 nm. An additional concern is the need to compensate for or control pulse dispersion, which can accumulate over long fiber distances to limit signal transmission speed. Care must also be taken to avoid the buildup of amplified noise.

Electro-Optic Repeaters and Regenerators

●

Electro-optic
repeaters are
designed to operate
at a specific signal
speed.

You saw earlier that electro-optic repeaters and regenerators are essentially a receiver and transmitter placed back to back and packaged in a single unit. The input end performs the usual receiver functions; the output end performs standard transmitter functions.

Because they contain sensitive electronic components, repeaters and regenerators require controlled operating conditions. Telephone companies have placed many repeaters inside switching offices or other telephone facilities. Repeaters installed outside require sealed cases for environmental protection, including temperature and humidity control. Repeater cases for submarine cables must withstand high pressures and immersion in salt water.

I won't go through the details of transmitters and receivers described in Chapters 7 and 8. However, a few points deserve special emphasis.

- Repeaters and regenerators are designed to operate at a specific transmission speed and format. They may contain timing circuits that generate clock signals at a specific rate, and their electronics may be optimized to operate at that rate. Circuits made to operate at 45 Mbit/s may contain components too slow to operate at 150 Mbit/s or 400 Mbit/s. Likewise, circuits may require specific signals. In short, repeaters and regenerators are not "transparent" to signal format.

● Regeneration can be important in long systems where noise and pulse dispersion can accumulate to obscure the signal. Pulse dispersion is the more important problem in many cases, but careful design to control dispersion should allow the use of optical amplifiers.

● True regeneration is usually limited to digital systems, where the digital structure of pulses makes it possible to discriminate between signals and noise. As long as the pulses can be recognized as pulses, the regenerator can produce new ones with the same timing and signal information. However, receivers have no way to remove noise from analog signals, so repeaters merely amplify it and pass it along. (Because they generate a new signal, repeaters can control the accumulation of pulse dispersion that can limit transmission bandwidth of a long analog system.)

Optical Amplifiers

Optical amplifiers boost optical signal strength internally, without first converting the signal into electrical form. They work on the stimulated emission principle that is basic to the laser, described in Chapter 6, and are essentially special-purpose lasers designed to amplify signals from an external source rather than to generate their own light. They amplify a weak signal beam that enters one side, to produce a stronger signal emerging from the other side. The light makes a single pass through the amplifier, which—unlike a laser—lacks mirrors.

Two basic types of optical amplifiers have been developed. Optical fibers doped with elements that amplify light at certain wavelengths have come the furthest and are finding system applications. The operating wavelength depends largely on the dopant. Semiconductor optical amplifiers are essentially semiconductor lasers with their ends coated to suppress reflections back into the chip, so they can amplify light passing through them. Researchers are investigating other possibilities.

A critical concern for the practical use of optical amplifiers is their operating wavelength, which determines their compatibility with existing systems and how well they fit into new systems. Most interest has centered on the two windows where silica fibers have their lowest attenuation: 1300 and 1550 nm. Most current systems operate at 1300 nm, but the best optical amplifiers operate at 1550 nm. As a result, designers of high-performance, long-distance systems are switching to 1550 nm for new systems.

Doped-Fiber Amplifiers

Optical-fiber amplifiers are fibers with cores doped with elements that can amplify light at certain wavelengths. The wavelength depends primarily on the dopant, and secondarily on the fiber composition. The best-developed fiber amplifiers are types doped with the rare earth erbium, which has good gain between about 1520 and 1560 nm. Praseodymium-doped

● Regeneration can prevent dispersion limits on digital transmission.

● Optical amplifiers include fibers doped with rare earths and special types of semiconductor lasers.

● The best-developed optical amplifiers are erbium-doped fiber amplifiers, which work at 1550 nm.

fibers are the best amplifiers so far available in the 1300 nm window, but their performance is not as good as that of erbium-doped fibers. Developers hope to improve praseodymium fibers enough that they could be used to upgrade existing 1300 nm systems.

Figure 9.3 shows the basic workings of a fiber amplifier. A weak optical signal enters from the left, passing through an optical isolator (which blocks scattered light that could cause noise) and a filter that transmits the signal wavelength. It then enters a coil of doped fiber, typically several meters (or several yards) long. The fiber amplifier is illuminated from the other end by a stronger beam at a shorter wavelength from an external laser, usually a semiconductor laser. Light from the pump laser excites the dopant atoms, raising them to a higher energy level. Light at the signal wavelength can stimulate these excited atoms to emit their excess energy as light at the signal wavelength, and in phase with the signal pulses. A coupler at the end of the fiber amplifier (at right) routes the amplified signal to the output fiber, separating it from the pump light.

FIGURE 9.3.

A fiber amplifier.

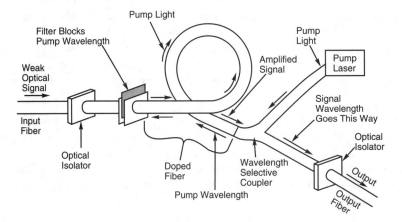

I should point out a few key features of this generic fiber amplifier. The pump laser operates continuously and has to supply photons with more energy than those the dopant atoms emit. Thus, the pump wavelength is shorter than the signal wavelength. The fiber amplifier generates light at the signal wavelength only when an input signal is passing through it. When the input level is zero, it should produce no light. (It may actually emit a little light, which represents noise.) Optical amplifiers work by stimulated emission, so the amplified light is in phase with the input signal. Add it all together and you have an efficient low-noise amplifier of optical signals that looks simpler than an electro-optic regenerator.

Details depend on the type of fiber amplifier. Standard erbium-doped fiber amplifiers are pumped by semiconductor lasers that emit light at 980 or 1480 nm and amplify light between about 1520 and 1560 nm. Ytterbium can be added to the fiber to absorb light at a broad range of other wavelengths, including the 1064 nm output of neodymium-YAG lasers; the ytterbium can transfer the energy it absorbs to the erbium atoms, exciting them so the fiber amplifier can be pumped with other wavelengths. Praseodymium-doped fibers can amplify light at 1280 to 1330 nm when pumped with a laser at 1017 nm.

I mentioned earlier that performance of a fiber amplifier depends on the fiber material as well as on the dopant, because the host material affects the energy-level structure of the dopant. As a result, the degree of amplification at different wavelengths can differ greatly among materials. One big advantage of erbium is that it works fine when doped into the silica glass used in standard optical fibers. Unfortunately, praseodymium does not. It works best in fluoride glasses, like those covered in Chapter 4. This means that it can't be spliced to silica fibers and would require special enclosures in harsh environments. (Even more frustrating is the case of neodymium. It has a laser line that is strong near 1320 nm in some bulk materials but that when doped into fibers does not amplify light at the 1300 to 1310 nm wavelength of most fiber systems.)

● Erbium is doped into silica glass fibers; praseodymium must be doped into non-silica fibers.

The actual degree of amplification depends on the input power as well as on the wavelength. You can measure amplification in two ways. One is to measure the ratio of output to input signal in decibels—for a small input signal. For example, if an input signal of 1 µW is amplified to 1 mW, the gain is 30 dB. That, in fact, is a typical small-signal gain.

The other measurement is saturated output power—the maximum output power available from the amplifier. If you increased the input power, you would find that the amplification factor dropped. The amplifier would be saturated and unable to crank out any more power. For example, with a 1 mW input, the amplifier might generate a 20 mW output, corresponding to a gain of only 13 dB.

Figure 9.4 shows how the amplification (in dB) changes with wavelength and input power for an erbium-doped fiber amplifier. Note that at low input powers the gain is highest at 1530 nm, but that at higher input powers it becomes more uniform with wavelength. This has important consequences in designing systems that carry multiple wavelengths in the 1550 nm window. If you want to keep signals at 1530 and 1550 nm the same strength through the entire system (usually desirable), you have to recognize the differences in gain at the two wavelengths. The details become quite complex because overall system gain depends on the number of amplifiers, the input signal, and the wavelength; I won't deal with them here. However, I should stress that the problem is not insurmountable; it just requires careful design.

● Gain depends on wavelength and input power.

FIGURE 9.4.

Erbium-fiber amplifier gain versus wavelength at different input powers. (Courtesy of Corning Inc.)

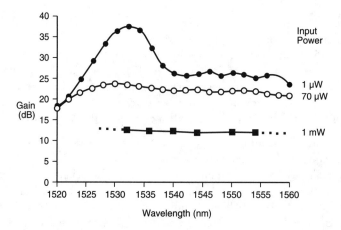

Erbium-doped fiber amplifiers are very good, but they are not perfect. They generate small amounts of noise by a process called amplified spontaneous emission. Earlier I showed how an external laser excited dopant atoms like erbium and left them waiting for an input signal to release their energy. Some of the dopant atoms don't wait but instead spontaneously emit light in the 1550 nm range. Some of that spontaneous emission travels along the fiber and is amplified or is simply transmitted along the communication fiber to the next amplifier, which dutifully amplifies it. Most spontaneous emission is at other wavelengths, but some is at the signal wavelength. The result after several amplifiers is a curve like the one in Figure 9.5. Good digital receivers can discriminate against this noise.

FIGURE 9.5.

Noise and signal in a fiber amplifier.

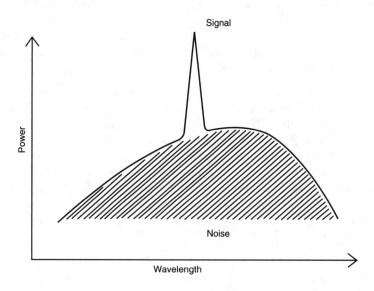

Fiber amplifiers also cannot compensate for dispersion in the transmission fiber. If there is dispersion, the dispersed signal is amplified, dispersion and all. Thus it becomes important to control dispersion for long transmission distances. One consequence is that optical-amplifier systems typically use dispersion-shifted fiber to give low dispersion at 1550 nm. Developers are working on ways to compensate for dispersion by including fiber segments with different types of dispersion, so total dispersion adds to a low level.

● Fiber amplifiers cannot compensate for dispersion or pulse jitter.

Pulse jitter, as well as dispersion, can accumulate over a long system with many amplifiers, potentially introducing noise. Developers are working on optical regenerators, which—like electro-optic regenerators—would retime and regenerate the received signal. Some concepts have been demonstrated in the laboratory, but at this writing they are not ready for practical use.

Semiconductor Laser Amplifiers

In principle, any laser can serve as an optical amplifier. Just remove the mirrors and send light from an external source through it, as you send light through a fiber amplifier. That led to early interest in semiconductor laser amplifiers. The idea is attractive because diode laser technology is well developed, and diode lasers are available at a wide range of wavelengths, including 1300 and 1550 nm. In practice, however, semiconductor laser amplifiers suffer serious problems that may limit them to certain applications.

● Semiconductor lasers can serve as optical amplifiers.

Figure 9.6 shows the basic idea of a semiconductor laser amplifier. The ends of the laser chip are coated with materials that reduce the reflection of light from the surface back into the semiconductor. Light from a single-mode fiber enters one side of the chip, is amplified in the active layer, and emerges from the other side. There, the output signal is collected by a second fiber. Ideally, none of the signal light is reflected back into the laser.

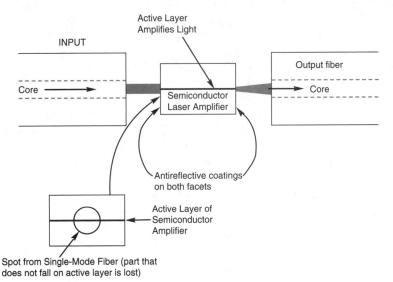

FIGURE 9.6.

Semiconductor laser amplifier.

Semiconductor laser amplifiers, like fiber amplifiers, have a wide operating bandwidth, so they can amplify signals over a range of wavelengths. The wavelengths where they have gain are controlled by composition, so they can be made to amplify at 1300 as well as 1550 nm. However, they suffer from serious practical problems that have limited their applications.

●
Semiconductor laser amplifiers suffer from high coupling loss and polarization sensitivity.

Look carefully at the expanded circle in the figure and you can see one big problem. The light spot in a single-mode fiber is a circle about 9 micrometers across. The light-emitting active layer in a semiconductor laser is several micrometers across, but usually under a micrometer thick. This means that a large fraction of the input light falls outside the active layer and is lost. (Losses in transferring the output into an optical fiber are much smaller.) The high transfer losses offset much of the gain possible in a semiconductor laser amplifier, if the input light comes from an optical fiber.

Another problem is more subtle. The gain of a semiconductor laser amplifier depends on the polarization of the input light; it may differ by 5-8 decibels for the two orthogonal polarizations. This is undesirable because standard single-mode fibers do not control polarization, so it fluctuates along the length of the fiber. That means that the signal gain depends on an uncontrollable factor—hence so does the output power. It is possible to reduce this loss by using pairs of semiconductor laser amplifiers, in series or parallel, but this gets cumbersome and leads to additional losses.

●
Semiconductor amplifiers can be integrated with diode lasers or other waveguide components.

High coupling losses can be avoided by integrating a semiconductor laser amplifier on a chip with other optical components. One possibility is to fabricate the semiconductor optical amplifier on the same chip as a semiconductor laser oscillator. The two can have the same active layer, with a waveguide transferring light efficiently between them—avoiding excessive losses.

Another possibility is to integrate a semiconductor amplifier with an optical switch or coupler fabricated in a semiconductor waveguide. Such optical components typically have high losses, but the semiconductor amplifier could offset them. Figure 9.7 shows one possible arrangement. Several narrow-line semiconductor lasers generate light at different wavelengths, which is combined in an optical coupler integrated on the same substrate, and split into several identical outputs to be sent to different places. Each coupler output is then amplified (at all wavelengths) by a semiconductor laser amplifier. In this example, the initial 1 mW output of the laser oscillator is reduced 20 dB by coupler losses, then that 0.01 mW signal (at each wavelength) is amplified 20 dB by the semiconductor amplifier—giving 1 mW output at each wavelength. Thus the gain of semiconductor laser amplifiers compensates for coupler loss, which would otherwise leave too little signal power for transmission.

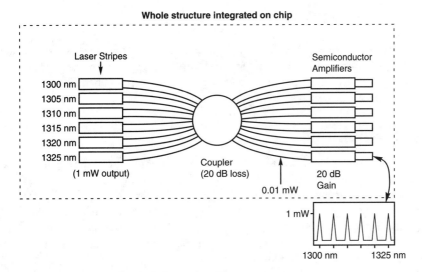

FIGURE 9.7.

Semiconductor amplifier integrated with laser array and waveguide coupler.

Optical Amplifier Systems

Optical amplifiers can serve various system roles, as shown in Figure 9.8. The most obvious is as in-line amplifiers in the middle of a system, to boost power level after transmission through a length of fiber so the signal can pass through another fiber segment. On each end, they can serve as post-amplifiers, raising signal strength at the transmitter, or as preamplifiers, which increase optical power to a receiver. In addition, they can be connected to components such as couplers and switches to compensate for their inherent losses. I'll take a brief look at each of these applications.

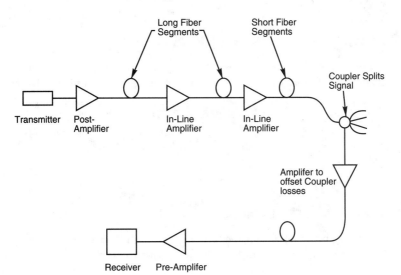

FIGURE 9.8.

Roles for optical amplifiers.

In-Line Amplifiers

In-line optical
amplifiers extend
transmission
distance.

In-line optical amplifiers compensate for signal attenuation in long stretches of optical fiber. They serve the same essential function as an electro-optic repeater, boosting the input signal level to generate an output that can drive the next segment of fiber. They are typically used in long systems, such as submarine cables.

Post-Amplifiers

Post-amplifiers boost
transmitter output.

Post-amplifiers are placed immediately after a transmitter to increase the signal being sent through a length of fiber. They are desirable in high-performance systems where a clean signal is critical, because raising transmitter output may degrade signal quality. (For example, by spreading the laser emission bandwidth, which would increase dispersion.) The optical amplifier takes the clean low-power signal from the transmitter (after external modulation, if it is used), then generates a higher-power signal that can travel further through the fiber without amplification.

Post-amplification is desirable in point-to-point links where it can provide enough power to avoid the need for repeaters. It can also be useful in generating signals for distribution to many subscribers, because the higher power lets the same signal be sent to more terminals.

Preamplifiers

Preamplifiers boost
receiver sensitivity.

As you saw in the last chapter, a preamplifier can increase receiver sensitivity by raising the optical power that reaches the detector. This makes it possible to stretch transmission distances and improve signal quality.

Offsetting Component Losses

Optical amplifiers
can offset compo-
nent losses.

Optical amplifiers offer an attractive way to overcome the high losses inherent in optical components such as couplers and switches. Optical couplers must physically divide the signal among different terminals, so they reduce its strength. Splitting the signal in half causes a 3 dB loss, whereas dividing it among 20 terminals causes an 11 dB loss. As you saw in Figure 9.8, an optical amplifier can raise the signal level enough to compensate for these losses, making the components appear "lossless." The same can be done with optical switches.

What Have You Learned?

1. Signals require amplification because they fade with distance in fibers as well as wires. Fibers can carry high-speed signals much farther.

2. Repeaters amplify signals and repeat them for transmission through another length of cable. Regenerators clean up signals as well as amplify them. Many "repeaters" are actually regenerators.

3. Electro-optic repeaters must convert an optical signal into electronic form for amplification. Optical amplifiers directly amplify the optical signal. They are simpler than electro-optic repeaters or regenerators.

4. Electro-optic repeaters are designed to operate at a specific signal speed. Optical amplifiers can accommodate different signal speeds.

5. Regeneration can prevent dispersion from accumulating to limit digital transmission.

6. Optical amplifiers include doped fiber amplifiers and semiconductor laser amplifiers.

7. The best-developed optical amplifiers are silica fibers doped with the rare earth erbium, which operate at 1550 nm. Praseodymium-doped fibers are the most promising fiber amplifiers for use at 1300 nm.

8. The gain of fiber amplifiers depends on wavelength and input power.

9. Fiber amplifiers cannot compensate for dispersion or pulse jitter.

10. Semiconductor lasers with facets coated to prevent reflection can amplify light from an external source. They work at any semiconductor laser wavelength, but when used with fibers suffer from high transfer losses and polarization sensitivity.

11. Semiconductor optical amplifiers can be integrated on a chip with other semiconductor optical components, without high transfer losses.

12. Optical amplifiers can be used in-line (between cable segments), as post-amplifiers to increase transmitter output, as preamplifiers to increase receiver sensitivity, or combined with other components to offset high losses.

What's Next?

In Chapter 10, I move to the connectors that bridge the gaps among optical fibers, transmitters, receivers, and other components.

Quiz for Chapter 9

1. Amplifiers are needed

    a. to overcome the threshold for driving an optical fiber.

    b. to compensate for fiber attenuation.

    c. only with copper-wire systems.

    d. to convert optical signals into electronic form.

2. What is the difference between amplification and regeneration?

    a. Regeneration retimes and cleans up the signal as well as amplifies it.

    b. Regeneration does not increase signal power.

    c. There is no difference.

    d. Regeneration is done electronically; amplification is done optically.

3. What can optical amplifiers do that electro-optic repeaters cannot?

    a. Compensate for fiber dispersion.

    b. Retime signals.

    c. Operate at a wide range of signal speeds without adjustment.

    d. Convert signal wavelength.

4. What can electro-optic repeaters do that optical amplifiers cannot?

    a. Compensate for fiber dispersion.

    b. Retime signals.

    c. Operate at a wide range of signal speeds without adjustment.

    d. a and b.

    e. None of the above.

5. At what wavelengths does an erbium-doped fiber amplifier operate?

    a. 1520 to 1560 nm.

    b. 1280 to 1330 nm.

    c. 750 to 900 nm.

    d. At all important fiber windows.

6. At what wavelengths does a praseodymium-doped fiber amplifier operate?

    a. 1520 to 1560 nm.

    b. 1280 to 1330 nm.

    c. 750 to 900 nm.

    d. At all important fiber windows.

7. A fiber amplifier has 30 dB small-signal gain and saturated output power of 50 mW. The input signal level is 1 μW. What is the output power?

    a. 50 μW.

    b. 1 mW.

    c. 10 mW.

    d. 30 mW.

    e. 50 mW.

8. Semiconductor laser amplifiers can operate at which of the following wavelengths?

    a. 1520 to 1560 nm.

    b. 1280 to 1330 nm.

    c. 750 to 900 nm.

    d. None of the above.

    e. a, b, and c.

9. Why are fiber amplifiers preferred over semiconductor laser amplifiers for boosting signals after transmission through an optical fiber?

    a. Fiber amplifiers are sensitive to polarization.

    b. Fiber amplifiers have much lower coupling losses.

    c. Fiber amplifiers operate at more wavelengths.

    d. Fiber amplifiers are smaller.

10. Why are optical amplifiers used?

    a. To amplify signals attenuated by fiber transmission.

    b. To increase receiver sensitivity.

    c. To raise transmitter output.

    d. To compensate for coupling losses.

    e. All of the above.

Connectors

About This Chapter

In the world of fiber optics, connectors are not the only way to make connections. The term "connector" has a specific meaning: a device that can be mated (and unmated) repeatedly with similar devices to transfer light between two fiber ends or between a fiber end and a transmitter or receiver. The connector is mounted on the end of a cable or on a device package. A permanent junction between two fibers is called a splice, which you'll learn about in Chapter 11. A device to interconnect three or more fiber ends or devices is called a coupler, which I will describe in Chapter 12.

This chapter first explains connectors and how they work, starting with the basic concepts behind fiber-optic connectors and the mechanisms causing their inherent loss or attenuation. Then it discusses important types of connectors.

Why Connectors Are Needed

Electrical connectors are common in modular electronic, audio, or telephone equipment, although you may think of them as plugs and jacks. Their purpose is to connect two devices electrically and mechanically, such as a cable and a stereo receiver. A plug on the cable goes into a socket in the back of the receiver, making electrical contact and holding the cable in place. Both the electrical and mechanical junctions are important. If the cable falls out, it can't carry signals; if the electrical connection is bad, the mechanical connection doesn't do any good. (You'll understand the problem all too well if you've ever tried to find an intermittent fault in electronic connectors.)

● **Connectors make temporary connections among equipment that may need to be rearranged.**

Fiber-optic connectors are intended to do the same job, but the signal being transmitted is light through an optical fiber, not electricity through a wire. That's an important difference because, as you learned in Chapter 4, the way light is guided through a fiber is fundamentally different from the way current travels in a wire. Electrons can follow a convoluted path through electrical conductors (wires) if the wires make good electric contact. However, fiber cores must be precisely aligned with each other. Just how precisely you'll see later.

Electrical connectors are used for audio equipment and telephones because the connections are not supposed to be permanent. You use fiber connectors for the same reason. For permanent connection, you splice or solder wires, and you splice optical fibers. Permanent connections have some advantages, including better mechanical stability and—especially for fiber optics—lower signal loss. However, those advantages come at a cost in flexibility; you don't want to unmake a splice each time you move a computer terminal or telephone.

● **Splices and connectors are used in different places.**

Fiber-optic connectors and splices are far from interchangeable. Connectors are normally used at the ends of systems to join cables to transmitters and receivers. Connectors are used in patch panels where outdoor cables enter a building and have their junctions with cables that distribute signals within the building. They are used where configurations are likely to be changed, such as at telecommunication closets, equipment rooms, and telecommunication outlets. Examples include the following:

- Interfaces between devices and local area networks
- Connections with short intrabuilding data links
- Patch panels where signals are routed in a building
- The point where a telecommunication system enters a building
- Connections between networks and terminal equipment
- Temporary connections between remote mobile video cameras and recording equipment or temporary studios
- Portable military systems

Splices are used where junctions are permanent or where the lower loss of splices is critical. For example, splices are made in long cable runs because there is no need to disconnect the cable segments and because connector losses would reduce maximum transmission distance. (Splices offer better mechanical characteristics for outdoor locations.)

● **Distinctions between splices and connectors are not always sharp.**

The distinctions between connectors and splices are not always as sharp as they might seem. The most common fiber connectors, which superficially resemble those for metal coaxial cables, differ obviously from the most common splices—the welding, fusion, or gluing of two fiber ends together. However, between these extremes are such hybrids as demountable splices, which nominally bond fibers together permanently but can be removed. Some of these approaches will be covered in Chapter 11. In addition, some connectors are installed by splicing them to the cable.

Connector Attenuation

The key optical parameter of fiber connectors is attenuation—the fraction of the signal lost within the connector. This loss is measured in decibels for a mated pair—that is, the loss in going from one fiber (or other device) to the other. (Light actually passes through two connectors, but the loss of one connector is not meaningful because the signal isn't going anywhere.) Typical attenuation is a fraction of a decibel, but some connectors have loss in the 1 dB range. Manufacturers specify loss for specific fiber types; as you will see in the next section, mismatched fibers almost always have higher loss. Note that in most of the discussion that follows I assume that a connector is joining the ends of two fibers. Connectors can be mounted on transmitters and receivers as well, and although details differ, the principles are the same.

Connector attenuation is the sum of losses caused by several factors, which are easier to isolate in theory than in practice. These factors stem from the way light is guided in fibers. The major ones are as follows:

● Overlap of fiber cores

● Alignment of fiber axes

● Fiber numerical aperture

● Fiber spacing

● Reflection at fiber ends

These factors interact to some degree. One—overlap of fiber cores—really is the sum of many different effects, including variation in core diameter, concentricity of the core within the cladding, eccentricity of the core, and lateral alignment of the two fibers. The fiber geometry can affect not only loss but also the geometry of the connector ferrule, both the inside core and the outside surface.

Overlap of Fiber Cores

To see how core overlap affects loss, look at Figure 10.1, where the end of one fiber is offset from the end of the other. For simplicity, assume that light is distributed uniformly in the cores of identical fibers and that the two fiber ends are next to each other and are otherwise well aligned. The loss then equals the fraction of the input-fiber core area that does not overlap with that of the output fiber. If the offset is 10% of the core diameter, the excess loss is about 0.6 dB.

The most important optical characteristic of connectors is loss, measured in decibels for a mated pair.

Offset of fiber cores by 10% of their diameter can cause a 0.6 dB loss.

FIGURE 10.1.

*Offset fibers can
cause loss.*

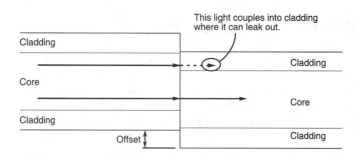

Mismatches of emitting and collecting area also occur if core diameters differ. Suppose that the fibers were perfectly aligned but that the 50 μm nominal fiber core diameter varied within tolerance of ±3 μm, as specified on a typical commercial graded-index fiber. With simple geometry, you can calculate the loss for going from fiber with core radius r_1 to core radius r_2. The relative difference in area is

$$(0.5r_1^2 - 0.5r_2^2)/0.5r_1^2$$

For the worst case of going from fiber with 53 μm core to one with 47 μm core, this is a factor of 0.21. If light was distributed uniformly through the core, that fraction of the light would be lost, about a 1 dB loss. Fortunately, things are rarely that bad—light is not distributed uniformly through the core, and core diameter rarely varies as much as the maximum allowed by the specifications. The same principles apply for single-mode fiber, but in that case, the critical dimension is mode-field diameter, which is typically slightly larger than core diameter. Because single-mode fibers have much smaller cores, the tolerances are much tighter.

Significant losses can occur if fiber types are mismatched and signals are going from a large-core fiber into one with a smaller core. (There are virtually no extra losses when transferring light from a small-core fiber to one with a larger core that can collect all its output.) This makes it important to be sure you know the fiber type. Excess loss of 1.9 dB occurs when going from a 62.5 μm to a 50 μm core multimode fiber. It's very bad news going from a 62.5 μm graded-index fiber to a single-mode fiber with 10 μm core—loss amounts to 97.9% of the light, or 17 dB.

Mismatches in area can arise from other factors, too, as shown in Figure 10.2. The fiber core might be slightly elliptical, or the core might be slightly off center in the fiber. (These problems have been exaggerated in Figure 10.2.) Variations in cladding dimensions can throw off alignment in connectors that hold the fiber in position by gripping its outside.

FIGURE 10.2.

Losses arise when cores are elliptical or off-center.

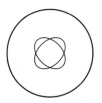 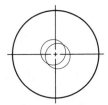

a. Elliptical Cores   b. Off-Center Cores

Alignment of Fiber Axes

The importance of aligning fiber axes is shown in Figure 10.3. As the fibers tilt out of alignment and the angle q increases, the light enters the second fiber at increasingly steeper angles, so some rays are not confined. The severity of this loss decreases as numerical aperture increases because the larger the NA, the larger the collection angle.

● Angular misalignment of fiber ends can cause significant losses.

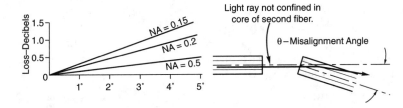

FIGURE 10.3.

Misaligned fiber axes cause losses.

Fiber Numerical Aperture

Differences in NA between fibers can also contribute to connector losses. If the fiber receiving the light has a smaller NA than the one delivering the light, some light will enter it in modes that are not confined in the core. That light will quickly leak out of the fiber, as shown in Figure 10.4. In this case, the loss can be defined with a simple formula:

● Differences in NA can contribute to connector losses.

$$\text{LOSS(dB)} = 10 \log_{10}(NA_2/NA_1)^2$$

where NA_2 is the numerical aperture of the fiber receiving the signal and NA_1 is the NA of the fiber from which light is transmitted. The NA must be the measured value for the segment of fiber used (which for multimode fibers is a function of length, light sources, and other factors), rather than the theoretical NA. Note also that there is no NA-related loss if the fiber receiving the light has a larger NA than the transmitting fiber.

FIGURE 10.4.

Mating fibers with different NAs can cause losses.

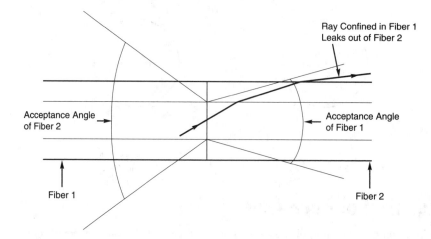

Spacing Between Fibers

Numerical aperture also influences the loss caused by separation of fiber ends in a connector. Light exits a fiber in a cone, with the spreading angle—like the acceptance angle—dependent on numerical aperture. The more the cone of light spreads out, the less light the other fiber can collect, as shown in Figure 10.5. This is one case where transfer losses increase with numerical aperture because the larger the NA of the output fiber, the faster the light spreads out. The following formula gives values for end-separation loss, assuming identical transmitting and receiving fibers:

$$\text{LOSS(dB)} = 10 \log_{10} [(d/2)/((d/2)+(S \times \tan[\arcsin(NA/n_0)]))]^2$$

where d is core diameter, S is the fiber spacing, and n_0 is the refractive index of the material between them.

FIGURE 10.5.

End-separation loss.

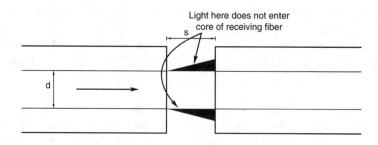

An additional spacing loss is Fresnel reflection, which occurs whenever light passes between two materials with different refractive indexes. It occurs for all transparent optical materials, even ordinary window glass. (It causes the reflections you see on windows when looking from a lighted room out into darkness.) Fresnel loss depends on the difference in refractive index between the fiber core and the material in the gap. For uncoated glass fiber ends in air, it is about 0.32 dB. This loss can be reduced by putting the fiber ends together (carefully, to avoid damage), by adding antireflective coatings, which have a refractive index between that of glass and air, or by filling the gap with a transparent material called index-matching gel, which has a refractive index closer to that of glass.

● Fresnel reflection losses occur when light enters material of a different refractive index.

Other End Losses

Other effects can add to connector loss. So far, I have assumed that the fiber ends are cut and polished cleanly and perfectly perpendicular to the fiber axis. However, extra losses can occur if the ends are cut and polished at a slight angle and the angles do not match in the connector. Other losses can arise if the ends are not smooth or if dirt gets into the connector.

With all these loss mechanisms, it is no wonder early fiber-optic developers were very worried about connectors. Tremendous progress has been made, but connector losses can still be significant in designing fiber-optic systems, as you will see in Chapter 16. Typical losses of good connectors are 0.2 to 0.7 dB for either single- or multimode fiber, but in practice all the factors discussed earlier can cause considerable variation. Connector installation is a critical variable. Repeated matings and unmatings of the connector can change attenuation. Generally a typical connector loss is specified, along with changes that both mechanical and environmental factors are expected to cause during use.

Internal Reflections

Losses are not the only potentially harmful things that happen within connectors. Strong back-reflections from fiber ends can cause problems with laser light sources. As mentioned in Chapter 6, the operation of a semiconductor laser relies on optical feedback, reflection from front and rear facets of the semiconductor cavity. Reflection from fiber ends in a connector can provide additional feedback that in some cases effectively adds a spurious modulation signal to the laser input, increasing noise levels.

● Internal connector reflections can cause spurious modulation and noise in laser light sources.

Internal reflections can be suppressed by reducing refractive-index differences at fiber ends. One possibility is to butt fiber ends together, so they appear (to the light) to be a single continuous piece of glass. Another is to fill the inside of the connector with an index-matching fluid or gel having a refractive index close to that of the glass. However, wet connectors are subject to contamination or leakage that can increase losses.

● Index-matching fluids reduce reflections in connectors.

Reflection noise is an important concern for cable television systems, which are more vulnerable because they use analog transmission. Some cable systems use angled-fiber connectors, in which the fiber ends are cut at an angle so any reflected light is lost from the fiber (as in the case of misaligned fibers, described earlier). However, angled connectors must be mated carefully to avoid excess losses if the angles don't match.

Mechanical Considerations

So far, I have concentrated on the optical characteristics of a fiber connector. However, mechanical characteristics are also important, and in some cases critical. Virtually all fiber connectors are designed well enough that they will stay in place under normal conditions. However, connectors must withstand physical stress applied during their use, from the normal forces in mating and unmating them to the sudden stress applied by a person tripping over a cable. Connectors must also prevent contamination of the optical interface, with dirt and moisture the main threats.

Durability

Durability is a concern with any kind of connector. Repeated mating and unmating of fiber connectors can wear mechanical components, introduce dirt into the optics, strain the fiber and other cable components, and even damage exposed fiber ends. Typical connectors for indoor use are specified for 500 to 1000 mating cycles, which should be adequate for most use. Few types of equipment are connected and disconnected daily. Specifications typically call for attenuation to change no more than 0.2 dB over that lifetime.

Connectors are attached to cables by forming mechanical and/or epoxy bonds to the fiber, cable sheath, and strength members. (Usually the fiber is epoxied, and the other bonds are crimped.) That physical connection is adequate for normal wear and tear but not for sudden sharp forces, such as those produced when someone trips over a cable. That sharp tug can detach a cable from a mounted connector, because the bond between connector and fiber is the weakest point. The same is true for electrical cords, and the best way to address the problem is to be careful with the cables.

Because sharp bends can increase losses and damage fibers, care should be taken to avoid sharp kinks in cables at the connector (e.g., when a cable mates with a connector on a patch panel). Fibers are particularly vulnerable if they have been nicked during connector installation. Care should also be taken to be certain that fiber ends do not protrude from the ends of connectors. If fiber ends hit each other or other objects, they can easily be damaged, increasing attenuation.

● Typical fiber connectors are specified for 500 to 1000 matings. Most can be torn from cable ends by a sharp tug.

Environmental Considerations

Most fiber-optic connectors are designed for use indoors, protected from environmental extremes. Keeping them free from contaminants is even more important than it is for electrical connectors. Dirt or dust on fiber ends or within the connector can scatter or absorb light, causing excessive connector loss and poor system performance. This makes it unwise to leave fiber-optic connectors open to the air, even inside. Many connectors and patch panels come with protective caps for use when they are not mated. These caps are the sort of things that are easily lost, but they should not be.

Special hermetically sealed connectors are required for outdoor use. As you might expect, those designed for military field use, such as the one shown in Figure 10.6, are by far the most durable. Military field connectors are bulky and expensive, but when sealed they can be left on the ground, exposed to mud and moisture. They are designed to operate even after having one end stuck in mud and wiped out with a rag! Normally, non-military users will avoid outdoor connectors, or house them in enclosures that are sealed against dirt and moisture.

●
Fiber ends must be kept free of contaminants to avoid excess losses.

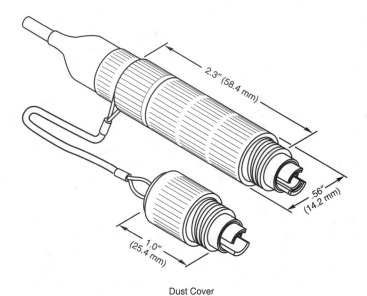

2.3" (58.4 mm)

.56" (14.2 mm)

1.0" (25.4 mm)

Dust Cover

FIGURE 10.6.
A military field connector. (Courtesy of AT&T Military Systems)

Connector Structures

I have talked about fiber connectors in fairly general terms so far for an important practical reason—many different types have been developed. This was a logical response to the difficult technical problems faced in developing durable, low-loss optical connectors. However, order is emerging from the chaos as users insist on standard types to simplify their logistics.

Most connectors in common use today have some common elements, shown in simplified form in Figure 10.7. The fiber is mounted in a long, thin cylinder called a ferrule, with a hole sized to match the fiber cladding diameter. The ferrule centers and aligns the fiber and protects it from mechanical damage. The end of the fiber is at the end of the ferrule. The ferrule is mounted in the connector body, which is attached to the cable structure. A strain-relief boot shields the junction of the connector body and the cable.

FIGURE 10.7.

A simplified generic fiber connector, with coupling receptacle or adapter.

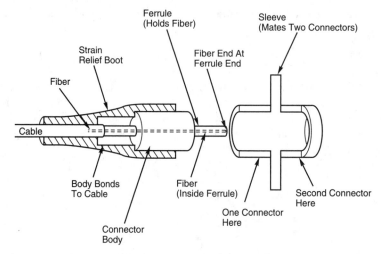

Standard fiber connectors lack the male-female polarity common in electronic connectors. Instead, fiber connectors mate in adapters (often called "coupling receptacles" or "sleeves") that fit between two fiber connectors. Similar adapters are mounted on devices such as transmitters and receivers, to mate with fiber connectors. Although this approach requires the use of separate adapters, which must be kept in stock to link cables, it otherwise reduces inventory requirements. The adapters can also be designed to mate different types of connectors.

Ferrules are typically made of metal or ceramic, but some are made of plastics. The hole through the ferrule must be large enough to fit the clad fiber and tight enough to hold it in a fixed position. Standard bore diameters are 126 +1/-0 μm for single-mode connectors and 127 +2/-0 μm for multimode connectors, but some manufacturers supply a range of sizes (e.g., 124, 125, 126, and 127 μm) to accommodate the natural variation in fiber diameter. Adhesive is typically put in the hole before the fiber is pushed in, to hold the fiber in place. The fiber end may be pushed slightly past the end of the ferrule, then polished to a smooth face.

The ferrule may be slipped inside another hollow cylinder (also called a sleeve) before it is mounted in the connector body. The body, typically made of metal or plastic, includes one or more pieces that are assembled to hold the cable and fiber in place. Details of assembly vary among connectors; cable bonding is usually to strengthen members and the jacket. The end of the ferrule protrudes beyond the connector body to slip into the mating receptacle. A strain-relief boot is slipped over the cable end of the connector to protect the cable-connector junction.

Connector Installation

Fiber users face important trade-offs in deciding where and how to install connectors. The tight tolerances needed for low loss are easier to reproduce in a factory environment. However, field installation gives much more flexibility in meeting system requirements, and allows on-the-spot repairs. Each approach has its advantages, and connector manufacturers have taken some steps to offer users the best of both worlds.

●
Fiber connectors may be installed in the factory or in the field.

A big advantage of factory installation is that it's the cable supplier's responsibility to do it right. Trained technicians mount and test the connectors, with all the equipment they need in a controlled environment. Generally, they can mass-produce standard lengths of connectorized cable economically. That's fine for short jumper cables used in patch panels, but it's more difficult to supply the many different lengths needed for intrabuilding cable.

An intermediate step is to supply cable segments with factory-mounted connectors on one end and fiber pigtails on the other. These pigtails can be spliced to cables in the field, using mechanical or fusion splices (described in Chapter 11). This is a quick and easy approach using splicing equipment that many field technicians already have. Factory-polishing makes connector losses low, and many types of connectors can be used. However, it requires additional splicing hardware and can add to costs. It works best for many-fiber loose-tube cables.

Field installation of the complete connector enhances flexibility and has low consumable costs. Labor costs may be low, depending on the location, but installation results depend on both the skill of the technician and the forgivingness of the connector design. It generally takes more time and skill on the part of the technician than splicing a premounted connector, and it requires some special tools. It works best for tight-buffered cables. Some manufacturers supply field connectorization kits with some of the most sensitive alignments already done.

Connector manufacturers continue efforts to simplify connector installation. One recent advance is the introduction of a field-installable connector with a built-in mechanical splice. This allows the incoming fiber to be spliced to a factory-polished and -installed fiber stub only an inch long—an extremely short pigtail.

Connecting Single- and Multifiber Cables

Connector installation is simplest for single-fiber cables. However, most cables contain two or more fibers, complicating matters.

The simplest case is the duplex connector, connecting a cable with two fibers. It is often made from two single-fiber connectors arranged side by side in a single housing. This is not quite enough for most practical applications, because fiber polarity is important. To ensure that the proper fibers are connected, most duplex connectors are keyed, so they can be inserted in only one way.

Multifiber cables are often broken out into separate fibers at patch panels or junction boxes, with single connectors on each fiber, as shown in Figure 10.8. This is often the simplest approach for multifiber cables, especially where they come into buildings and must be split to distribute signals to different areas. It is also the standard approach for multifiber cables within buildings.

Multifiber connectors can also be used, particularly in military field applications, where simplicity of use, and physical and environmental integrity are the paramount concerns. However, such connectors must be custom-designed for a particular cable and may not be available for nonstandard cables.

> ● Connectors for multifiber cables are more complex than those for single-fiber cables. Often multifiber cables may be broken out at each end into many single-fiber connectors.

> ● Multifiber cables may be broken out into separate fibers at patch panels.

FIGURE 10.8.

Fan-out of multifiber cable in a junction box.

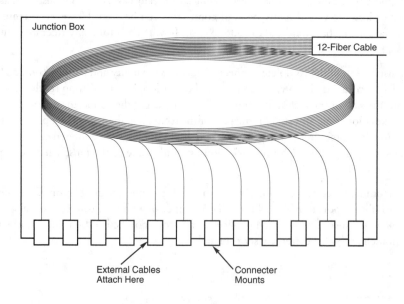

Standard Connector Types

During the 1980s, almost every company that made fiber connectors seemed to have its own designs. Some remain in production, but the recent actions of standards groups have narrowed the field considerably. The International Standards Organization, the International Electrotechnical Commission, and the Telecommunications Industry Association all endorsed one type, called the SC connector. A second type, the ST, is widely used in data communications and the telephone industry. A duplex connector developed for a standard local area network, the Fiber Distributed Data Interface (FDDI) has also become an accepted standard. Several other types remain in use, but many users are expected to shift to the standard SC and ST types. Those two connectors offer quite similar optical attenuation and back-reflection. Both work with single- and multimode fiber. Their major differences are in mechanical characteristics that affect ease of installation, packing density, and durability.

In the rest of this chapter, I'll look at the most common connector types and how they work.

> ●
>
> The square-format SC connector is specified by several standards. The round ST is also widely used.

SC

The SC connector, developed by Nippon Telegraph and Telephone in Japan, is shown in Figure 10.9. Like most modern connectors, it is built around a cylindrical ferrule that holds the fiber, and it mates with an interconnection adapter or coupling receptacle. It has a square cross-section that allows high packing density on patch panels. It can easily be packaged in duplex or four-fiber (quad) versions, unlike screw-on or bayonet-type connectors. The polarized duplex version has been adopted as a standard by ISO and TIA. Pushing the connector latches it in place, without any need to turn the connector in a tight space. It is better able to withstand pulling forces than ST connectors, which can lose optical contact when a force greater than that applied by the internal spring (typically two pounds) is applied to them.

ST

The older ST connector, long used in data communications, is shown in Figure 10.10. Like the SC, it is built around a cylindrical ferrule and mates with an interconnection adapter or coupling receptacle. However, it has a round cross-section and is latched into place by twisting it to engage a spring-loaded bayonet socket. (Some variants on the ST connector can be latched into place by pushing them.)

FDDI

The FDDI-standard duplex connector, shown in Figure 10.11, has gained acceptance for duplex connections. Like the ST and SC, it is a ferrule-based connector that mates with a coupling receptacle or adapter. It is keyed so that it can be installed in only one polarity, which is critical when dealing with input and output fibers. The above-mentioned polarized duplex SC connector has been adopted for the low-cost FDDI standard.

FIGURE 10.9.

SC connector, expanded and assembled. (Courtesy of AMP Inc.)

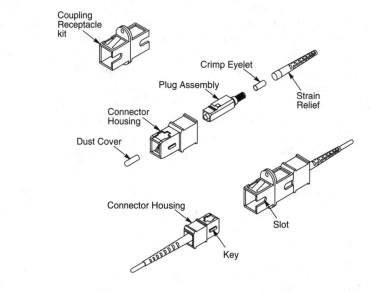

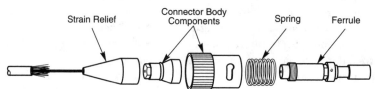

FIGURE 10.10.

ST connector, expanded and assembled. (Courtesy of Siecor Corp., Hickory, NC)

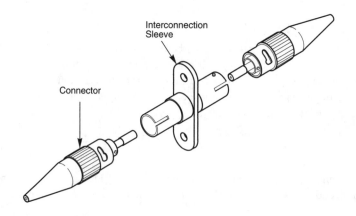

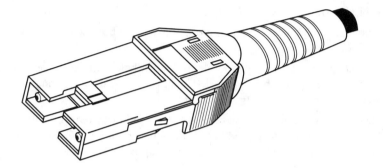

FIGURE 10.11.
FDDI-standard connector. (Courtesy of Siecor Corp., Hickory, NC)

FC

The FC connector is a screw-on type developed in the early 1980s that uses the same 2.5-millimeter ferrule size as the ST and SC connectors. Its optical losses are similar to the other two types. Like the SC connector, it can resist pulling forces. However, the screw-on design cannot be mounted as easily and cannot be used as a module in duplex connectors.

Other Connector Formats

The catalogs of many fiber-optic connector makers list many other styles of connectors. Some, like the SMA, Mini-BNC, and D4, look like the ST—but that doesn't mean you can mate the two together. Others, like biconic connectors, have their own distinct appearance. The use of coupling receptacles or adapters makes it possible to connect these connectors with other types.

There are also special-purpose connectors that don't meet general standards because they are designed for specific applications. Some are used in military environments, as mentioned earlier. Rotary connectors are used in places where signals have to be transmitted through a pivoting junction, but they are rare.

What Have You Learned?

1. Connectors make temporary connections between fiber ends.
2. The most important specification of connectors is attenuation, which is always given for a pair of connectors, measuring the loss in transferring a signal between two fibers.
3. Causes of connector loss include mismatch of fiber cores, misalignment of fiber axes, differences in numerical aperture, spacing between fibers, and reflection at fiber ends. Tolerances are tighter for small-core single-mode fibers than for larger-core multimode fibers.

4. Typical connector losses are 0.2 to 0.7 dB, but some have higher loss.

5. Connector back-reflection is an important parameter because it can cause noise in laser transmitters.

6. Most connectors contain cylindrical ferrules that hold the fiber inside a connector body. Most connectors lack male-female polarity and mate through interconnection adapters or coupling receptacles.

7. Many types of connectors have been marketed, and many remain available. Those widely accepted as standards are the square push-pull SC connector, the round bayonet-latch ST connector, and the duplex FDDI connector.

8. Fiber connectors can be installed in the field or in the factory. Some field-installable connectors are factory-mounted on cable segments that are spliced into place in the field.

What's Next?

In Chapter 11, I will look at splices—the permanent connections between two fiber ends.

Quiz for Chapter 10

1. Connectors are used
   a. to permanently join two fiber ends.
   b. to make temporary connections between two fiber ends or devices.
   c. to transmit light in only one direction.
   d. to merge signals coming from many devices.

2. Which of the following effects affect connector attenuation?
   a. Fiber core overlap.
   b. Alignment of fiber axes.
   c. Numerical apertures.
   d. End-to-end spacing of fibers.
   e. All of the above.

3. What does index-matching gel do in a connector?
   a. Holds the fibers in place.
   b. Keeps dirt out of the space between fiber ends.
   c. Prevents reflections at fiber ends.
   d. Eliminates effects of numerical aperture mismatch.

4. What will be the excess loss caused by the mismatch in core diameters when a connector transmits light from a 62.5/125 multimode fiber into a 50/125 fiber?
   a. 0 dB.
   b. 0.1 dB.
   c. 1 dB.
   d. 1.9 dB.
   e. 12.5 dB.

5. The largest excess loss will occur in which of the following cases?

a. Transfer of light from a single-mode to a multimode fiber.

b. Transfer of light from a fiber with a high NA to one with a low NA.

c. Transfer of light from a fiber with a low NA to one with a high NA.

d. Transfer of light through a connector filled with index-matching gel.

6. Back-reflections from fiber ends

a. can cause laser noise in analog systems.

b. are lower in high-loss connectors.

c. are not a significant problem in practical systems.

d. do not occur if fiber ends are separated in connectors.

7. Which of the following factors tends to favor field installation of connectors?

a. Ease of measuring connector performance.

b. Ease of matching precise cabling requirements.

c. Installation is more accurate and repeatable.

d. Minimizing time required to mount each connector.

8. What are functions of coupling adapters?

a. To provide an interface between two connectors.

b. To allow mating of two different connector types.

c. To attach connectorized cables to terminal equipment.

d. The same as interconnection adapters.

e. All of the above.

9. Ferrules do what in a fiber-optic connector?

a. Relieve strain on the cable.

b. Allow adjustment of attenuation.

c. Hold the fiber precisely in place.

d. Prevent back-reflection.

e. Screw the connector into place.

10. Which connector type has been chosen by the most standards organizations?

a. SC

b. ST

c. FDDI

d. Biconic

e. D4

Splicing

About This Chapter

Splices, unlike the connectors discussed in the last chapter, are permanent connections between fibers. Splices weld, glue, or otherwise bond together the ends of two fibers. Like fiber-optic connectors, fiber-optic splices are functionally similar to their wire counterparts. However, as with connectors, there are important differences between splicing wires and optical fibers.

In this chapter you will learn when and why optical fibers are spliced, the major considerations in fiber splicing, the types of splices, and the special equipment used in splicing.

Applications of Fiber Splices

Splices and connectors both join fiber ends, but they do so in different ways, so they are used for different purposes. (If you're familiar with electronics, you can compare splicing to soldering, and connectors to plugs.) Splices have lower loss, and they bond fibers together permanently. You use them in places where you never expect to make any more changes. Connectors have higher loss, but you can move them around, changing the way they connect fibers. To complicate things somewhat, there are different types of splices, although they are more similar to each other than to connectors.

UNDERSTANDING

●
Splices are low-loss, permanent connections between fiber ends.

Typically, splices are used to join lengths of cable outside buildings, whereas connectors are used at the ends of cables inside buildings. Splices may be incorporated in lengths of cable or housed in indoor or outdoor splice boxes; connectors are typically in patch panels or attached to equipment at cable interfaces. The decision isn't always that simple; it depends on the advantages of splices and connectors, listed in Table 11.1.

Table 11.1. Comparison of splice and connector advantages.

Connectors	Splices
Non-permanent	Permanent
Simple to use once mounted	Lower attenuation
Factory installable on cables	Lower back-reflection
Allow easy reconfiguration	Easier to seal hermetically
Provide standard interfaces	Usually less expensive per splice
	More compact

●
Permanent and non-permanent junctions are needed in different situations.

It might seem strange to see "Permanent" listed as an advantage of splices and "Non-permanent" as an advantage of connectors. However, each characteristic is desirable in certain applications. For example, splices to fix a broken underground cable should be permanent. However, you don't want to make permanent junctions between a local area network and terminals that may be moved about within a building.

●
The lower loss of splices allows long-haul cables to be spliced together as the cable is installed.

The lower attenuation of splices simplifies installation of long-haul fiber-optic systems. Bare fiber normally comes on reels in standard lengths from 1 to 25 km. Cables are much bulkier than fibers—particularly the heavy-duty types intended for outdoor use—and normally come in lengths of 12 kilometers or less.

Longer cable runs are made by splicing cable segments together. If the cables are installed in underground ducts, the splices are made and installed in manholes, with cable-segment length dependent on manhole spacing. Overhead cables are spliced in the field, from segments typically well over a kilometer long.

●
Enclosures are needed to protect splices.

The physical characteristics of splices are important in many long-distance applications. The spliced cables must be capable of withstanding the hostile outdoor environment, so splices are housed in protective enclosures. Although many splice enclosures are designed to be re-opened if repairs or changes are needed, they can be hermetically sealed to protect against moisture and temperature extremes. This combines with the low loss of splices to make them the preferred way to join lengths of fiber in long-haul telecommunication systems. These considerations are less important in shorter systems and systems in more controlled environments, where connectors are often used.

Types of Splicing

There are two basic approaches to fiber splicing: fusion and mechanical. Fusion splicing melts the ends of two fibers together so they fuse, like welding metal. Mechanical splicing holds two fiber ends together without welding them, using a mechanical clamp and/or glue. Each approach has its distinct advantages. Fusion splicers are expensive, but they require almost no consumable costs, and fusion splices have slightly better optical characteristics. Mechanical splicing requires less equipment (and no costly fusion splicer), but consumable costs per splice are much higher. I'll discuss each approach in turn before looking at considerations of splice performance.

Fusion Splicing

Fusion splicing is performed by butting the tips of two fibers together and heating them so they melt together. This is normally done with a fusion splicer, which mechanically aligns the two fiber ends, then applies a spark across the tips to fuse them together. Typical splicers also include instruments to test splice quality and optics to help the technician align the fibers for splicing. Typical splice losses are 0.05 to 0.2 dB, with more than half below 0.1 dB. The basic arrangement of a fiber splicer is shown in Figure 11.1.

●
Fusion splicing welds fiber ends together.

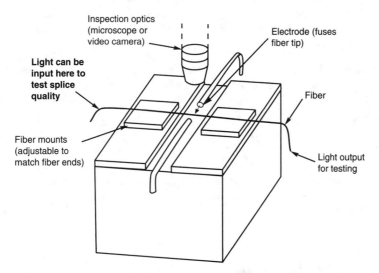

FIGURE 11.1.

Key components of a fusion splicer.

Individual fiber splicers are designed differently, but all have the common goal of producing good splices reliably. Many are automated to assist the operator. They are expensive instruments, with prices starting at thousands of dollars and reaching tens of thousands of dollars for the most sophisticated models. Major differences center on the degree of automation and the amount of instrumentation included. Most models share the following key elements and functions:

● A fusion welder, typically an electric arc, with electrode spacing and timing of the arc adjustable by the user. The discharge heats the fiber junction. Flames and the infrared beams from carbon-dioxide lasers have also been used, but virtually all commercial splicers use electric arcs. Portable versions are operated by batteries that carry enough charge for a few hundred splices before recharging. Factory versions operate from power lines or batteries.

● Mechanisms for mechanically aligning fibers with respect to the arc and each other. These include mounts that hold the fibers in place, as well as adjust their position. More expensive splicers automate alignment and measurement functions.

● A video camera or microscope (generally a binocular model) with magnification of 50 power or more so the operator can see the fibers while aligning them.

● Instruments to check optical power transmitted through the fibers both before and after splicing. Typically, light is coupled into a bent portion of the fiber on one side of the splice and coupled out of a bent portion on the other side. With proper calibration, this can measure the excess loss caused by the splice. (This may be missing from inexpensive field splicers.)

● **Before fusion splicing, plastic coatings must be removed from the fiber, and the end must be cleaved to have a face within a few degrees of perpendicular to the fiber axis.**

Fusion splicing involves a series of steps. First, the fiber must be exposed by cutting open the cable. Then the protective plastic jacket must be stripped from a few millimeters to a few centimeters of fiber at the ends to be spliced. The fiber ends must be cleaved to produce faces that are within 1°-3° of being perpendicular to the fiber axis. The ends must be kept clean until they are fused.

The next step is alignment of the fibers, which may be done manually or automatically by different splicer models. After preliminary alignment, the ends may be "prefused" for about a second with a moderate arc that cleans their ends and rounds their edges. These ends are then pushed together, allowing power transmission to be tested to see how accurately they are aligned. After results are satisfactory, the arc is fired to weld the two fiber ends together. Care must be taken to ensure proper timing of the arc so the fiber ends are heated to the right temperature. After the joint cools, it can be recoated with a plastic material to protect against environmental degradation. The spliced area can also be enclosed in a plastic jacket. The entire splice assembly is then enclosed mechanically for protection, which in turn is mounted in a splice enclosure. The case around the individual splice provides strain relief.

Mechanical Splicing

● **Mechanical splicing gives higher losses but requires simpler equipment than fusion splicing.**

Mechanical splices join two fiber ends either by clamping them within a structure or by gluing them together. A variety of approaches have been used in the past, and many are still in use. The extremely tight tolerances in splicing single-mode fiber often require special equipment not needed for splicing multimode fiber. Those extra requirements typically make single-mode splicing more expensive.

In general, mechanical splicing requires less costly capital equipment but has higher consumable costs than fusion splicing. This can tilt the balance toward mechanical splicing for organizations that don't perform much splicing, or for emergency on-site kits for temporary repairs. Mechanical splices tend to have slightly higher loss than fusion splices, but the difference is not dramatic. Back-reflections can occur in mechanical splices, but they can be reduced by using epoxy to connect the fibers, or by inserting into the splice a fluid or gel with a refractive index close to that of glass. This index-matching gel suppresses the reflections that can occur at a glass-air interface.

We will look briefly at important types of splices.

CAPILLARY SPLICE

One of the simplest types of splices relies on inserting two fiber ends into a thin capillary tube, as shown in Figure 11.2. The plastic coating is stripped from the fiber to expose the cladding, which is inserted into a tube with an inner diameter that matches the outer diameter of the clad fiber. The two fiber ends are then pushed into the capillary until they meet (typically with index-matching gel inserted to reduce reflections). Compression or friction usually holds the fiber in place, although epoxy may also be used.

A capillary splice holds two fiber ends in a thin tube.

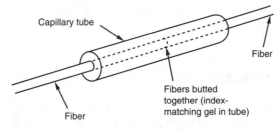

Capillary tube

Fiber

Fibers butted together (index-matching gel in tube)

Fiber

FIGURE 11.2.

Capillary splice joins two fibers.

Alignment of the fiber ends depends on mechanical alignment of the outside of the fibers. The result is a simple splice that is easy to install and can compensate for differences in the outer diameters of fibers. However, it is not designed to compensate for other differences between the fibers being joined.

ROTARY OR POLISHED-FERRULE SPLICE

A more elaborate arrangement is needed to compensate for subtle differences in the fibers being spliced. An older approach is the polished-ferrule splice, often called a rotary splice, shown in Figure 11.3. As with other splices, the plastic coating is first removed from the fiber. Then each fiber end is inserted into a separate ferrule, and its end is cleaved and polished to a smooth surface. The two polished ferrules then are mated within a jacket or tube.

FIGURE 11.3.

Rotary or polished-ferrule splice.

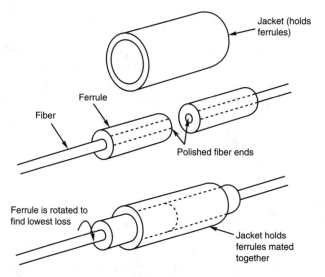

The ferrules are designed to mount the fibers slightly off-center. After they are inserted into the tube, the ferrules are rotated while splice loss is monitored. The ferrules are fixed in place at the angle where splice loss is at a minimum. Although this technique is more complex and time-consuming than capillary splicing, it offers a more precise way of mating fibers. Its sensitivity to rotation of the fiber around its axis makes it suitable for splicing polarization-sensitive fibers.

V-GROOVE SPLICING

V-groove splices are valuable for multifiber splicing.

Fibers can also be held in V-shaped grooves in a plate. V-groove splices can take various forms. The two fiber ends can be slipped into the same groove, and a matching plate applied on top. The fiber ends can be put into separate grooved plates, and the ends polished before they are mated and aligned with another plate.

The V-groove splice is particularly amenable for multifiber splicing, as shown in Figure 11.4 for ribbon cable. Each fiber in the cable slips into a separate groove, and mating the top and bottom plates automatically aligns all fibers with respect to one another. A practical limitation is the need for tight tolerances in the grooves. Field splicing of multifiber ribbon cables is aided by assembling end mounts on each cable in the factory and mating them in the field.

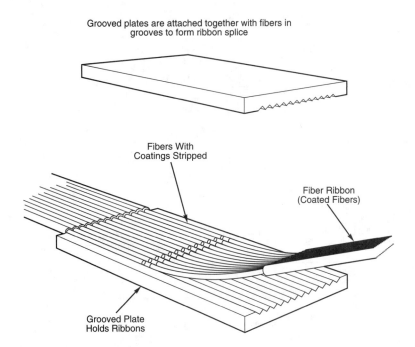

Grooved plates are attached together with fibers in grooves to form ribbon splice

Fibers With Coatings Stripped

Fiber Ribbon (Coated Fibers)

Grooved Plate Holds Ribbons

FIGURE 11.4.

Mass splicing of 12-fiber ribbons.

ELASTOMERIC SPLICE

Another type of mechanical splice is the elastomeric splice shown in Figure 11.5. The internal structure is similar to the V-groove splice, but the plates are made of a flexible plastic material. An index-matching gel or epoxy is first inserted into the hole through the splice. Then one fiber end is inserted until it reaches about halfway through the splice. Finally, the second fiber end is inserted from the other end until it can be felt pushing against the first.

The elastomeric splice is often used in field kits for emergency restoration of service over a broken cable. Typical losses are about 0.25 dB, good for such applications.

●
Elastomeric splices align fibers in a hole in a flexible plastic.

FIGURE 11.5.

*Elastomeric splice
developed for field use.*
(Courtesy of GTE)

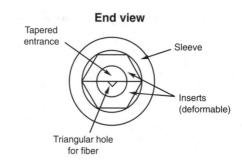

End view

Tapered entrance

Sleeve

Inserts (deformable)

Triangular hole for fiber

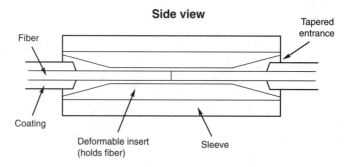

Side view

Tapered entrance

Fiber

Coating

Deformable insert (holds fiber)

Sleeve

The Connector-Splice Borderland

As you saw in the last chapter, the borderline between connectors and splices can be hazy. Some splices can be opened and reused a few times, but they are not designed for use as connectors. The factory-mounted splice elements on ribbon cables have some resemblance to connector components, and the splices can be opened and remated, but these devices still are not true connectors.

Splicing Issues

Loss, durability, and ease of operation are major concerns in splicing.

There are three principal concerns in splicing: the optical characteristics of the finished splice, its physical durability, and ease of splicing. Trade-offs among them enter into the choice of splice technology.

Attenuation

The attenuation mechanisms described for connectors in the last chapter also affect fiber-optic splices. However, splicing tends to align fibers more accurately, giving lower attenuation. Some sources of connection loss are essentially eliminated in splices; others are greatly reduced.

In a splice, the two fiber ends are bonded together by melting (fusing) them, gluing them, or mechanically holding them in a tightly confined structure. Bonding the two fiber ends together with no intervening air space reduces or eliminates losses from fiber spacing, as well as back-reflections.

Splice losses fall into two categories: intrinsic and extrinsic. Many are analogous to those encountered in connectors.

Intrinsic losses arise from differences in the fibers being connected. Mechanisms include variations in fiber core and outer diameter, differences in index profile and in ellipticity and eccentricity of the core. They can occur even in fibers with nominally identical specifications, because of inevitable variations in the manufacturing process.

Extrinsic losses arise from the nature of the splice itself. They depend on fiber end alignment, end quality, contamination, refractive-index matching between ends, spacing between ends, waveguide imperfections at the junction, and angular misalignment of bonded fibers.

Typically these two loss mechanisms are comparable in magnitude for well-made splices. Fortunately, the two types of loss combine to give a total splice loss that may be less than their arithmetic sum. (Note, however, that modal distribution effects can give misleading results in multimode fibers, as described in Chapter 14.) Losses can be very low—near 0.05 dB—in properly made splices, but imperfect junctions can suffer from high loss. A single 10 μm dust particle in the wrong place can block the entire core of a single-mode fiber. With proper tools and procedures, splice attenuation is comparable for single-mode and multimode fibers.

Fiber Curl

If you've ever tried to splice together broken cassette tape, you've discovered the phenomenon of end curl. The ragged ends refuse to lie flat and straight, but instead curl, making splicing a frustrating task. Optical fibers suffer from a much smaller degree of end curl, but even a few micrometers can cause problems in splicing single-mode fibers. This makes it vital to design fibers and splicers to control fiber end curl.

Mismatched Fibers

All specifications are written with the implicit assumption that the two fibers being spliced are identical. If they are not, losses can be significant. For single-mode fibers, the most important losses come from mismatches in mode field diameter—the effective size of the mode transmitted by the fiber—and offset of the fiber cores. Slight variations occur in these characteristics even for fibers with nominally identical specifications, contributing to average splice loss of 0.05 to 0.1 dB. Note, however, that individual splices may have higher losses because of natural variations in fiber characteristics.

● Fiber ends are fused, glued, or mechanically held together.

● Average splice loss is comparable for multimode and single-mode fibers.

● Specifications assume splicing of identical fibers.

Most telecommunication fibers have the same 125 μm cladding diameter, making it physically possible to splice different fiber types. The loss of such splices depends on transmission direction. It would be negligibly small in going from a 10 μm core single-mode fiber to a 62.5/125 multimode fiber, but very large—about 20 dB—in the opposite direction. The mechanism involved, the difference in core areas, is the same as for connectors.

Back-Reflection

Reflections from within a splice can also affect optical performance. A good fusion or mechanical splice should have low reflection, but high reflections could occur if the splice is not made properly. Problems are more likely in mechanical splices because the ends can be physically separated, allowing reflection.

Strength

Fusion splices are most likely to break near the fused zone.

If you pull hard on a spliced metal wire, you expect it to part at the splice long before the wire itself fails. Optical fibers likewise are more vulnerable at splices. Different mechanisms affect fusion and mechanical splices.

Stripping the coating from fibers can damage them before splicing, causing microcracks that can cause failure later. In fusion splicing, contaminants can weaken the fusion zone itself, and thermal cycling can weaken the surrounding area. When fusion splices fail, they typically do so close to, but not at, the splice interface. The lifetime of the splice can be enhanced by claddings or jackets that protect the spliced zone from mechanical and environmental stresses.

Mechanical splices likewise can fail because the fibers were damaged in preparation. Other failures can occur in bonding the splice to the two fibers.

Ease of Splicing

Specialized equipment is used in fiber splicing.

Because splices are often installed in the field, the ease with which they can be made is an important concern. This has led to development of specialized equipment for field as well as factory use.

The need for a fusion splicer is the main drawback to fusion splicing in the field. Standard factory splicers are boxes that measure up to a foot (30 centimeters) on all three sides, weigh 20 pounds (10 kilograms) and up, and may cost $20,000 to $40,000. Special portable models cost several thousand dollars, weigh only a few pounds (a couple of kilograms), and measure 4 to 6 inches (10 to 15 centimeters) on a side. The differences are similar to those between laptop and desktop computers; the smaller splicers have fewer features and less battery power (so they require more frequent recharging).

Installation of mechanical splices requires special splice housings and tools. However, the housings are small and the tools are much less bulky and expensive than those needed for fusion splicing. This makes mechanical splicing practical on a smaller scale than fusion splicing, and emergency cable repair kits come with mechanical splicing equipment.

Mass versus Individual Splicing

So far I've implicitly assumed that fibers are spliced one at a time. However, cables generally contain two or more fibers that must be spliced. Often the simplest approach is to splice the fibers individually and house the splices in a splice enclosure of the sort described in the next section. However, this approach becomes time-consuming and costly for many-fiber cables.

⬤ Cabled fibers can be spliced individually or—in some special cables—en masse.

An alternative is mass splicing many fibers at once. This can be done simply for ribbon cables, which contain many fibers in a flat line. Figure 11.4 showed mechanical splicing of a 12-fiber ribbon by aligning the fibers in parallel grooves. Similar alignment methods can be used to arrange fibers for mass fusion splicers. Mass splicing of multifiber cables can be considerably faster than individual splicing, and developers claim that losses are comparable or only slightly higher.

Splice Housings

Fiber-optic splices require protection from the environment, whether they are indoors or outdoors. Splice enclosures help organize spliced fibers in multifiber cables. They also protect the splices from strain and contamination.

⬤ Splice housings organize splices in multifiber cables, protect them from strain, and isolate them from hostile environments.

Splice housings typically contain a rack such as the one shown in Figure 11.6, which contains an array of individual splices. This rack is mounted inside a case that provides environmental protection. Individual fibers broken out from a cable lead to and from the splices. To provide a safety margin in case further splices are needed, an excess length of fiber is left in the splice case. Like splice enclosures for telephone wires, fiber-optic splice cases are placed in strategic locations where splices are necessary (e.g., in manholes, on utility poles, or at points where fiber cables enter buildings).

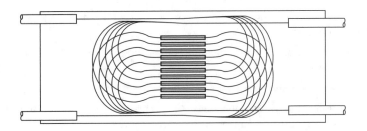

FIGURE 11.6.

Splices arrayed inside housing.

Fiber splice enclosures should be designed to

- Hold cable strength member tightly
- Block entrance of water
- Provide redundant seals in case one level fails
- Electrically bond and ground any metal elements in the cable (e.g., strength members and armor)
- Be re-enterable if the splice must be changed or repaired
- Organize splices and fibers so they can be readily identified
- Provide room for initial splicing and future modifications
- Leave large enough bend radii for fibers and cables to avoid losses and physical damage

Splice Testing

Where high performance is crucial, attenuation can be measured as the splice is being made. Many fusion splicers come with equipment that measures loss as the fibers are being aligned, then test loss after the splice is completed. As will be described in more detail in Chapter 14, precise measurement requires passing light through the splice from the remote end of one fiber being spliced to the remote end of the other—in the same direction that light will travel in the system. This is awkward in the field, where the remote ends may be several kilometers away.

A simpler approach is possible as long as the person making a splice needs to know only relative attenuation as the fibers are aligned. In practice, this is the needed information, because a peak in transmitted power indicates that the fiber alignment is best and the fibers are ready to be spliced. Such relative measurements can be made by bending the fibers near the splice point so light can be coupled into (and out of) them. Although this does not measure actual splice loss, it can make field fiber splicing a one-person job.

What Have You Learned?

1. Splices are permanent connections between fibers. They have lower attenuation than connectors.

2. Major issues in splices are attenuation, physical durability, and ease of installation.

3. Splices are normally made in the field.

4. Fusion splicing melts two fiber ends together; typical loss of fusion splices is under 0.1 dB. It requires an expensive fusion splicer.

5. Mechanical splices hold fiber ends together mechanically or with glue. Losses tend to be slightly higher than with fusion splices, but they do not require expensive equipment to install.

6. The factors influencing attenuation of splices are very similar to those influencing connector attenuation. Mismatched fibers can cause high attenuation.

7. Splices are mounted in enclosures for protection against stress and the environment. The enclosures may be indoors or outdoors.

8. Fibers in multifiber cables may be spliced individually or in mass splices.

What's Next?

In Chapter 12, I will look at the issues involved in coupling three or more fiber ends together, a job for fiber-optic couplers.

Quiz for Chapter 11

1. Which of the following is not an advantage of a splice?
   a. Permanent junction between two fibers.
   b. Ease of making changes.
   c. Low attenuation.
   d. Ease of installing in the field.
   e. Strength.

2. Which place is a splice most likely to be used?
   a. In the middle of a long-distance cable.
   b. To connect a computer terminal with a local area network.
   c. To couple light from an LED to a short-distance fiber system.
   d. To join an intrabuilding cable to a patch panel.

3. Splice loss typically is around
   a. 0.01 dB.
   b. 0.1 dB.
   c. 0.5 dB.
   d. 1.0 dB.
   e. 0 dB.

4. Which of the following mechanisms could not cause loss in a fusion splice?
   a. Differences in fiber core diameter.
   b. Dirt in the splice zone.
   c. Misalignment of fiber ends.
   d. Separation between fiber ends.

5. What would happen in a splice between fibers of identical outer diameters but different size cores?

    a. The splice would fail mechanically.

    b. Loss would be high in both directions.

    c. Loss would be high going from one large-core fiber to the small-core fibers, and low in the opposite direction.

    d. Loss would be high going from the small-core fiber to the large-core fibers, and low in the opposite direction.

6. Where is a fusion splice most likely to fail under mechanical stress?

    a. At the exact splice point.

    b. About a millimeter from the splice point.

    c. Far from the spliced point.

    d. At the end being pulled.

    e. Failure is unpredictable.

7. What is the major advantage of mechanical splices over fusion splices?

    a. Lower attenuation.

    b. Higher mechanical strength.

    c. Elaborate fusion-splicing system not required.

    d. Physically smaller.

8. Which of the following is not required in a fusion splicer?

    a. A welder to heat fiber ends.

    b. A microscope to view fiber ends.

    c. A mechanical alignment system.

    d. A device to cut fibers.

    e. A glue-delivery system.

9. What is the primary advantage of mass splicing?

    a. Lower attenuation.

    b. Requires less costly splicing equipment.

    c. Reduces labor requirements and installation costs for multifiber cables.

    d. Reduces operator errors in matching fiber ends.

10. Splice housings are important because they

    a. reduce splice attenuation.

    b. protect splices from physical and environmental stresses.

    c. prevent hydrogen from escaping from splices.

    d. allow measurement of splice attenuation.

Couplers and Switches

About This Chapter

The term "coupler" has a special meaning in fiber optics. A coupler connects three or more fiber ends (or optical devices such as detectors and transmitters). As such, it is distinct from connectors and splices, which join two fiber ends, or a fiber with a light emitter or detector. The distinction is much more important in fiber optics than in electronics because of the way signals travel in fibers. Good fiber couplers are very hard to make, and even good ones suffer higher losses than their electronic counterparts.

Although they serve different functions, many optical switches are based on the same technology as couplers. In fact, some switches are specialized versions of couplers, with extra features added so they can switch the light passing through them.

In this chapter, you will see what fiber-optic couplers and switches are and how they are used. You will also examine the various types of couplers and switches, as well as their strengths and weaknesses.

What Are Couplers?

Dividing Signals

Connectors and splices join two fiber ends together. That's fine for sending signals between two devices. However, many applications require connecting more than two devices. Connecting both a telephone and an answering machine on the same phone line, for example, requires a coupler (the other end of that phone line—the third point connected—goes to the rest of the world). That coupler

●
Couplers connect three or more points.

must take the signal from one set of telephone wires and split it between two modular phone sockets, one for the telephone and one for the answering machine.

That's not a problem with a telephone or other electrical equipment because of the way electricity flows in a circuit. If you hook one, two, or twenty identical resistors across an ideal voltage source, each will see the same voltage signal. Of course, that's only an approximation of what happens in a telecommunication system. Transmission line resistance and other effects can cause voltage across the load to drop somewhat as the load resistance drops (because more devices are put in parallel across the signal source). Nonetheless, in a carefully designed system, many loads in parallel will all see voltages close to what they would see individually. Likewise, if the signal is carried by variations in current from a current source, many devices in series will all see currents comparable to what an individual device would see.

●
An optical signal must be divided among output ports, reducing its magnitude.

Optical signals are different from electrical signals and, thus, are transmitted and coupled differently. An optical signal is not a potential, like an electrical voltage, but a flow of signal carriers (photons), similar in some ways to an electric current. However, unlike a current, an optical signal does not flow through the receiver on its way to ground. It stops there, absorbed by the detector. That means multiple fiber-optic receivers cannot be put in series optically, because the first would absorb all the signal. If an optical signal is to be divided between two or more output ports, the ports must be in parallel. However, because the signal is not a potential, the whole signal cannot be delivered to all the ports. Instead, it must be divided between them in some way, reducing its magnitude.

This limits the number of terminals that can be connected to a passive fiber-optic coupler that merely splits up the input signal. After some maximum number of output ports is exceeded, there is not enough signal to go around (i.e., to be detected reliably with low enough bit-error rate or high enough signal-to-noise ratio for the application). This division of power typically limits one transmitter to sending signals to tens of receivers, although amplifiers or repeaters can boost that number. To further complicate matters, couplers are not easy to build and usually suffer losses above the theoretical lower limit from simply dividing power.

Coupler Applications

●
Couplers are needed for the many systems that interconnect more than two points.

Couplers were not needed for the first fiber-optic systems, which carried signals between pairs of points. However, many communication applications require connecting many terminals, such as the local area network shown in Figure 12.1. At each point where a device is connected to the network, the signal must be split into two parts—one to be passed along the network, and the other sent to the device. This can be done in various ways. One of the simplest to envision is dividing the optical signal at each connection, with part of the light going to the device and the rest going around the network. (This turns out to be rather inefficient, because the coupler losses accumulate around the ring.) Another is to send signals to a central multiport coupler, which distributes output to all terminals. Details differ, but the need for couplers arises any time optical signals are split and sent to two or more places. The only way to avoid the need for couplers is to convert the optical signal to electronic form and use it to drive multiple transmitters, which is done in some networks.

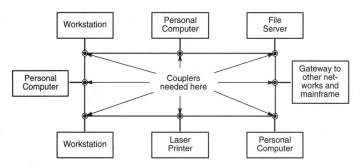

FIGURE 12.1.
A local area network.

Couplers are also needed to separate or combine signals, usually at different wavelengths, being sent through the same fiber. Figure 12.2 shows how three signals at different wavelengths can be carried through the same fibers, a technique called wavelength-division multiplexing (WDM). This is possible because light of different wavelengths traveling through the same fiber does not interact strongly enough to affect signal transmission. Couplers are needed to combine light signals from different sources at the input and separate them at the output. The separation (at least) requires couplers that are wavelength-selective, to direct different wavelengths along different paths. The transmission direction is important because some couplers transmit light differently in different directions. In Figure 12.2, signals are sent to the right at λ_1 and λ_2 and to the left at λ_3.

●
Couplers are needed to combine or divide signals being sent through the same fiber.

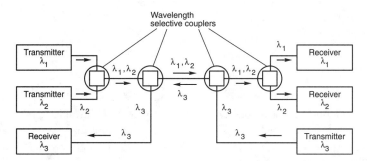

FIGURE 12.2.
Wavelength-division multiplexing.

Other components can be used to separate signals at different wavelengths, but they require couplers to divide the signal before they can select the desired wavelength. Although this process is less efficient, it can separate closely spaced wavelengths.

Wavelength-selective couplers are also used to separate the pump and signal wavelengths in optical amplifiers, as described in Chapter 9.

Coupler Issues

Many issues enter
into the choice and
design of couplers.

Important issues in the design and choice of couplers include choices among fundamental types and performance factors. The major issues are as follows:

- Number of ports
- Sensitivity to light-transmission direction
- Wavelength selectivity
- Type of fiber (single- or multimode)
- Signal attenuation
- Cost

These considerations have contributed to the development of several different types of fiber couplers. The labels that follow are all attached to some types:

- Passive coupler
- Star coupler
- T (or tee) coupler
- Y coupler
- Tree coupler
- Directional coupler
- Bidirectional coupler
- Waveguide coupler
- Fiber coupler
- Splitters
- Combiners
- Wavelength-selective coupler
- Wavelength-independent coupler
- n×m coupler (where n and m are two integers, e.g., 2 and 3 or 4 and 4)

This proliferation of labels is confusing, particularly because the same coupler can wear two or more labels. For example, a T coupler can also be a directional coupler and a passive coupler (i.e., a passive, directional T coupler). I'll break this confusing mass into smaller chunks that should be more digestible.

The Basics of Couplers

Total output power
from a passive
coupler can be no
more than the input
power.

Most couplers are passive optical devices that divide signals among two or more output ports. The fundamental fact of life for passive couplers is that the total output power can be no more than the input power. From the viewpoint of each output device, the coupler has a characteristic loss, equal to the ratio (in decibels) of output to that device to total input power. The equal division of an input signal between two output ports causes a 3 dB loss. Any additional loss above that theoretical minimum is called excess loss.

In the general case of a coupler with one input and many outputs, the total output, summed over all ports, equals input power minus excess loss. Note that in general, power need not be divided equally among output ports. A two-port coupler can be designed so 10% of the light emerges from one port and 90% from the other.

T, Y, Tree, and Star Couplers

Three major types of passive couplers are the star, tree, and T types shown in Figure 12.3. They get their names from their geometry. The T coupler has three ports and is analogous to electrical taps that take a signal from a passing cable to a terminal. Thus, this type of coupler is normally shown as one fiber coming off a fiber passing by, in a T configuration, as in Figure 12.3, although some designs have a Y-shaped geometry and are sometimes called Y couplers. The input light need not be divided equally between the output ports, and T couplers are made with various power-splitting ratios.

●
A T coupler has three ports; a star or tree coupler has more than three.

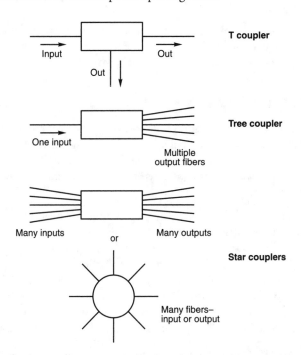

FIGURE 12.3.

T, tree, and star couplers.

Tree couplers have a single input and more than two outputs (or can be inverted to have multiple inputs and a single output). Star couplers have multiple inputs and multiple outputs. Some star couplers are directional, with separate groups of input and output fibers. In other star couplers, any fiber can serve as either an input or output, and signals are distributed to all fibers attached to the coupler.

The different couplers are used in different ways. The standard T coupler is used in a network such as that shown in Figure 12.1, where individual devices are connected to a data bus or ring that carries signals. A star coupler is used where all signals pass through a central point (where the coupler is inserted). A tree coupler is used to distribute signals from a single source to multiple points. In each case, the number of devices that can be connected is limited by coupler loss and receiver sensitivity. In a T-coupler system, the limit is reached when losses added along the ring or data bus reduce signal level below receiver requirements. With a star or tree coupler, the limit is imposed by the number of output ports. For example, if a 1 mW signal is divided equally among 100 ports without excess loss, each receives 10 μW from the coupler, a 20 dB loss. On the other hand, total loss would be 21 dB if a signal passed in series through seven T couplers that divided signals equally between two outputs (i.e., with 3 dB loss). Even with unequal T couplers that have loss of only 1 dB for signals sent along the main fiber, 20 dB loss would accumulate after 20 couplers. This means that networks must be designed carefully to avoid excess coupling losses.

Most couplers can be used with multimode fibers because their large cores can collect light reasonably efficiently. However, some types cannot be used with single-mode fibers because the small cores do not collect enough light from the mixing area.

Directional Couplers

Some couplers work only if light is going in one direction; others work if light is going in either direction.

Another variable in coupler design is how they operate on light going in different directions. Some couplers operate in different ways depending on which port the light enters and which direction it travels; these are called directional couplers. Others operate in the same way no matter which port light enters or which direction it travels; these are called bidirectional couplers. Figure 12.4 shows samples of the two types.

In the directional coupler, light must be input via the left-hand fiber (A) to be split between the two output fibers (B and C). If light enters via B, essentially all of it emerges through A and virtually none goes to fiber C. Such a coupler can also take inputs from B and C and combine them as a single output in A, but it cannot combine A and B in C, or A and C in B. In this example, directionality is caused by the way light is guided. The simple bidirectional coupler shown in Figure 12.4 uses reflection in a mixing area to distribute signals input from any of three fibers to the other two. For purposes of signal splitting, it does not matter which is the input fiber and which are the outputs.

Tree and Y couplers are directional couplers; star couplers are bidirectional. As mentioned previously, the directionality of T couplers depends on the design. Both directional and bidirectional couplers may be used to split or combine signals in fiber-optic systems. The choice depends on the application, but it is vital to know which is which.

FIGURE 12.4.

Directional and bidirectional T couplers.

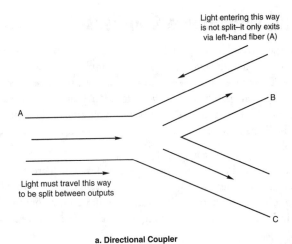

Light entering this way
is not split–it only exits
via left-hand fiber (A)

A

B

Light must travel this way
to be split between outputs

C

a. Directional Coupler

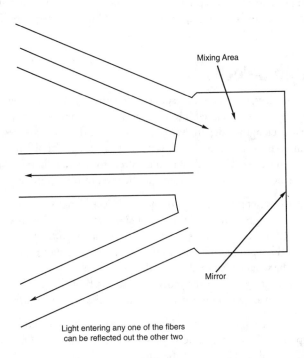

Mixing Area

Mirror

Light entering any one of the fibers
can be reflected out the other two

b. Bidirectional Coupler

Fiber and Waveguide Couplers

Couplers can be made by merging fibers or planar waveguides.

Look back at the Y coupler in Figure 12.4, and you can see that it looks like a pair of fibers merging together. That, in fact, is one way that couplers are made—from a pair of fibers. The fibers may be melted or pressed together, usually after some of the cladding has been removed, so light can pass between their light-carrying cores. This actually produces a coupler with two inputs and two outputs, but often one of the inputs is cut off to give a Y coupler. The fibers may be drawn so they are narrow at the junction, in what is called a biconic coupler.

Similar Y couplers can be made in planar waveguides, which are described in more detail later in this chapter. These waveguides are essentially stripes formed in a thin layer of material on a substrate. Differences in refractive index confine light within the stripes just as in an optical fiber. Splitting the waveguide divides the light between the two output waveguides.

Waveguide couplers have both advantages and disadvantages compared to fiber couplers. Waveguide components are conceptually simple, and fairly easy to fabricate. However, there are higher losses in coupling light from a fiber into a waveguide than in a fiber coupler.

Wavelength Selectivity

Wavelength-selective couplers transmit light of different wavelengths in different ways. Applications are in wavelength-division multiplexing.

Like other optical devices, couplers may respond differently to light of different wavelengths. In general, the ratio of light coupled into two (or more) output ports can depend on the wavelength. In wavelength-independent couplers, the differences are negligibly small. However, the differences can be made very large to separate light of different wavelengths. Many optical systems use devices called beamsplitters to separate light of different wavelengths in a single beam. The same concept can be used in fiber-optic systems.

The major use of wavelength-selective couplers is in wavelength-division multiplexing, where two (or more) signals are sent through a single fiber at different wavelengths, as shown in Figure 12.5. Here, a single fiber carries signals in different directions at two wavelengths, λ_1 and λ_2, produced by light sources on different ends of the fiber. The two wavelengths must be separated at each end of the fiber. The top wavelength-selective coupler transmits λ_1, entering it from the transmitter and reflects light at λ_2 to the receiver. The lower coupler reflects λ_1 to the receiver and transmits λ_2 to the fiber.

Wavelength-selective couplers can be designed to separate wavelengths that are closely spaced or far apart. Some can separate signals at the 1300 and 1550 nm windows, and others can isolate signals at 1293 and 1308 nm.

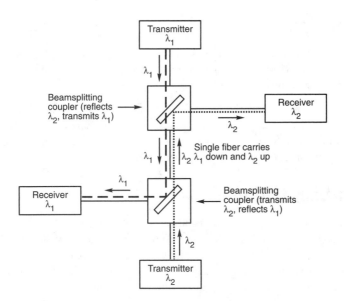

FIGURE 12.5.
Use of wavelength-selective beamsplitting couplers.

Splitters and Combiners

Couplers can serve two distinct functions in signal transmission. Sometimes they divide one input into two or more output signals to drive multiple devices. In other cases, they combine two or more inputs from separate devices to provide input to another device. These functions are inherently directional, depending on which way the signal is being transmitted. If signals are transmitted in both directions through the system, the same bidirectional coupler can serve both as a splitter and combiner. Sometimes couplers are called splitters or combiners.

●

Couplers can split one input into two or more outputs, or combine multiple inputs into a single output.

Number of Ports

Each coupler has a characteristic number of input and output ports. Those numbers are normally identified by describing the coupler as an n×m type, where n is the number of inputs and m the number of outputs. For directional couplers, the order of the numbers is fixed; for bidirectional couplers, it depends on how the coupler is installed. A bidirectional three-port coupler, for example, could be either a 1×2 or 2×1 coupler, depending on its use in a network.

●

An n×m coupler has n input ports and m output ports.

The classic T coupler is a 1×2 or 2×1 device. Closely related are 2×2 couplers, which have two input and two output fibers and sometimes function as T couplers with one input or output port unused. In star couplers, n and m are larger, and often equal, because in practice the inputs and outputs are connected to the same terminal devices. Tree couplers are 1×N (or N×1) devices.

Type of Fiber

The type of fiber used is critical in design and selection of couplers. Individual couplers are designed for use with specific fiber types and normally come with fiber pigtails or connectors that mate to those fibers.

Some coupling techniques work only with single-mode fibers, such as waveguide devices, because the light has to be concentrated in a small area. Others work much better with large-core multimode fibers. Many work with either single- or multimode fibers, but as mentioned previously, different devices must be built for each.

Power Distribution

Power distribution among output ports differs among couplers. Some T or 2×2 couplers divide power equally, but others split it in various fractions, such as 30%-70%, 20%-80%, or 10%-90%. The choice of the splitting ratio depends on system design. For multiport star or tree couplers, the goal is generally to distribute power as equally as possible among output ports, so the smaller the port-to-port power variations, the better the coupler.

In some systems, signals should be isolated from each other. For example, in a 2×2 directional coupler, the signal entering through one input fiber ideally should not be transmitted down the other input fiber. In practice, the isolation is not perfect, but the signal attenuation is usually strong—40 dB or more.

Note that some couplers, particularly single-mode types, may have different effects on light of different polarizations. This may be intentional or unintentional, but you should be aware that putting the nominally unpolarized output of a fiber through some couplers can generate two polarized outputs.

Excess Loss

Splitting an optical signal among two or more outputs in a passive coupler means that each output has less power than the input. In a perfect coupler, those would be the only losses experienced by the signal, and the sum of the outputs would equal the input. In real couplers, an excess loss is given by taking the ratio of the total output to the input. You can consider it power wasted in the coupler. It is normally given in decibels.

$$\text{EXCESS LOSS (dB)} = -10 \log (\Sigma \text{OUTPUT/INPUT})$$

Cost

Coupler prices have come down from the early days of fiber optics, but those used to working with electronic connectors may be surprised to find that fiber couplers cost much more than connectors. Simple three- or four-port couplers are the least expensive; prices increase with the number of ports.

Practical Couplers

The basic concepts described earlier have been used in many different coupler designs. I can't cover them all, so I will concentrate on a few important examples that illustrate the internal workings of couplers.

Connectors and Couplers

I stressed earlier that couplers are not connectors, but you should realize that connectors may be attached to couplers. Standard connectors are often mounted on the cases that house couplers, to make it easy for fiber-optic cables to deliver signals to the coupler. Alternatively, some couplers are spliced to other fibers. Those are really the only options for connections, so the terminology makes technical sense—but it can be confusing.

⬤ Connectors are mounted on the ends of many couplers.

Fused Fiber Couplers

As I mentioned previously, many couplers rely on the transfer of light signals between a pair of fibers. Their cores must be placed very close together, so light can transfer between them. First, the plastic buffer is removed entirely from the coupling zone. Then the two fibers are melted together and pulled to create a tapered region where light can be transferred between the fiber cores, shown in Figure 12.6. For single-mode fibers, where the core is much smaller than the fiber's outer diameter, some of the cladding may be removed before the fibers are heated and fused together, but this is not necessary for multimode fibers. These couplers are called fused fiber or biconic couplers.

⬤ Fiber couplers rely on light transfer between fibers.

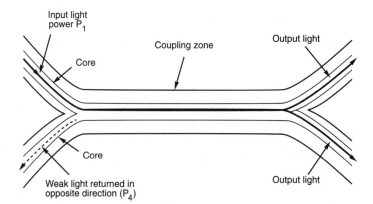

Input light power P_1
Coupling zone
Output light
Core
Core
Weak light returned in opposite direction (P_4)
Output light

FIGURE 12.6.

A 2×2 coupler made by fusing two fibers.

EVANESCENT WAVE COUPLING

The cores themselves need not be fused together. If the distance between them is small enough—say, a few micrometers—a phenomenon called evanescent wave coupling will transfer light between them. Some light, called evanescent waves, actually travels in the inner part of the cladding rather than in the core, and this light can leak between cores.

Fusing two fibers produces a 2×2 coupler with two inputs and two outputs. (To make a three-port coupler, one fiber end is ignored or cut off.) The design makes the 2×2 coupler inherently directional, with the signal reflected in the wrong direction down the fiber end parallel to the input typically 40-45 dB below the input power. Directivity is measured by comparing the input power entering through a single fiber (P_1) to the power reflected back to the other input fiber in the direction opposite the input (P_4 in Figure 12.6). Mathematically,

$$\text{DIRECTIVITY (dB)} = -10 \log (P_4/P_1)$$

OPTICAL BEAMSPLITTERS

Wavelength-selective T couplers can be built around a beamsplitter that separates wavelengths.

Another approach, often used in wavelength-selective couplers, relies on optical beam-splitters to divide one input or combine two inputs into a single beam. The beamsplitter reflects some incident light and transmits the rest. Some beamsplitters are made to reflect one wavelength and transmit others (for example, reflecting 1300 nm but transmitting 1550 nm). When the beamsplitter plate is set at 45° to the incoming light, one wavelength is transmitted but others are reflected, as shown in Figure 12.7. Light entering through the two output fibers would be directed out the fiber at right, because the beamsplitter would still transmit and reflect the same wavelengths.

FIGURE 12.7.

A wavelength-selective T coupler.

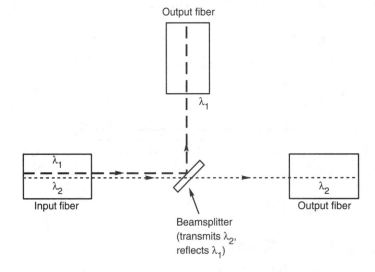

Star Couplers

A star coupler has different functional requirements than a T or 2×2 coupler. When light is transferred between only two fibers, it can be concentrated in a small region (e.g., the core of a single-mode fiber). However, if light is to be distributed among many fibers, it must be spread out reasonably uniformly over the region where the fibers collect light. Then the cores of the output fibers must collect this light efficiently. This can lead to large excess losses unless the output fibers are bunched tightly together.

One way to make star couplers is to fuse many fibers together, forming a mixing region where the signals combine. These are called transmissive stars because light travels from the input fibers through the mixing region to output fibers. This approach can be used for coupling up to about 64 fibers. Their efficiency depends on how well the output fibers can collect light.

Reflective star couplers are also possible. These are made like transmissive couplers, but the mixing region is cut in half and coated with a reflective material so light entering through one fiber returns to be distributed among all the fibers entering the coupler. Unlike a transmissive star, a reflective star does not discriminate between input and output fibers. It sends light into all fibers, even those that provided the input.

The choice between reflective and transmissive stars depends on network architecture. Reflective stars are needed if signals are transmitted and received through the same fiber. However, if input and output channels are separate, transmissive stars are a better choice because no light is lost into input fibers, reducing excess loss at least 3 dB.

Most star couplers work with multimode rather than single-mode fibers. This is largely because the larger-type multimode fibers collect light much more efficiently. Single-mode star couplers with only a few input and output fibers can be reasonably efficient, but it is very difficult to produce multiport star couplers for single-mode fibers. Fortunately, that is not a serious practical problem, because single-mode fibers are rarely used in networks that require multiport couplers.

> ●
> Couplers can be made by fusing many fibers to create a mixing region.

> ●
> Star couplers can be transmissive or reflective. Transmissive stars separate input and output fibers and, hence, have lower excess loss.

Tree Couplers

Tree couplers have a single input and more than two outputs, making a branching structure that gives them their name. Many are functionally similar to star couplers, but have one instead of multiple inputs. (They could be fabricated by cutting off the extra input fibers from a star coupler.) Others are more related to Y couplers, or may even be made from a series of branching Y couplers.

> ●
> Tree couplers branch out from a single input to many output fibers.

Active Couplers

So far, I've talked entirely about passive couplers, which transmit and reflect light but don't generate any new signal. There is also another entire family of active devices that serve the same function as couplers but do generate or amplify light.

●
Active couplers are
essentially special-
purpose repeaters
that drive both a
terminal device and
an output fiber.

Figure 12.8 shows the basic concept of one type of active coupler, which drives both a terminal device and an output fiber. A receiver detects the input light, generating an electronic signal that goes to decoding electronics. The decoder separates signals intended for that terminal from those intended for the rest of the network and generates two electronic outputs—one for the terminal device, and the second for an optical transmitter. The transmitter then produces a signal that drives the next fiber segment. This approach is used in some local area networks like the one shown in Figure 12.1, including the Fiber Distributed Data Interface (FDDI) network covered in Chapter 20. You can think of the whole package as essentially a special-purpose repeater.

FIGURE 12.8.

*An active coupler links
a terminal to a
network.*

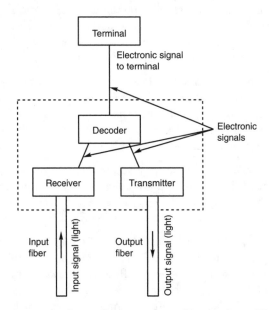

Another way to enhance coupler performance is to add optical amplifiers either before or after the coupler splits the signal. The amplifier can make up for lost power, raising signal strength to meet receiver requirements. For example, if the outputs of a tree coupler/splitter did not meet receiver requirements, an optical amplifier could be put before the coupler to boost input power.

Planar Waveguide Technology

●
Planar waveguides
are flat structures
that guide light in
couplers and other
components.

I mentioned earlier that couplers can be made from planar waveguides. That technology is important enough to deserve a closer look, particularly because of its importance in optical switches and modulators. The field is often called integrated optics, because it allows many optical devices to be integrated on a single substrate.

As their name implies, planar waveguides are flat structures that guide light. They are actually stripes fabricated in a thin layer on top of a substrate, as shown in Figure 12.9. The refractive index of the waveguide is higher than that of the substrate or thin-film layer, so light is confined by total internal reflection, as in an optical fiber. The waveguides are typically formed by diffusing a dopant from above. If the waveguides are made in suitable shapes, they can split input light signals. Special structures can be made to control the degree of splitting dynamically, so waveguide couplers can serve as optical switches or modulators, as described below.

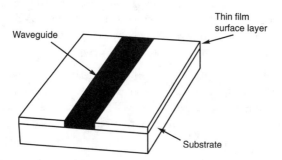

FIGURE 12.9.

Embedded thin-film waveguide.

Waveguide Structures

The planar waveguide of Figure 12.9 differs in some subtle ways from the core of an optical fiber besides its rectangular cross-section. The waveguide is not completely surrounded by identical material of slightly lower refractive index, because the air layer on top has an index of 1, much lower than the waveguide. In practice, the waveguide layer is made with index slightly higher than the substrate and the thin-film layer in which it is embedded. The embedded waveguide shown in Figure 12.9 is the most common type for couplers. However, other structures are possible, some with ridges on top of the substrate.

Planar waveguides are made by depositing materials on a substrate or diffusing material into a substrate in patterns formed by standard semiconductor fabrication techniques. Processing details vary, depending on the materials and waveguide structure. The technology allows fabrication of many patterns, in addition to simple linear waveguide, opening possibilities for new types of components.

Waveguide Materials

Many materials can be used for planar waveguides, but developers have settled on a few. Glass is an attractive material because the technology is well-developed and its attenuation is low. Photolithography and diffusion techniques can form waveguides along the surface that are thick enough to carry light from multimode or single-mode fibers. This makes it a good choice for couplers and other passive waveguide components.

●

Waveguides can be made in glass, semiconductors, and lithium niobate.

Semiconductors such as gallium arsenide are attractive because lasers, detectors, and electronic devices can be made from them as well. This lets researchers make complex integrated devices, such as a single chip that includes multiple lasers with different wavelengths, waveguides to carry their outputs, and couplers to mix the outputs. Active optical components can also be made in semiconductors, making possible waveguide switches and modulators as well as couplers.

Lithium niobate ($LiNbO_3$) is often used for planar waveguides because its refractive index changes with the electric field applied to it. Because an optical waveguide's transmission properties depend on refractive index, applying a voltage across a lithium niobate waveguide can change how it guides light. To go a step farther, changing the voltage can change the waveguide properties enough to switch light between two optical waveguides. That is the basis of the optical switches and modulators that I describe later.

Many other materials also display the electro-optic effect—the variation of refractive index with ambient electric field. However, it is particularly strong and well characterized in lithium niobate. Gallium arsenide and similar III-V semiconductors, silicon, and certain glasses also display the electro-optic effect, so they can also be used in optical switches and modulators.

Couplers

●

Simple waveguide couplers can be made by making a waveguide branch.

The simplest type of waveguide coupler is a Y-shaped structure like the directional coupler in Figure 12.4a. (Actual split angles are much smaller than shown.) If the output waveguides go off at equal angles, the light is split evenly between the two. This approach can be extended to more than two output waveguides either by making more than two output branches or by putting multiple Y junctions in series. The division of power depends on junction angle and the way successive junctions are arranged.

●

Light can leak between two adjacent waveguides on the same substrate, forming a coupler.

Another approach is to rely on coupling of the evanescent wave that leaks out of the high-index waveguide into the surrounding lower-index material. If two waveguides are close enough on the same substrate, light can be transferred between them, as shown in Figure 12.10.

The power in each waveguide varies along the region where they are close enough to couple. As light travels along the upper waveguide in Figure 12.10, more and more of it transfers to the lower one. After a certain distance (sometimes called the transfer length), which depends on the optical characteristics of the waveguide, all the light shifts to the lower waveguide. Then the light starts transferring from the lower waveguide to the upper one. Thus, as the light travels along the two coupled waveguides, the distribution of light energy oscillates back and forth between them. The oscillation stops at the end of the region where the two are coupled, determining the final distribution of light. (This simplified case ignores losses, but the real world is not so kind. Planar waveguides have much higher losses than fibers.) Designers can select lengths and optical properties of the two waveguides to divide the light into the desired proportions (e.g., 50-50 or 75-25).

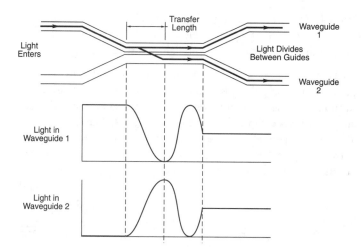

FIGURE 12.10.

Transfer of light energy between two coupled waveguides.

Waveguide Limitations

Integrated waveguide optics may sound great, but they do have some important drawbacks. One is the loss inevitable when transferring light from the circular core of an optical fiber to a planar waveguide with a thin rectangular cross-section. Losses are higher for multimode fibers but can be significant even for single-mode fibers. Another is that the waveguides themselves are lossy compared to fibers. In addition, the whole field of integrated optics suffers from the comparative weakness of optical interactions. Electrons interact strongly, so electronic devices can be made small; the weakness of optical effects means that waveguides have to be much longer than electronic integrated circuits if they are to do anything to an optical signal passing through.

⬤

Some light is lost when transferring light from a fiber to a waveguide.

Waveguide Switches and Modulators

Why bother with the limitations of integrated optics and the complexity of evanescent-wave couplers? Because they can be more than static, passive devices. Remember that waveguides are made in materials that display the electro-optic effect. Changing the voltage applied across them changes their refractive index, which, in turn, affects light coupling between adjacent waveguides. So change the voltage and you can control how much light emerges from each waveguide. That lets you modulate or switch the output.

⬤

Voltages applied to coupled waveguides can make them switch or modulate optical signals.

Start by looking at the evanescent-wave coupler in Figure 12.10. I said that the transfer length—the distance light has to travel along the two coupled waveguides before it is transferred completely from one to the other—depends on their optical characteristics. (The details are the type of complexity that you don't want to worry about.) To convert the coupler into a switch or modulator, you can apply electrodes to produce an electric field across the two waveguides, as shown in Figure 12.11. This changes their optical properties (specifically the refractive index)—and light transfer.

Suppose that at zero voltage, all the energy entering Waveguide 1 in Figure 12.11 emerges in Waveguide 2. Increasing the voltage applied across the waveguides changes the refractive index, and at a voltage V, light first shifts from Waveguide 1 to Waveguide 2, then at higher voltages switches completely back to Waveguide 1. The result is an optical switch, controlled by the voltage applied to the electrodes. Further increases in voltage might cause additional transfers, but that is not normally done in practice. Similar effects can be used to convert other waveguide couplers into switches.

FIGURE 12.11.

Transfer of light energy between two coupled waveguides.

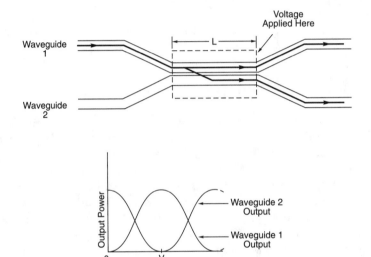

Similar concepts can be used to make waveguide light modulators. From the outside, a light modulator seems to change its transparency, at times transmitting nearly all light entering it, at others transmitting almost none. However, inside a waveguide modulator is a special type of optical switch or coupler. Like a two-port switch, a waveguide modulator divides input light between two waveguides, with the degree of division depending on the electrical modulation voltage. One waveguide delivers the (modulated) output light signal. Light entering the other waveguide is thrown away.

●
Waveguide modulators change the fraction of light coupled out of the two outputs.

It is important to note that modulation can be continuous, whereas switching is supposed to be either on or off. Thus, modulators must be capable of transmitting not just 100% or 0% of the input light but also intermediate values, and they must be capable of responding quickly to changes in the modulation signals. Switches need to respond quickly, but they should not have gradations between off and on. (In practice, optical switches are not ideal switches, so a little light may leak out the "off" side.)

Other Optical Switches

Other types of optical switches can be made, but optical switching in general is difficult. There is no simple, compact optical equivalent of the electronic logic circuits that switch electrical signals. In many cases, optical signals have to be converted into electronic form before they can be switched.

● Optical switching is difficult.

Opto-Mechanical Switches

Optical signals can be switched by mechanically moving fiber ends or other components, but this is slow. A fiber can be moved between two slots to make connections with two different fibers, or a focusing lens can be moved to direct light to different points in the switch. However, switching speeds are measured in milliseconds, much too slow to handle routine signals in the telecommunication network, and moving fibers are vulnerable to breakage. Thus, low-performance opto-mechanical switches are limited to low-duty applications, such as bypassing failed network nodes.

Bistable Optics

Over the years, optical scientists have developed and tested many different concepts for optical switching. Most have not proved practical, but some are still in development. One interesting possibility is a family of "bistable" optical devices that have two stable states—one in which they transmit light well, and one in which they transmit light poorly.

● Bistable devices switch between high- and low-transparency states.

The basic idea of a bistable optical device is shown in Figure 12.12, which plots light transmitted by a bistable device versus light entering it. In this example, only a small fraction of input light is transmitted at low intensities. Transmission remains small until the input intensity passes a threshold value, above which it increases rapidly.

If light transmission simply retraced its path as you reduced input intensity, the device would not be truly bistable (although it still might be useful for optical switching). In a true bistable device, the transmission remains high after the input power is reduced below the turn-on threshold value, and drops only at a lower turn-off threshold, as shown in Figure 12.12. If you're familiar with electronics, that plot looks familiar, because it contains a hysteresis loop like those of some electronic devices. The implications in optics are the same as they are in electronics. A device with such a loop can be in two stable states at one input level, depending on whether it started from higher or lower on the input plot. This means that it can be "latched" into an "on" or "off" state, like a transistor circuit, by applying the proper inputs. And that opens the door to optical logic as well as switching.

FIGURE 12.12.

Light transmission by a bistable optical device.

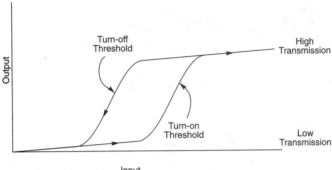

Turn-off Threshold

High Transmission

Output

Turn-on Threshold

Low Transmission

Input

● Bistable optical devices can be used like switching transistors.

Some bistable optical devices lack true hysteresis loops, but they can still be used as switches because their input-output curves have a sharp turn-on like a switching transistor. That sounds like great news for the designers of fiber-optic transmission systems, but bistable optics is still in its infancy, and may not prove practical.

What Have You Learned?

1. A coupler connects three or more fiber ends.

2. Dividing an optical signal among output ports reduces its strength.

3. Couplers are needed for wavelength-division multiplexing if multiple signals are sent at different wavelengths through the same fiber.

4. Major elements in coupler design include number of ports, directional sensitivity, wavelength selectivity, type of fiber, cost, and power requirements.

5. Total output power from a passive coupler can be no more than the input power.

6. A T or Y coupler has three ports. A tree coupler has one input and more than two outputs. A star coupler has multiple inputs and outputs.

7. Some couplers work only if light is going in one direction; others work if light is going in either direction.

8. An n×m coupler has n input ports and m output ports.

9. Excess loss is reduction in signal strength above that needed to divide the signal.

10. Couplers can be made by merging fibers or planar waveguides.

11. Wavelength-selective T couplers can be built around a beamsplitter that separates wavelengths.

12. Couplers can distribute power evenly or unevenly among output ports.

13. Star couplers can be made by fusing many fibers to create a mixing region.

14. Active couplers are essentially special-purpose repeaters that drive both a terminal device and an output fiber.

15. Planar waveguides are flat structures that guide light in couplers and other components. Waveguides can be made in glass, semiconductors, and lithium niobate. They can serve as couplers, optical switches, or modulators.

What's Next?

In Chapter 13, I will look at the other components, tools, and accessories used in fiber optics.

Quiz for Chapter 12

1. A fiber-optic signal can be split among two or more devices by
   a. putting them in parallel across an input fiber.
   b. passing the signal through them in series.
   c. splitting the signal so some light goes to each.
   d. none of the above.

2. What is not required for wavelength-division multiplexing?
   a. Wavelength-selective couplers.
   b. Transmission of signals at different wavelengths through the same fiber.
   c. A fiber capable of transmitting more than one wavelength.
   d. Transmission of signals in different directions.

3. If a T coupler with no excess loss equally divides input power between two output ports, how does the power at one output port compare with the input power?
   a. The same.
   b. 3 dB higher.
   c. 3 dB lower.
   d. 0.5 dB lower.

4. A star coupler divides output equally among 100 ports, with excess loss of 3 dB. The total drop in power between input port and one of the outputs is
   a. 3 dB.
   b. 10 dB.
   c. 20 dB.
   d. 23 dB.
   e. 100 dB.

5. How many fibers can be connected to a 64×64 star coupler?
   a. 64.
   b. 128.
   c. 1024.
   d. 4096.

6. A 2×2 coupler can be made from a pair of fibers by

   a. fusing them together so light transfers between their cores.

   b. enclosing them in a common plastic coating.

   c. putting them in the same loose tube in a cable.

   d. attaching them end to end.

   e. all of the above.

7. A fused 2×2 fiber coupler is inherently

   a. directional.

   b. bidirectional.

   c. wavelength-selective.

   d. polarizing.

8. What is a planar waveguide?

   a. An optical fiber squashed flat by a truck.

   b. A high-index stripe embedded in a substrate with lower refractive index.

   c. A low-index region in a block of semiconductor.

   d. A thin-film layer that completely covers a flat surface.

9. How can light be switched between a pair of planar waveguides?

   a. By adding a third waveguide to connect them.

   b. By changing the optical wavelength.

   c. By changing the voltage applied across a material like lithium niobate.

   d. By changing the spacing between them.

10. Which technology can be used to make both optical switches and modulators?

   a. Fusion splicing of fibers.

   b. Thin-film waveguides in lithium niobate.

   c. Star couplers.

   d. Opto-mechanical movement of fibers.

Other Components, Tools, and Accessories

About This Chapter

So far I've covered the major components of fiber-optic systems. However, an assortment of other equipment is also used with optical fibers. Some are optical components that operate on light in the fiber, such as filters, attenuators, and optical isolators. Others are specialized tools used to manipulate optical fibers. And some are simply things I need to mention that don't fit elsewhere. (I will cover measurement tools in Chapter 14.)

This chapter will give you a basic introduction to this specialized fiber-optic equipment, describing what it is and how it is used.

Optical Components

Optical fibers are versatile tools, but they cannot do everything that is needed in a fiber-optic system. In many places, the light may have to be focused, or part of the light may have to be blocked or redirected. These are the jobs of other optical components. Many have equivalents in larger-scale optics, but most versions used with fiber optics are designed expressly for that purpose. I can't go into a lot of detail, but I will cover each briefly.

Microlenses

A microlens is a very small lens that focuses light from an optical fiber or light source.

My earlier descriptions of transmitters and receivers glossed over one optical component common in such devices—the microlens. As shown in Figure 13.1, a microlens is a very small lens that focuses light to improve light transfer between fibers and other devices. They are important because the light emerging from fibers and light sources forms diverging cone-like beams that spread out rapidly with distance. Microlenses bend those light rays to make a more nearly parallel beam. They can also focus parallel light rays onto a small spot, such as a fiber core.

FIGURE 13.1.

A microlens focuses light.

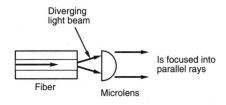

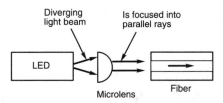

Most microlenses are discrete components. However, a microlens can be formed at the tip of a fiber by melting the fiber end. The surface tension of the molten glass rounds the tip into a hemisphere that can act like a lens and focus light. Small segments of special graded-index fiber can also serve as microlenses. They will be described in Chapter 23.

Optical Filters

Filters selectively transmit some wavelengths and reflect others.

If you've ever done much photography, you have probably encountered filters. They block some light and transmit the rest. The most common in photography are color filters, which transmit only the desired colors, and background filters, which compensate for differences between the lighting and color response of film. There are also other types, including narrow-line filters, which transmit (or block) only a narrow range of wavelengths, and neutral-density filters, which block a certain fraction of light at all wavelengths.

Wavelength-selective filters are important in fiber-optic systems that carry more than one wavelength. The wavelengths do not interfere with each other in the fiber, but they can interfere at the detector. To prevent this, they must be separated at the end of the fiber. Typically this is done by placing a filter at the fiber end that reflects certain wavelengths and transmits others. The most common of these filters are small pieces of glass coated with many thin layers that control transmission and reflection.

Filters are important for wavelength-division multiplexing. They are also used to isolate signals from background noise, and to block stray pump light when using optical amplifiers. They are typically included within couplers, receivers, and equipment for wavelength-division multiplexing.

Attenuators

Fiber-optic systems need a certain amount of optical power to work properly, but too much power can be a bad thing. You saw in Chapter 8 that fiber-optic receivers can be overloaded by power beyond their dynamic range. There are also other cases where power may need to be restricted, perhaps because couplers do not distribute signals evenly, or to protect sensitive measurement instruments. In other cases, technicians need a way to vary signal level to test system response. All these situations are places where attenuators can be used.

Attenuators discard surplus optical power. They can reduce signal levels in communication systems to those that the receivers can handle best. This makes it possible for all terminals in a network to use the same transmitters and receivers, even though light traveling between them suffers different losses. Adding attenuators makes it possible to use the same terminal equipment in all parts of a local area network. The terminal next to the distribution node might receive a signal 20 dB higher than one on the opposite corner of the building, but an attenuator could balance the power levels.

The simplest and least costly attenuators are filters that block a fixed portion of the incoming light. These are installed in communication systems to balance power levels, which normally will not need to be changed again.

Variable attenuators are more costly but more versatile. They contain internal optics that can be adjusted to control the degree of attenuation, typically over a large range. For example, the gap loss attenuator is adjusted by changing the distance between a pair of fiber ends, altering the overall loss. Variable attenuators are typically used in research and engineering laboratories, and for test and measurement.

Polarizers

You saw in Chapter 4 that light can be divided into two polarizations, depending on the orientation of the light wave. Polarizers block light waves of one polarization and transmit those with the perpendicular polarization. This divides the light roughly in half. They are the polarization-sensitive equivalent of filters and are often called polarizing filters. (Polarizing sunglasses use them both to reduce light intensity and to cut down on glare, which is polarized by the way it reflects from certain surfaces.)

Although most fibers do not maintain the polarization of input light, a few do. More importantly, polarization can be used in other ways to divide and control light.

● Attenuators reduce optical power to a desired power range.

● Variable attenuators are used for testing; fixed attenuators are installed in systems.

● Polarizers control the polarization of light.

Optical Isolators

Optical isolators block noise from reaching sensitive lasers and other components.

One important use of polarizers and other polarizing components is in optical isolators, which prevent back-reflections and other noise from reaching sensitive optical components. You can think of them as optical one-way streets with their own traffic cops. Figure 13.2 shows the basic idea. Light traveling to the right first passes through a polarizer that transmits vertically polarized light. Then it goes through a device called a Faraday rotator, which rotates the polarization 45° (as shown by the polarization vectors in the circles). The rotated light can travel straight through the second polarizer, which is oriented at a 45° angle to the first, with essentially no losses.

FIGURE 13.2.

An optical isolator transmits light in only one direction.

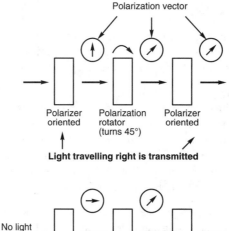

Light going to the left first passes through the second polarizer, then through the Faraday rotator, which rotates it another 45° in the same direction. (Polarization is always rotated in the same direction, no matter which direction light travels through the device.) This leaves it rotated 90° from the first polarizer, which blocks it. Thus the optical isolator transmits light to the right but blocks light going in the other direction.

Optical isolators are important components in high-performance systems because of the way they can block noise traveling in the wrong direction through the fiber. Typically they attenuate light passing in the wrong direction by 40 dB or more, enough to greatly reduce noise.

Fiber Components

So far I have assumed that optical components are separate from optical fibers. However, optical fibers can be made to serve some functions of discrete components. Most work is developmental.

Single-polarization fibers can be used as polarizers, because they carry light in only one polarization direction. However, long lengths are required to attenuate the undesired polarization to low levels. A more practical alternative is to modify a short segment of fiber so it strongly attenuates one polarization.

Recent research points to other possibilities. Ultraviolet light can permanently change the refractive index of fibers rich in germanium. With proper illumination, the ultraviolet light can form alternating high- and low-index regions in a fiber segment. Depending on the spacing and other fiber properties, these regions may strongly reflect or transmit certain wavelengths—making them into wavelength-selective filters. The technology is still in the laboratory stage, but it is promising for future applications.

● Optical components, including filters and polarizers, can be made from fibers.

Index-Matching Materials

Earlier I mentioned that index-matching gels and other index-matching materials can limit internal reflections in connectors and in mechanical splices. These deserve a bit more explanation because their function is unique to optical systems.

Reflection at the interface between two optical materials depends on the difference in their refractive indexes and the angle at which light strikes the interface. The larger the difference in refractive index, the bigger the reflective (Fresnel) losses, and the larger the angle over which they occur. If you make the simplifying assumption that all the light is going straight through, these losses can be calculated for optical fibers from the formula

$$\text{LOSS (dB)} = 10 \log_{10} (1\text{-}p)$$

where p is a reflection coefficient given by

$$p = [(n_1\text{-}n_0)/(n_1+n_0)]^2$$

where n_1 and n_0 are respectively the refractive index of the fiber and the medium from which light enters it.

● Index-matching gels reduce surface reflection losses.

This formula gives the loss per surface. Where two surfaces are involved (as in the glass to air to glass transition for light passing through an air gap between two fibers), total loss is the sum of reflections at two interfaces. If the two interfaces are between the same materials, total loss is double the loss at one (because the difference in refractive index is squared, the signs and direction of travel doesn't matter).

The glass used in fibers has a refractive index of about 1.5, much higher than the 1.0 index of air, so the loss at an air-glass interface is about 0.3 dB. (Some of that light becomes undesired back-reflections that can cause noise in the fiber system.) That loss can be reduced by filling the space between fiber ends with a transparent gel or solid with refractive index close to that of the fiber core. Transparent epoxies fill that role in glued joints. Thick, viscous liquids such as glycerine or silicone grease can serve as index-matching gels in unglued joints. They fill the interior of the connector so no air spaces remain between the fiber ends. The index match need not be precise. A refractive index of 1.4 would cut Fresnel loss down to 0.005 dB per surface, or 0.01 dB overall.

Mode Strippers or Filters

Mode strippers or filters remove undesired modes from multimode fibers to improve measurement precision.

As you saw in Chapter 4, multimode fibers can transmit light in many distinct waveguide modes. This has advantages for the transmission of light from large-area, high-divergence sources. But it makes the measurement of fiber characteristics difficult because of the way light is distributed among fiber modes. Much light enters the fiber in high-order modes, which are barely confined in the fiber. As the light travels along the fiber, it leaks out of these modes, and eventually—after traveling a few kilometers—the light assumes an equilibrium distribution among fiber modes. However, if light hasn't traveled far enough, the distribution of light among high-order modes is uneven and very hard to predict. That complicates attenuation measurements because they depend strongly on the distribution of light among fiber modes.

To avoid this problem, light can be passed through a mode stripper or mode filter after it enters the fiber. This is a length of fiber bent or otherwise treated so high-order core modes propagating in the cladding will escape.

Microscopes

Microscopes are needed to examine or visually align tiny fiber-optic components.

To examine or visually align tiny fiber-optic components, you need a microscope. Specialized microscopes come with virtually all fusion splicers. Most are binocular types, with eyepieces for both eyes and magnification of about 50 power. Those microscopes stare down onto an illuminated stage built to align two fibers end-to-end. They make fibers easily visible even to the severely nearsighted (as I can testify from trying several at trade shows). Some fusion splicers come with video cameras that project a magnified image on a video monitor.

To inspect connector terminations, you need another type of microscope to look at fiber ends. The usual microscopes for that purpose are monocular (one-eyed) types with higher magnification—100-400 power. That magnification is large enough to make a 50 μm fiber core appear several millimeters in diameter. Most such microscopes are hand-held or have small stages to hold the connector or fiber end in place.

Fiber Tools

Specialized tools help prepare optical fibers and cables for use in systems. Conventional cable and wire cutters can remove the outer jackets of fiber-optic cables, and ordinary household scissors or tin snips can cut strength members. However, care must be taken in the delicate jobs of removing fiber coatings and cutting fibers to avoid weakening the fiber and/or causing excess loss at the connector or splice being installed.

Mechanical strippers can remove plastic coatings from fibers. Standard versions look like wire strippers but are made to match the standard dimensions of fibers. Special strippers are needed for fibers because it is vital to avoid nicking the glass. Microscopic surface nicks can be starting points for fiber breaks. Other alternatives for all-glass fibers include using chemical solvents to dissolve the plastic but not the glass, or heating to soften the low-melting-point plastic but not the fiber. Solvents and heat cannot be used for plastic fibers.

Cutting glass fibers is a delicate process. Normally a diamond- or carbide-tipped tool scribes the fiber, forming a weak point where the fiber breaks when bent under tension. Special tools and small jigs are built to perform the whole cleaving operation. For most applications, the ends should be close to perpendicular to the fiber axis. Cleanly cleaved fiber ends are needed for low-loss fusion splices. Some plastic fibers can be cut with scissors.

An additional step—polishing—is needed for high-performance optical connectors and mechanical splices. Typically, the fiber is inserted and glued into the connector ferrule with a small tip left protruding. This tip is ground and polished to a flat, smooth surface by rubbing it against polishing pads. Grinding and polishing is a multistage process similar to sanding down a rough wood surface. You start grinding with a coarse grit to remove rough edges, then polish with successively smaller grits until you produce an optically smooth surface.

This equipment often comes in tool kits that come with other accessories needed to install connectors, such as a microscope to view the fiber, a crimping tool to mount the connector, and components to assemble the connector.

Mounting and Positioning Equipment

Because optical fibers are small, special equipment is needed to mount and position them precisely, both for testing and for field use. I can't cover this equipment in any detail, but you should have a general idea of what it is.

Fiber splicers and connector-installation equipment include the required positioning tools. In a fusion splicer, the stage where the fibers are mounted includes precision micromanipulators to match the fiber ends, with micrometer tolerances. The motion may be controlled manually by threaded screws or electronically by digitized controls. Connectors and the equipment used in mounting them appear less elaborate. However, they have internal alignment mechanisms that provide the same tight tolerances.

● Conventional cable cutters and scissors can cut cable. Fiber coatings can be removed carefully with mechanical strippers or special solvents.

● Fiber cleavers cut fibers; the ends are usually polished for mechanical splices or connectors.

● Micromanipulators are used to align fibers.

Special micropositioning equipment is made for handling loose fibers in the laboratory. Figure 13.3 shows a mount that aligns a fiber to collect light from a laser or other source. Such mounts come with various levels of precision adjustments. Some simply hold the fiber steady, others come with adjustments in two dimensions, and others allow adjustments in three linear dimensions (X, Y, and Z axes) and two angles.

FIGURE 13.3.

Mount for positioning optical fibers in the laboratory. (Courtesy of Newport Corp)

Optical laboratories are usually equipped with special optical tables or benches designed for sensitive experiments. An optical table has a precisely flat surface, usually with an array of threaded holes on which accessories can be mounted. An optical bench is a linear rail-like structure on which you can place mounts to hold optical devices. Optical benches and tables provide the rigid, stable work areas needed in laboratory work with components as small as discrete fibers, light sources, and detectors.

Fiber Reels

The way fibers are wound onto reels affects their transmission while on the reel.

Optical fibers fresh from the factory are wound around reels, just as you would expect. However, the reels are not simply another piece of throw-away packaging for everyone but the janitor to ignore. The reel can have a surprisingly large effect on fiber characteristics while the fiber is on the reel—and fibers are often tested or used in experiments while still on their reels. The way the fiber is wound on the reel can lead to microbending losses affecting fiber attenuation. Variations in attenuation can arise from changes in winding tension, the way fiber is laid down on the reel, or the structure of the reel itself. Some fiber manufacturers take these effects seriously enough to write technical notes about the reels they use and expend extensive effort in developing the best possible reel.

Enclosures and Mounting Hardware

Many catalogs of fiber-optic equipment come with thick sections labeled "hardware" or "distribution equipment" that show lots of boxes that don't look very interesting at first glance. These are important parts of any fiber-optic installation, used like junction boxes are in electrical wiring.

This hardware includes splice enclosures, connector panels, interconnect centers, housings, wall outlets, splice trays, and assorted other equipment. The precise terminology depends on the vendor. Their functions are essentially similar: they are places to put fibers, splices, connectors, and other equipment. The housings protect the fiber equipment from mechanical stresses, dirt, and environmental stress. They also organize the fiber equipment to simplify further servicing.

⬤
Enclosures and
mounting equipment
protect fibers from
the environment.

What Have You Learned?

1. A microlens is a very small lens that focuses light from an optical fiber or light source.

2. Filters selectively transmit some wavelengths and reflect others.

3. Attenuators reduce optical power to a desired power range. Variable attenuators are used for testing; fixed attenuators are installed in systems.

4. Optical isolators block noise from reaching sensitive lasers and other components.

5. Index-matching gels reduce surface reflection losses. They need not precisely match the index of glass, but they should be close.

6. Mode strippers or filters remove undesired modes from multimode fibers to improve measurement precision.

7. Microscopes are needed to examine or visually align tiny fiber-optic components.

8. Conventional cable cutters and scissors can cut cable. Fiber coatings can be removed carefully with mechanical strippers or special solvents.

9. Fibers are scribed and broken to make smooth ends, which are polished for use in connectors.

10. Micromanipulators are used to align fibers.

11. The way fibers are wound onto reels affects their transmission while on the reel.

12. Enclosures and mounting equipment protect fibers from the environment.

What's Next?

Your next step in exploring the world of fiber optics is to learn about the specialized optical techniques used for fiber-optic measurements.

Quiz for Chapter 13

1. How can properties of a fiber change when it is wound onto a reel?

    a. Adhesion of the fiber to the reel may weaken it.

    b. Microbending can increase fiber attenuation.

    c. There are no significant changes in any fiber property.

    d. Pressure on the fiber inside the reel changes its attenuation.

2. The plastic coating on a fiber can be removed by

    a. mechanical stripping.

    b. solvents.

    c. peeling with fingers.

    d. a and b.

    e. none of the above.

3. Surface reflection losses when light passes between a glass fiber and air and back into the glass fiber are

    a. negligible.

    b. under 0.1 dB.

    c. 0.3 dB.

    d. 1 dB.

    e. over 1 dB.

4. Index-matching gel reduces attenuation when light is transferred between fibers by

    a. reducing Fresnel reflection losses caused by the large difference in refractive index between glass and air.

    b. preventing light leakage from the connection.

    c. guiding light through the junction.

    d. changing the refractive index of the fiber.

5. A microlens

    a. is a very small lens that helps transfer light between a source and fiber.

    b. can be formed by melting the end of a fiber.

    c. can be made from a segment of graded-index optical fiber.

    d. is found inside some transmitters.

    e. all of the above.

6. Attenuators are used to

    a. limit output power from a laser.

    b. prevent reflected light from being scattered into a fiber and inducing noise in a laser.

    c. reduce power level to match a receiver's dynamic range.

    d. prevent excess optical power from damaging fibers.

7. A mode stripper is used to

    a. discard excess modes in coupling from multimode to single-mode fiber.

    b. discard modes that might cause errors in measuring properties of a multimode fiber.

    c. prevent modal dispersion.

    d. match transmission characteristics of two fibers.

8. Optical isolators do which of the following?

    a. Limit output power from a laser.

    b. Prevent reflected light from being scattered into a fiber and inducing noise in a laser.

    c. Reduce power level to match a receiver's dynamic range.

    d. Prevent excess optical power from damaging fibers.

9. Optical filters do which of the following?

    a. Separate wavelengths by transmitting some and reflecting others.

    b. Prevent excess optical power from damaging fibers.

    c. Limit output power from a laser.

    d. Reduce system noise levels.

10. Fiber-optic enclosures serve what functions?

    a. Protect fibers from mechanical damage.

    b. Organize fibers, cables, connectors, and splices.

    c. Protect fibers from environmental damage.

    d. Provide a place to put the fibers.

    e. All of the above.

Measurements

About This Chapter

Measurement capability is as vital for fiber optics as it is for other technologies, but optical measurements differ in important ways from other ones. As with electronic systems, some measurements are done in the factory or laboratory, and others are done in the field, to monitor system performance or make repairs. However, the quantities being measured are different, as are the techniques and (often) the terminology.

In this chapter, I discuss the basics of optical and fiber-optic measurements, the quantities that are measured, and the capabilities of important measurement tools. I also examine some pitfalls of fiber-optic measurements and discuss misconceptions that you might carry over from other fields. I emphasize optical measurements because they are what make fiber optics different from other fields.

About Measurement

You can't start making measurements by pulling out the fiber-optic version of a yardstick and measuring away. Proper measurements require careful definition and control so that everyone knows exactly what's being measured and what it means. Otherwise, you end up with units like the cubit. Though the word "cubit" recurs throughout the Bible and other ancient literature, nobody is sure how long a cubit was. In fact, scholars believe that the cubit had many different definitions, just as the foot would if each of you defined the unit as the length of your own foot. To avoid such problems, specialists develop standard techniques and definitions so measurements made in one place yield the same results as those made elsewhere.

Specialists at organizations such as the National Institute of Standards and Technology devote their lives to developing, perfecting, and standardizing measurement techniques. The Department of Defense and industrial groups such as the

●
Measurement terms and techniques must be defined carefully so everyone knows what's being measured.

Electronic Industries Association and the American National Standards Institute have programs to establish standard measurement techniques. To outsiders, much of this work may seem like the splitting of technological hairs. Indeed, the level of detail involved is far more than you need to know at this stage of your involvement in fiber optics; I won't delve into it deeply in this chapter.

Nonetheless, such exhaustive standardization does have practical importance. Without precise and generally accepted measurement standards, you can't be sure what performance specifications and other quantities mean, or that they are measured in the same way. If you get involved in fiber measurements, you will find that many have highly detailed and specific procedures. However, I will concentrate on basic concepts here.

Basics of Optical Measurement

●
Fiber-optic measurements involve light and other quantities, such as the variation of light with time.

Most important fiber-optic measurements involve light, in the same way that important electronic measurements involve electric fields and currents. There are some exceptions, such as the length and diameter of optical fibers and cables, the sizes of other components, and the electrical characteristics of transmitter and receiver components. Because this is a book about fiber optics, I will mention such measurements only in passing. However, I will talk about measuring light in conjunction with other things, because you cannot qualify the properties of optical fibers if you consider only light. For example, to measure the dispersion of light pulses traveling through an optical fiber, you must observe how light intensity varies as a function of time, which requires measuring time as well as light.

When you're working with light, you need to know what can be measured. The most obvious quantity is optical power, which like electrical voltage is a fundamental measuring stick. However, that is rarely enough; it usually must be measured as a function of other things, such as time, position, and wavelength. Wavelength itself is important because optical properties of optical components, materials, light sources, and detectors all depend on wavelength. Other quantities that are sometimes important are the phase and polarization of the light wave. You need to learn a little more about these concepts before getting into more detail on measurement types and procedures.

Optical Power
DEFINED

●
Specialized terminology makes fine distinctions about quantities related to optical power.

People have an intuitive feeling for the idea of optical power (measured in watts) as the intensity of light. However, a closer look shows that optical power and light intensity are rather complex quantities and that you need to be careful what you talk about. Table 14.1 lists the most important quantities, which are described in more detail in the following section.

Table 14.1. Measurable quantities related to optical power.

Quantity and Symbol	Meaning	Units
Energy (Q)	Amount of light energy	joules
Optical power (P or φ)	Flow of light energy past a point at a particular time (dQ/dt)	watts
Intensity (I)	Power per unit solid angle	watts per steradian
Irradiance (E)	Power incident per unit area	W/cm^2
Radiance (L)	Power per unit angle per unit projected area	W/steradian-m^2
Average power	Power averaged over time	watts
Peak power	Peak power in a pulse	watts

What I call power (P or φ) measures the rate at which electromagnetic waves transfer light energy. It is a function of time, because this rate can vary with time. Mathematically, it is expressed as

$$P = dQ/dt$$

or

$$POWER = d(ENERGY)/d(TIME)$$

Sometimes called radiant flux, this optical power is measured in watts, which is equivalent to joules (a measure of energy) per second. It is the same type of power that you measure electrically or thermally in watts. (Note, however, that the power ratings of light bulbs measure how much electrical power they use—not the amount of light output, which is much lower.)

Like electrical power, optical power is a measurable manifestation of more fundamental quantities. In the case of electrical power, those are voltage (V) and current (I)

$$P = VI$$

or power = volts × amperes. This relationship can also take other forms, using the relationship V = IR (voltage = current × resistance):

$$POWER = I^2R = VI = V^2/R$$

As you saw earlier, light and other electromagnetic radiation can be described as a wave comprised of oscillating electric and magnetic fields. The wave has a characteristic amplitude (A) and oscillates with a particular period or frequency ν, as shown in Figure 14.1.

•

Power measures the rate of flow of light energy.

The frequency, in turn, determines the wavelength λ, which equals the speed of light c divided by the frequency ν:

$$\lambda = c/\nu$$

FIGURE 14.1.

Properties of an electromagnetic wave.

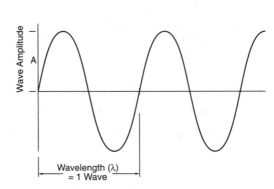

COMPARED TO ELECTRICAL POWER

●
Optical power is proportional to the square of the light wave amplitude and to the wave's frequency.

For light and other forms of electromagnetic radiation, the power is proportional to the square of the electromagnetic wave amplitude (A) at a given wavelength. Power is also proportional to the oscillation frequency of the electromagnetic wave n. Thus in physical terms, power in a light wave is proportional to

● The square of the field amplitude (A^2)
● The oscillation frequency of the electromagnetic wave (ν)
● The inverse of the wavelength (1/λ)

The latter two relationships should look familiar. Earlier, when I discussed the basic nature of light, I mentioned that the energy in a photon increases with frequency and decreases with wavelength. The optical power measures the flow of photons (energy) per unit time and thus depends on photon energy. (Photon energy is given by Planck's law, $E = h\nu$, where h is called Planck's constant.)

●
Electrical field and current can be measured directly, but in optics it is power that can be measured directly.

Look carefully and you can see some similarities between optical and electrical power. Optical power is proportional to the square of electromagnetic wave amplitude, which measures the electrical field in the wave. Electrical power is proportional to the square of the voltage across a resistance (R). That similarity should not surprise you, because optical and electrical power are just different versions of the same thing.

There are important differences from a measurement standpoint. Electrical field (voltage) and the flow of electrons (current) can be measured directly. However, there is no easy way to measure the amplitude of an electromagnetic wave directly. Instead, optical detectors directly measure power.

MEASUREMENT QUIRKS

Before I go deeper into measuring various forms of optical power, I'll warn you about a few potentially confusing measurement quirks. In electrical measurements, the decibel power ratio can be defined in terms of voltage and current. These are in the form

$$\text{POWER RATIO (dB)} = 20 \log (V_1/V_2) = 20 \log (I_1/I_2)$$

where the Vs and Is are voltages and currents, respectively.

● The definition of the decibel power ratio is different when power and voltage are measured.

That definition differs from the definition of power ratio in decibels for powers P_1 and P_2:

$$\text{POWER RATIO (dB)} = 10 \log (P_1/P_2)$$

Why the different factor preceding the log of the power ratio? Because electrical power is proportional to the square of voltage or current. If you measure the ratio of voltage or current, you have to square it to get the power ratio, which is the same as multiplying the log of the ratio by two. You don't have to do that if you measure power directly, either optically or electrically. Electrical measurements are usually in voltage or current, but optical measurements are in power, so it may seem that the difference is between optical and electrical. However, the real difference is between measuring power directly or indirectly. Both formulas are correct, but be careful to use the proper one for power measurements.

A second potentially confusing point is measurement of optical power in some peculiar-seeming units. Normally, power is measured in watts or one of the metric subdivisions of the watt—milliwatts, microwatts, or nanowatts. Sometimes, however, it is convenient to measure power in decibels to simplify calculations of power level using attenuation measured in decibels. The decibel is a dimensionless ratio, so it can't measure power directly. However, power can be measured in decibels relative to a defined power level. In fiber optics, the usual choices are decibels relative to 1 mW (dBm) or to 1 μW (dBμ). Negative numbers mean powers below the reference level; positive numbers mean higher powers. Thus, +10 dBm means 10 mW, but -10 dBm means 0.1 mW.

● Optical power can be measured in decibels relative to 1 milliwatt (dBm) or 1 microwatt (dBμ).

Such measurements will come in very handy when I talk about system design in Chapter 16. Suppose, for instance, that you start with a 1 mW LED, lose 3 dB coupling its output into a fiber, lose another 10 dB in the fiber, and lose 1 dB in each of three connectors. You can calculate that simply by converting 1 mW to 0 dBm and subtracting the losses:

INITIAL POWER	0. dBm
Fiber coupling loss	-3. dB
Fiber loss	-10. dB
Connector loss	-3. dB
FINAL POWER	-16. dBm

Convert the -16 dBm back to power, and you find that the signal is 0.025 mW; however, that often isn't necessary because many specifications are given in dBm. This ease of calculation and comparison is a major virtue of the decibel-based units.

Types of Power Measurement

As Table 14.1 indicates, optical power can be measured not just by itself but also in terms of its distribution angle or space. In many cases (e.g., measuring the brightness of illumination), it is important to know not just total power but also power per unit area. The main concern of fiber-optic measurements is with total power (in the fiber or emerging from it) or power as a function of time, but you should be aware of other light-measurement units to make sure you know what you're measuring.

● Total power is the most important type of power measurement in fiber optics.

LIGHT DETECTORS

Light detectors measure total power incident on their active (light-sensitive) areas—a value often given on data sheets. Fortunately, the light-carrying cores of most fibers are smaller than the active areas of most detectors. As long as the fiber is close enough to the detector, and the detector's active area is large enough, virtually all the light will reach the active region and generate an electrical output signal.

There are two important complications that arise from light detectors. One is that their response depends on wavelength. As you saw in Chapter 8, silicon detectors respond strongly to 800 nm, but not to 1300 nm, whereas InGaAs detectors respond strongly to 1300 nm but not to 800 nm. The spectral response of detectors has to be taken into account for accurate measurements.

In addition, individual detectors give linear response over only a limited range. Powers in fiber-optic systems can range from over 100 mW near powerful transmitters used to drive many terminals to below 1 μW at the receiver ends of other systems. Special detectors are needed for accurate measurements at the high end of the power range.

IRRADIANCE AND INTENSITY

● Irradiance (E) is power per unit area. Intensity (I) is power per unit solid angle.

Things are more complicated when measuring optical power distributed over a large area; all the power may not be collected by the detector. Then another parameter becomes important, irradiance (E), the power density per unit area (e.g., watts per square centimeter). You cannot assume that irradiance is evenly distributed over a given area unless the light source meets certain conditions (e.g., that it is a distant point source such as the sun and that the entire area is at the same angle relative to the source). Total power (P) from a light source is the irradiance (E) collected over area (A). This can be expressed as an integral:

$$P = \int E \, dA$$

over the entire illuminated surface. If the irradiance (E) is uniform over the entire area, this becomes

$$P = EA$$

where A is the area.

The term intensity (I) has a special meaning in light measurement—the power per unit solid angle (steradian), with the light source defined as being at the center of the solid angle. This is a measure of how rapidly light is spreading out from its source. Unfortunately, the term is often misused (used in place of "irradiance") to indicate power per unit area, so it is wise to be certain what is intended.

PEAK AND AVERAGE POWER

When output of a light source or optical fiber varies with time, measured power can differ with time. Power is an instantaneous measurement. The highest power level reached in an optical pulse is called the peak power, as shown in Figure 14.2. The average level of optical power received over a comparatively long period (say over a minute), is average power. For digitally modulated fiber optic light sources operating at a 50% duty cycle (i.e., on half the time), average power is half peak power. However, the peak power may be thousands of times the average power for a laser that emits only a few very short (but quite powerful) pulses per second. Most meters made specifically for fiber optics measure average power.

⬤
Power is an instantaneous measurement that varies with time.

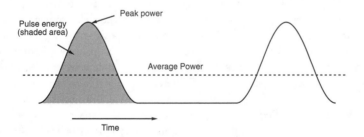

FIGURE 14.2.

Peak and average power, and total pulse energy.

ENERGY

At the beginning of this section, you saw that power measures the flow of energy (which in light measurement is usually symbolized by the letter Q, not E). Most fiber-optic measurements are of power rather than energy, but you should recognize the relationship of the two quantities. Energy can be measured in joules or (equivalently) watt-seconds. Total energy (Q) delivered is the area under the power curve in Figure 14.2, or mathematically the integral of power P over time:

$$Q = \int P(t) \, dt$$

where P(t) is in general a function that varies with time. If the average power over an interval t is P, this can be simplified to

$$Q = P \times t$$

Thus, if average power in a pulse that lasts t seconds is P watts; the pulse energy is Q = Pt joules. This approximation is useful for digital signals, for which the power is roughly constant while a pulse is on.

As this formula indicates, each light pulse contains an amount of energy (Q). If you go deeply into communication theory, you will find that the ultimate limits on detection of pulses and communication capacity are stated as the minimum pulse energy required to deliver a bit of information. Energy per pulse is measured in some experiments with ultra-high performance laboratory systems. However, in practical fiber-optic systems, the emphasis is on measuring power.

Radiometry and Photometry

Fiber-optic measurements are only a small part of light measurement. I have only skimmed the surface of that broader field because not much of it is critical if you're working only with fiber optics. However, you should appreciate the critical difference between the related fields of radiometry and photometry.

Radiometry and photometry are often used interchangeably, but they are not synonyms. Strictly speaking, photometry is the science of measuring light visible to the human eye. If the light isn't visible by the human eye, it doesn't count in photometric measurements. Radiometry, in contrast, is measurement of light in the whole electromagnetic spectrum, whether or not people can see it. Radiometry measures power in watts. The corresponding unit in photometry (power visible to the human eye) is the lumen. Radiometric measurements give equal weight to light at visible and invisible wavelengths. Photometric measurements weigh the contributions of different wavelengths according to eye sensitivity. Thus, 550 nm light, where the eye is most sensitive, counts much more on a photometric scale than light of equal power at a wavelength where the eye is less sensitive (e.g., 450 nm in the blue or 650 nm in the red). Photometry ignores invisible ultraviolet and infrared light.

Fiber-optic measurements are made on a radiometric scale. The only time to use a photometric scale is when measuring light visible to the human eye. Some people in the industry are inexcusably sloppy in their terminology and call instruments photometers even though they measure radiometric units. Perhaps they'll learn better if you insist you want only radiometers. (A radiometer-photometer is a common instrument calibrated in both radiometric and photometric units.)

So far I've talked about ideal photometers and radiometers, but the real world doesn't work quite that nicely. Detectors do not have uniform response across the electromagnetic spectrum. As shown in Figure 14.3, their response varies markedly with wavelength and is limited in range. This makes it essential to calibrate radiometers to account for detector response at the wavelengths being measured.

In practice, fiber-optic power meters are calibrated specifically for the wavelengths used in fiber systems, at 850, 1300, and 1550 nm. (This means that in the strict sense, they are not true radiometers because they work only at certain wavelengths.) You should also recognize that fiber-optic power meters, radiometers, and photometers all measure average power. Their response times are much slower than signal speeds, and their displays cannot track instantaneous power fluctuations.

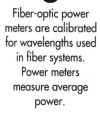

Fiber-optic power meters are calibrated for wavelengths used in fiber systems. Power meters measure average power.

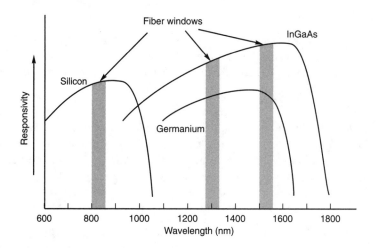

FIGURE 14.3.

Detector response at different wavelengths.

Wavelength

The major importance of wavelength is not its effect on power, as described earlier, but its more visible effects on the interaction between light and matter. Because detector response depends on wavelength, as shown in Figure 14.3, you need to know wavelength to calibrate power measurements (except with a radiometer that has uniform spectral response over a wide range). Wavelength also affects light transmission by fibers and the performance of optical components.

This means that wavelength must be known to characterize light waves. The most straightforward ways to determine wavelength are by measuring the angle at which light is bent or scattered by prisms or diffraction gratings, because that angle depends on wavelength, as shown in Figure 14.4. Prisms refract light of different wavelengths at different angles,

Wavelength affects the interaction between light and matter, including detector response. It can be measured directly.

because the refractive index of glass varies with wavelength. Diffraction gratings scatter light at different angles, depending on its wavelength and the spacing of lines on the grating. The most precise measurements are made by comparing the light being studied with light of a known wavelength.

FIGURE 14.4.

Prisms and diffraction gratings spread out light by wavelength.

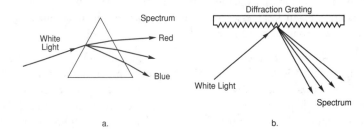

Phase

●

Phase measures a light wave's progress in its oscillation cycle.

Like other types of waves, electromagnetic waves have a property called phase. Envision a light wave as a sine wave, which goes through a cycle of 360° or 2π radians before repeating itself, as shown in Figure 14.5. The phase of a light wave is a measure of its progress in its oscillation cycle.

Phase of a light wave can be measured only by comparing it to other light waves of the same wavelength. This measurement relies on the interference of two waves, a phenomenon shown in Figure 14.5. When two light waves come together at a point, their net amplitude at any instant is the sum of their amplitudes, which can be either positive or negative. If the peak of one wave arrives at the same time as the valley of an equal-amplitude wave, the two cancel and the net amplitude is zero. However, if the peaks arrive at the same time, the two waves add.

Normally, many light waves with different phases and wavelengths arrive at once, overwhelming interference effects. However, interference can be seen if light is coherent (i.e., in phase and of the same wavelength) like the light emitted by a laser. Its main practical effect in present fiber-optic systems is modal noise, which is caused by interference between light in different modes in a multimode fiber with a laser light source. The interference shows up as a pattern of bright and dark areas—zones of constructive and destructive interference—on the detector. Even slight disturbances to the fiber can change this pattern, adding noise to the system or increasing the bit-error rate. (Interference could be an important consideration in the use of coherent communication systems in development.)

Polarization

You saw earlier that light waves are made up of electric and magnetic fields oscillating perpendicular to each other and to the direction the wave is traveling. The polarization direction is defined as the alignment of the electric field, which automatically sets the direction of the perpendicular magnetic field. Light waves with their electric fields in the same plane are linearly polarized. If the field direction changes regularly along the light wave, it is elliptically or circularly polarized. And if the electric fields are not aligned with each other, the light is unpolarized.

You measure polarization by passing light through a polarizer, which transmits light only if its electric field is aligned in a particular direction. The fraction of light transmitted by the polarizer indicates the degree of polarization in the direction of the polarizer. Polarization measurements are important when using components such as optical isolators and couplers that are sensitive to polarization, for sensing applications, and for use with polarization-sensitive fibers.

Polarization is the alignment of the electric fields that make up light waves.

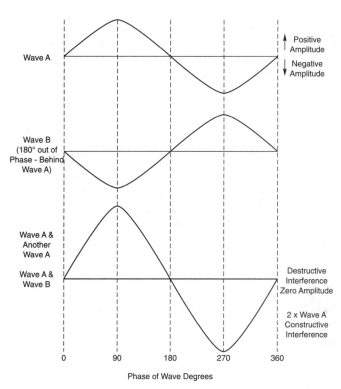

FIGURE 14.5.

Phase and interference of light waves.

Timing

As you saw in Chapter 8, the electrical output from a detector reproduces the input optical signal. Monitoring this electrical output lets you measure time variation of an optical signal.

In practice, detector output takes time to rise and fall—even in response to an instantaneous change in optical signal. Thus, the rise and fall times of detector output signals depend

Time variations in electrical output from a detector reproduce time variations in the input optical signal.

both on rise and fall times of the optical pulse itself and on detector response time. Thus, slow detectors cannot measure fast optical pulses.

Note that pulse repetition rates in even the slowest fiber-optic systems are extremely fast on a human scale—tens of thousands per second—so you cannot see variations in signal level in real time. They are recorded on an oscilloscope or other display in a way that lets people see events too fast to perceive otherwise.

Position

Optical detectors can measure distribution of light, but high resolution is needed for precise measurements.

Optical power varies with position as well as time, and suitable optical detectors can sense this differential. The cells in the retina of your eye can tell light regions from dark spots—otherwise, you couldn't read this page. Retinal cells have small areas, giving your eye a very fine resolution. Electro-optic light detectors can be made with large or small active areas. The detectors used with fiber optics generally have a single active cell a millimeter or more across. Other detectors, used to record images, may have many discrete active cells, like the retina of your eye. Some special types have electrical output that varies with the position of a spot of light on the sensitive area.

As mentioned previously, a single-element detector measures the total power incident on its active area. It can measure the distribution of optical power by scanning an illuminated surface. Alternatively, a multi-element array of detectors can measure power distribution over its surface. Use of either technique allows measurement of light distribution, for example, as it emerges from an optical fiber. However, you should realize that assumptions made about the pattern of light distribution can have an impact on the results. The coarser the resolution relative to the size of the measured structure, the larger the effect of the assumptions. And as you would expect, measurements with a detector with 500 μm active area can tell nothing about light distribution in a 50 μm fiber core.

Quantities To Be Measured in Fiber Optics

Some vital characteristics of fiber-optic systems can be measured only indirectly.

Some fundamental optical quantities I have talked about so far in this chapter are themselves important in fiber-optic systems. Optical power level is one example. However, other factors important in fiber-optic systems (e.g., transmission bandwidth and numerical aperture) can be measured only indirectly. You must measure other quantities, such as power distribution in time and space, then calculate the desired characteristics from those values.

Measurement Standards

When you start digging seriously into fiber-optic measurements, you quickly find references to cryptic-seeming codes, like EIA/TIA-455A. These codes identify standards that specify how certain measurements should be made, and which organizations have written them. They may sound befuddling, but they serve the vital function of making sure that measurements performed by different people in different places are comparable.

Measurement standards specify the techniques and equipment that should be used to measure various quantities. They often go into excruciating detail, but that detail is important. Seemingly small differences in measurement techniques can lead to large differences in results. As a simple example, output power from an optical fiber cannot be measured accurately if some light from the fiber misses the active area of the detector. Likewise, calculated results can be thrown off if pulse width is measured one way at the light source and another way at the output of a fiber. Standards make sure that measurements are repeatable.

This is not the place to go into detail on standards. As a starting point, you should check with manufacturers of test equipment, or with suppliers of other equipment. Table 14.2 lists major organizations that write standards for fiber measurements.

> ●
> The use of standard measurement techniques assures that results are comparable.

Table 14.2. Standards organizations.

ANSI	American National Standards Institute
ASTM	American Society for Testing and Materials
Bellcore	Bell Communications Research
CCITT	International Consultative Commission on Telephone and Telegraph
EIA	Electronic Industries Association
ICEA	Insulated Cable Engineers Association
IEEE	Institute of Electrical and Electronics Engineers
NEC	National Electrical Code
REA	Rural Electrification Administration
TIA	Telecommunications Industry Association
UL	Underwriters Laboratories

Measurement Assumptions

Making such indirect measurements requires some implicit and explicit assumptions, and unless you're careful, those assumptions can lead to the wrong results. The assumptions you think are absolutely essential to make a measurement task manageable may oversimplify the task so you throw away vital data.

One example is assumptions about light distribution in a fiber. It is reasonable to assume that light is distributed in the same way along the length of a single-mode fiber, but not in

> ●
> Unwarranted assumptions can cause you to discard essential data.

● Modal distribution changes along the length of a multimode fiber.

a multimode fiber. Light may take a kilometer or more to distribute itself stably among the modes a multimode fiber can transmit, and that mode distribution can be rearranged by a splice or connector. This mode distribution affects many things, including loss within the fiber, numerical aperture and light-acceptance angle, transfer of light between fibers, and the distribution of light emerging from the fiber. Standard test methods and techniques such as mode scrambling can control mode distribution to produce consistent results.

● Loss of a connector or splice can depend on the direction light is traveling.

Another logical (but sometimes false) assumption is that loss of a connector or splice is the same for light going in either direction. Suppose that light distribution in the cores is uniform, and the only difference between fibers is that one has a 49 μm core and the other has a core 51 μm across—within normal manufacturing tolerances. When light goes from the smaller fiber into the larger one, there are no geometrical losses caused by the difference in core size. However, if light were going in the opposite direction, the core-diameter difference would cause an added 0.3 dB loss.

Pitfalls become subtle in more sophisticated measurements. I can't go through them all here, but the important point to remember is to think measurement techniques through carefully and exercise care in making assumptions. Using standard measurement techniques is one way to minimize the risk.

Fiber Continuity

● Fiber continuity checks can verify system function. The simplest test is to see if light can pass through the fiber.

A major concern in installing and maintaining fiber-optic cables is system continuity. If something has gone wrong with the system, you need to check to see if the cable can transmit signals. If it can, you know you have another problem. If it can't, you need to find out where the break or discontinuity is. In some cases, the break may be obvious—a cable snapped by a falling tree limb or a hole dug by a careless contractor. However, such damage is not always obvious, and the cable route may not be readily accessible.

Early fiber technicians developed a quick-and-dirty test of fiber continuity that required no elaborate equipment. One shined a flashlight into the fiber, and a second on the other end looked to see if any light emerged. However, that simple approach has a number of problems. Flashlight beams do not couple efficiently into optical fibers, especially single-mode types, and faint light emerging from a fiber is hard to see in a brightly lit area. The measurement is hardly quantitative. Furthermore, it requires people at both ends of the fiber—one to send the light and the other to look for the transmitted light—and those people must be able to communicate with each other.

Other instruments are available that can do a much better job, often without requiring people on both ends. Optical fault indicators send pulses of light down the fiber and look for reflections that indicate a fault. They work somewhat like simplified versions of the optical time domain reflectometer (OTDR) described later, which can also locate faults. Optical test sets measure power transmission. Fiber identifiers can tell if exposed fibers are carrying signals (they work by bending the fiber and observing light that leaks out at the bend). Visible fault identifiers send visible red light through the fiber, and visual inspections show if any is leaking out.

In practice, these instruments may be used in different places. Visible fault indicators are best for tests where fibers are exposed in a small area close to the fiber end, but they are of little use in searching for faults in buried exterior cables. Conversely, fault indicators and OTDRs excel at finding faults in buried or aerial cables but may not detect breaks close to the fiber end.

Optical Power

Optical power is the quantity most often measured in fiber-optic systems. The power may be output from a light source, power emerging from a length of optical fiber, or power in some part of a system. The wavelength must be known so that the detector can be calibrated for that wavelength. Duty cycle—the fraction of the time the light source is on—should also be known to interpret properly measurements of average power. The usual assumption is 50% (half on, half off) for digital modulation, but under certain circumstances that may be far off (e.g., if a series of 1s are being transmitted in NRZ code so the transmitter is continually sending at its high level).

Normally, power is measured where the light emerges from a light source or fiber. Fiber-optic power meters collect the light from the fiber through an optical connector, which directs the light to a detector. Electronics process the detector output and drive a digital display that shows the power level in linear units (nanowatts to milliwatts) or in dB referenced to either 1 mW or 1 μW. Measurement ranges are automatically switched across the dynamic range, which is typically a factor of one million. Typical measurement accuracy is ±5%.

Measuring optical power stops the beam, because it's absorbed by the detector. If you want to sample the power level in a transmitted signal, you need a beamsplitter that will divert a calibrated fraction of the light to a detector and transmit the rest.

It is important to keep input power within the dynamic range of the power meter. Excessive powers won't be measured correctly, and in extreme cases, long-term exposure to excess power can damage some detectors.

Measurements of optical power require knowing the wavelength and duty cycle.

Attenuation

Attenuation is the most important property of passive optical components, because it determines what part of an optical signal is lost within the component and how much passes through. It is always a function of wavelength, although the wavelength sensitivity varies widely. In fibers, the variation with wavelength is large; in some other components, it is negligibly small.

Attenuation is $-10 \log(P_{out}/P_{in})$.

Attenuation is measured by comparing input and output levels, P_{in} and P_{out} respectively. It is given in decibels by

$$\text{ATTENUATION (dB)} = -10 \log (P_{out}/P_{in})$$

The negative sign is added to give attenuation a positive value, because the output power is always lower than the input power for passive devices.

The standard way to measure cable loss with an optical test set—which includes a light source and transmitter—is shown in Figure 14.6. First the light source is connected to the power, meter through a short launch cable, and the power is adjusted to a convenient level (-10 dBm in this case). Then a short receive cable is added between the launch cable and the power meter; a power change no more than 0.5 dB verifies the receive cable is good. Then the cable to be tested is connected between launch and receive cables and the power it transmits is read (-14.2 dBm in this case). The difference, 4.2 dBm, is taken as the total attenuation of the cable being tested, including fiber, connectors, and splices. For more precise measurements, the loss should be measured in both directions through the test cable. As I mentioned earlier, attenuation typically differs slightly in the two directions because of differences in light coupling at the connectors.

FIGURE 14.6.

Cable loss measurement.

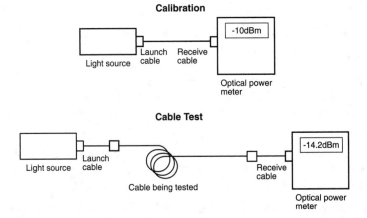

The same principles can be used to measure the attenuation of other components, such as couplers, or of segments of cable installed in a system. If cable ends are located at different places, the tests can be performed by technicians working at both ends, one with a light source and the other with a power meter, or by temporarily installing a "loop-back" cable to send the signal back to the origination point through a second fiber. Inevitably, small losses are measured less accurately than large ones.

Optical time domain reflectometers can also measure attenuation, as described later.

This simple comparison technique is adequate for most purposes, but it does not precisely measure pure fiber loss, because it includes loss within the connectors at each end. More precise measurements of fiber loss alone require the cut-back technique. First, power transmission is measured through the desired length of fiber. Then the fiber is cut to a short length (about a meter) and the power emerging from that segment is measured with the same light source and power meter. Taking the ratio of those power measurements eliminates input coupling losses (which occur in both measurements), while leaving the intrinsic fiber transmission loss (which is present only in the long-fiber measurement).

The cut-back method can be more accurate for single-mode fibers than for multimode fibers because of the way mode distribution changes along the fiber. Accurate measurement of long-distance attenuation of multimode fibers requires use of a mode filter to remove the higher-order modes that gradually leak out of the fiber. However, this won't accurately measure the loss of short multimode fibers, which depends on propagation of the high-order modes.

One special problem with single-mode fibers is that light can propagate short distances in the cladding, throwing off measurement results by systematically underestimating input coupling losses. To measure true single-mode transmission and coupling, fiber lengths should be at least 20 or 30 m.

Dispersion

Pulse dispersion is an important characteristic of fibers that can limit bandwidth and data rate. As you saw in Chapter 4, it is the sum of modal dispersion (present only in multimode fibers), material dispersion (intrinsic to the material from which the fiber is made), and waveguide dispersion (dependent on refractive-index profile). All three quantities depend on wavelength, although the dependence is very weak and indirect for modal dispersion. Only material and waveguide dispersion are present in single-mode fibers, where they add to make chromatic dispersion. That quantity depends not merely on operating wavelength but also on the range of wavelengths being transmitted.

Sophisticated laboratory instruments can isolate the components of fiber dispersion. However, in practice the main concern is total dispersion. That can be measured directly in the time domain by sending a very short optical pulse down the fiber and measuring how much it has spread at the output. Note that it is important to know the central wavelength and the range of wavelengths. For multimode fibers, it is important to know modal distribution launched into the fiber.

Bandwidth and Data Rate

Bandwidth in analog systems and data rate in digital systems are essentially the inverse of pulse dispersion in a fiber. Although these quantities can be measured indirectly as dispersion, they can also be measured directly. Frequency measurements can be made by comparing the strengths of signals at various frequencies. The result is a plot of signal strength versus frequency, such as the one shown in Figure 14.7. This plot, with level response over a broad range of frequencies, then a rapid decline past a certain level, is characteristic of

● Precise measurements of fiber attenuation rely on cutting back fibers to compare power emerging from short and long lengths.

● Pulse dispersion is determined by measuring changes in pulse length along a fiber.

● Bandwidth and data rate can be measured directly or calculated from dispersion.

fiber-optic systems. Typically, the upper limit on bandwidth is specified as the point at which signal strength has dropped by 3 dB.

FIGURE 14.7.

Received signal strength as a function of frequency.

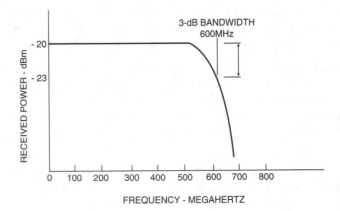

Maximum digital data rate can be extrapolated from analog bandwidth or pulse dispersion characteristics, or it can be estimated in other ways. One method is to study the shapes of output pulses in the eye pattern described later. Another is to define the maximum data rate as the highest possible with bit-error rate no more than a certain acceptable level, which is also described later.

Bit-Error Rate

● Digital transmission is evaluated by measuring the fraction of bits received incorrectly, the bit-error rate.

The bit-error rate is a straightforward concept used in assessing the performance of many digital systems, including fiber-optic as well as other types. It is based on generating a randomized bit pattern and comparing the original pattern with the signal emerging from the system being tested. Counting both total bits and the number of errors detected gives the bit-error rate—the fraction of bits received incorrectly. This gives a convenient and quantitative assessment of the performance of a digital communication system.

As might be expected, the bit-error rate increases as received power drops, as well as when the system approaches other performance limits, such as maximum transmission rate. The drop is generally quite steep and can be more than a factor of 100 for a 1 dBm increase in received power in certain power ranges. Other factors generally set a minimum bit-error rate when the receiver gets adequate output power. If too much power reaches the detector, the bit-error rate again increases because the receiver is saturated.

Different users have their own standards for bit-error rate. The limit is typically 10^{-9} (one bit in a billion) for a telephone voice transmission system, and 10^{-11} or 10^{-12} for data transmission.

Eye Pattern Analysis

One popular way to assess performance of a digital fiber-optic link in the laboratory is to superimpose the waveforms of a series of pulses on an oscilloscope display. This produces the "eye pattern" shown in Figure 14.8. Each pulse trace draws its own pattern on the screen, with noise added to the signal. If there wasn't any noise, each trace would follow exactly the same line. The more noise, the more the signal varies, and the thicker the line is vertically. Likewise, the larger the jitter (difference in pulse arrival times), the thicker the lines are horizontally.

●
An eye pattern is the superposition of the waveforms of successive bits; open eyes indicate good quality transmission.

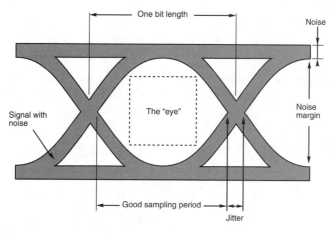

FIGURE 14.8.
An eye pattern.

In essence, the eye pattern measures the repeatability of pulses reaching the instrument. The better the transmission quality and the more uniform the received signal, the more open the eye will appear. If the eye starts to close, it indicates that transmission errors are likely as successive bits interfere with each other.

Careful interpretation of the eye pattern can yield important data on fiber link performance. Data signals can be sampled at any point within the eye, but the best point is in the middle. Some important points for interpreting eye patterns are as follows:

- Height of the central eye opening measures noise margin in receiver output.
- Width of the signal band at the corner of the middle of the eye measures the jitter (or variation in pulse timing) in the system.
- Thickness of the signal line at top and bottom of the eye is proportional to noise and distortion in the receiver output.
- Transitions between top and bottom of the eye pattern show the rise and fall times of the signal that can be measured on the eye pattern.

Mode Field and Core Diameter

Mode field diameter
is the region
occupied by light in
a single-mode fiber.

As you saw in Chapter 4, fiber core diameter can vary because of manufacturing toler-ances. In addition, mode field diameter—the diameter of the region occupied by light propagating in a single-mode fiber—is somewhat larger than the core diameter. These quantities can be measured.

Practical interest in the mode field and core diameters depends on the distribution of light, and measurements are therefore based on light distribution. One approach is to scan across the end of the fiber with another fiber of known small core diameter, observing variations in light power collected by the scanning fiber. Other approaches rely on observing the spatial distribution of light near to or far from the fiber—the near-field and far-field in-tensity patterns. Those distributions of optical power can be used to calculate the core diameter.

A related quantity important for both single- and multimode fibers is the refractive-index profile, the change in refractive index with distance from the center of the fiber. This is measured in the same way as mode field and core diameter.

Numerical Aperture and Acceptance Angle

Numerical aperture
is not measured
directly; it is
calculated from the
acceptance angle.

As you learned in Chapter 4, the numerical aperture measures how light is collected by an optical fiber and how it spreads out after leaving the fiber. It measures angles, but not di-rectly in degrees or radians. Although NA is widely used to characterize fiber, it isn't NA that is measured, but the fiber acceptance angle, from which NA can be deduced.

Numerical aperture and acceptance angle are most important for multimode fibers. As mentioned earlier, measured numerical aperture depends on how far light has traveled through the fiber, because high-order modes gradually leak out as light passes through a fiber. The measured numerical aperture can be larger for shorter fibers, which carry a larger complement of high-order modes, than it will be for long fiber segments. Measurements are made by observing the spread of light emerging from the fiber.

Cut-Off Wavelength

Cut-off wavelength
is measured by
observing where
stripping out the
second-order mode
causes an increase
in loss.

Cut-off wavelength, the wavelength at which the fiber begins to carry a second waveguide mode, is an important feature of single-mode fibers. The measured effective cut-off wave-length differs slightly from the theoretical cut-off wavelength calculated from the core diameter and refractive-index profile. As with core and mode-field diameter, cut-off wave-length is a laboratory rather than a field measurement.

Normally, the cut-off wavelength is measured by arranging the fiber in a test bed that bends the fiber a standard amount. Fiber attenuation as a function of wavelength is measured twice. First, the fiber is bent in a manner that causes the second-order mode to leak out almost completely. Second, the fiber is arranged so it transmits both first- and second-order modes. These two measurements are compared, giving a curve such as the one in Figure 14.9, which shows excess loss as a function of wavelength. In this case, λ_c is the effective cut-off wavelength, which is defined as the wavelength above which second-order mode power is at least a certain amount below the power in the fundamental mode. The measurement finds this value by locating the point where excess loss caused by stripping out the second-order mode is no more than 0.1 dB.

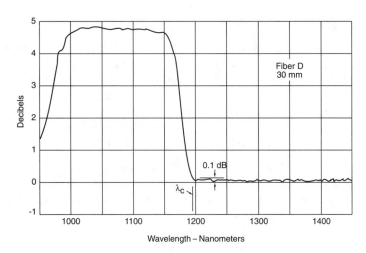

FIGURE 14.9.

Measurement of effective cut-off wavelength. (Courtesy of Douglas Franzen, National Institute of Standards and Technology)

Measurement Instruments

I started by describing fiber-optic measurements in general terms and by specifying the quantities that you needed to measure. Specialized instruments can perform these measurements easily. Many are analogous to instruments used for other optical or telecommunication systems. I'll look at the most important of these instruments in the following section.

Optical Power Meters

The simplest optical measurement instruments are optical power meters, which include a fiber connector, calibrated detectors, electronics to process the signal, and a digital display. Most are compact and portable, with autoranging digital readouts that show power on either decibel or watt scales. Priced at a few hundred dollars and up, they are invaluable tools that can be adapted for many measurements.

Power meters are calibrated for use at one or more of three standard wavelengths: 850, 1300, and 1550 nm. (Some are calibrated at extra, nonstandard wavelengths, such as 660 nm for plastic fiber transmission and 780 nm for CD laser systems.) Be careful that you find a meter usable at the wavelength you want. Many are calibrated for only one

Optical power meters are calibrated for wavelengths used in fiber-optic systems.

wavelength, and as you can see in Figure 14.3, the detectors used at 850 nm do not respond to light at 1300 and 1550 nm (and vice versa).

All measurements should be made at the calibrated wavelengths because they are the nominal transmission wavelengths of fiber systems. Newer versions can store measurements and come with computer interfaces.

Test Sources

Test sources provide output at fiber system wavelengths.

Fiber-optic test sources provide light for measurements of attenuation and other optical characteristics of fiber components and systems. The standard sources are LEDs or diode lasers emitting at 660 nm (for plastic fiber); 820, 850, and 870 nm (for short-wavelength systems); 1300 nm; or 1550 nm. The wavelength is critical for accurate loss measurements. Each LED or laser emits at only a single wavelength, so multi-wavelength test sets must include multiple sources. The source emits a continuous beam or can be modulated for certain tests. Power is normally stabilized but can often be adjusted.

LED sources are normally used only for multimode fiber tests; laser sources can be used with single-mode or multimode fibers. Matching the light source to the fiber is important, because a multimode source cannot deliver adequate power to a single-mode fiber. A connector on the source package mates with a short source cable, which delivers the light to the fiber, cable, or component being tested. Many types of connectors are in use, making it critical to match the source connector with the test cable. (Adapters are readily available for common connector types.)

Other sources may be used for laboratory measurements at nonstandard wavelengths. These typically use incoherent tungsten lamps, which emit a broad range of wavelengths. Internal optics select a narrow range of wavelengths for transmission. Such instruments are called monochromators and are used more for general optical measurements than for fiber optics.

Optical Loss Test Sets

An optical loss test set includes a light source and optical power meter calibrated for use together.

An optical power meter can be combined with a calibrated light source to serve as an optical loss test set. The power meter measures the drop in optical power from the level emitted by the source to the level received at the detector. Source and power meter wavelength must be matched for accurate results. Test sets are often packaged for specific applications, such as testing local area networks.

Optical loss test sets measure attenuation by comparing power levels measured with and without the component being tested. Figure 14.6 shows a typical arrangement. In practice, the source and power meter may be calibrated together but used in separate places in the field, with the source at one end of the cable and the power meter at the other. This simplifies measurements of installed cables.

Oscilloscopes

Oscilloscopes are useful in measuring waveforms transmitted through optical fibers, as they are in other waveform measurements. They are often used in measuring the eye pattern shown in Figure 14.8. They can also display the variations of optical power with time. However, optical probes are needed to convert the optical signals to electronic form.

Bit-Error Rate Meters

Specialized instruments are made for the bit-error rate measurements I discussed earlier. Some models are portable; others are designed for laboratory use. They are generally similar to instruments made for use in non-optical systems, although they have been adapted for use specifically with fiber optics.

Fiber-Optic Talk Sets

Strictly speaking, fiber-optic talk sets are simple communication systems, not measurement instruments. However, they are used by technicians performing measurements. The talk set includes a simple transmitter and receiver that lets it send voice signals through optical fibers. It lets pairs of technicians at opposite ends of an installed cable talk with each other to coordinate their activities.

Optical Time-Domain Reflectometer

The optical time-domain reflectometer (OTDR) is one of the most powerful (and unfortunately, one of the most costly) fiber-optic measurement instruments. It is a sort of optical radar that sends a short light pulse down a fiber and monitors the small fraction of that light scattered back to it. Plotting the returned light as a function of time can identify places where the fiber has excess loss, such as connectors, splices, or fiber breaks. Its slope indicates the loss of the fiber. Figure 14.10 shows typical features on an OTDR plot.

The optical time-domain reflectometer sends pulses down a fiber to measure its characteristics from one end.

The distance scale at the bottom is calculated from the time the light takes to return. Although the optical pulse is fast and the electronics respond rapidly, the signal received from the several meters to several tens of meters of fiber closest to the instrument is not useful; this is called the dead zone. The signal strength declines gradually through uninterrupted lengths of fiber. The slope of the decline indicates fiber loss.

Peaks in the slope indicate points where light is reflected back to the source. The largest peak in the plot in Figure 14.10 is reflection from the end of the fiber. The next largest peak is reflection from a connector. Look carefully and you can see that the signal just after the connector is slightly lower than it was before; this drop measures the connector loss. Mechanical splices likewise reflect some light back to the instrument and have some loss, although both loss and reflection are smaller in this example. The other discontinuity shows the loss from a fusion splice (or from a sharp fiber bend), which does not reflect light back to the OTDR. Breaks or other fiber flaws also appear in OTDR plots, much as connectors or the end of the fiber in this example.

FIGURE 14.10.

OTDR plot.

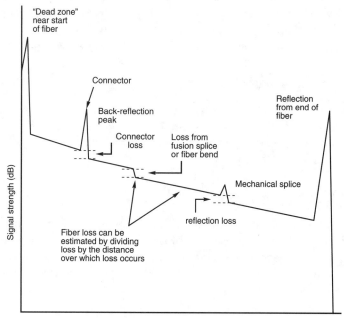

The major attraction of the OTDR is its convenience and ability to spot cable faults remotely. It requires access to only one end of the fiber in cable segments up to tens of kilometers long. Timing how long it takes light to travel from the instrument to a point in the fiber and back can locate flaws and junctions in the fiber. That's invaluable if you have to play detective and find where a fiber is broken in a long cable. Just plug your OTDR into one end of the cable and send a pulse of light down it. If you see a smoothly declining curve, the cable is okay. But if there is a large, sharp drop in the curve, the fiber is damaged. The instrument can locate a break to within a matter of meters, except in the dead zone. Because they are widely used in field measurements, OTDRs are normally packaged for field use. They also come with some internal computer power, which can add considerable information to the display and let you zoom in on areas of interest and measure losses directly. Many can interface directly with personal computers.

Care must be taken in interpreting OTDR loss measurements because they are not as accurate as direct measurements of attenuation. One problem is variations in the fraction of light backscattered toward the instrument. Splicing two fibers with different degrees of backscatter can give spurious results. In one direction, the splice becomes a "gainer," with the excess backscatter making it appear to have increased signal strength. In the other, the splice appears to have excess loss, because less of the signal is being returned to the instrument. OTDRs should also be matched to the fiber being tested, both in wavelength and core diameter, to enhance their accuracy.

OTDRs have proved invaluable for spotting fiber faults, optimizing splices, or inspecting fibers and cables for manufacturing flaws. However, it is important to understand their limitations.

Optical Fiber Analyzers

The instruments I've described so far are intended mostly for field use. Other measurements are performed in the laboratory or factory for quality control, research and development, and fiber characterization. Instruments called fiber analyzers perform many of these special measurements, such as mode-field diameter and numerical aperture. Specialized instruments also measure other quantities, such as spectral dispersion or characteristics of a fiber preform. However, few people will encounter such instruments.

●
Optical fiber analyzers perform many measurements required in factories or laboratories.

What Have You Learned?

1. Precision and standards define what is to be measured so that everyone makes the same measurements.

2. The basic quantity being measured in fiber optics is light. Measurements give information such as variations in optical power in time and space. Optical power is itself the rate of change in light energy with time.

3. Like electrical power, optical power is the square of a field amplitude, but the lightwave field (unlike voltage) is not easy to measure directly.

4. Optical power can be measured in decibels relative to 1 mW (dBm) or 1 μW (dBμ).

5. Pulse energy is the average power in the pulse multiplied by the length of the pulse.

6. Fiber-optic power meters are calibrated specifically for the wavelengths used in fiber systems. Power meters measure average power.

7. Wavelength is important because it affects the interaction between light and matter, including detector response. It can be measured directly.

8. Phase is a light wave's progress in its oscillation cycle of 360°. It can be measured only in relation to the phase of other light waves of the same wavelength.

9. Some vital characteristics of fiber-optic systems can be measured only indirectly by calculations based on more fundamental measurements.

10. Attenuation is measured by comparing input and output power levels. This can be done with optical loss test sets, by comparing power levels with and without the component being tested.

11. Pulse dispersion can be determined by sending a short pulse down a fiber and measuring how long it is at the end.

12. Bandwidth in analog systems and data rate in digital systems are essentially the inverse of pulse dispersion in a fiber. They can be measured indirectly as dispersion or directly as bandwidth or data rate.

13. An eye pattern is the superposition of the waveforms corresponding to a series of received bits. The clearer the eye, the more uniform the pulses, and the higher the transmission quality.

14. Numerical aperture is not measured directly; it is deduced from measurement of the acceptance angle.

15. Cut-off wavelength is measured by observing the wavelength above which power transmitted in the second waveguide mode is no more than a minimum level. This is done by measuring wavelength dependence of excess loss caused by stripping out the second-order mode.

16. Specialized instruments have been developed for the most common fiber-optic measurements.

17. Optical power meters are calibrated detectors with suitable electronics and readouts calibrated for particular wavelengths used in fiber-optic systems.

18. The optical time-domain reflectometer sends a short light pulse down a fiber and measures the light reflected back to it. In this way, it can spot discontinuities from one end of the fiber and indicate their magnitudes.

What's Next?

Now that you understand the nuts and bolts of fiber optics and the basics of measurement, the next few chapters will delve into fiber-optic systems and applications.

Quiz for Chapter 14

1. Optical power is

   a. the light intensity per square centimeter.

   b. the flow of energy.

   c. a unique form of energy.

   d. a constant quantity for each light source.

2. Optical power is proportional to

   a. the square of the electromagnetic field amplitude.

   b. the frequency of the electromagnetic wave.

   c. the inverse of the wavelength ($1/\lambda$).

   d. all of the above.

   e. none of the above.

3. Which measures power per unit area?

    a. Irradiance.

    b. Intensity.

    c. Average power.

    d. Radiant flux.

4. Which is not a directly measurable optical quantity?

    a. Wavelength.

    b. Field amplitude.

    c. Power variation in time.

    d. Phase of light waves.

    e. Polarization of light.

5. Attenuation is

    a. the difference in input and output powers measured in milliwatts.

    b. the ratio of output power divided by input power (in decibels).

    c. the increase in power provided by a repeater.

    d. a comparison of power levels at different wavelengths.

6. Which wavelengths are standard for attenuation measurements?

    a. 850 nm.

    b. 1300 nm.

    c. 1550 nm.

    d. All of the above.

    e. None of the above.

7. Why is bit-error rate measured?

    a. To calculate attenuation of an optical fiber.

    b. To measure pulse dispersion.

    c. To assess the quality of digital transmission.

    d. To locate broken fibers.

8. What does an open eye pattern indicate?

    a. Good-quality digital transmission.

    b. Only that a signal is reaching the receiver.

    c. That no signal is reaching the receiver.

    d. Poor-quality analog transmission.

9. Which of the following is not included in an optical loss test set?

    a. An optical power meter.

    b. An optical test source.

    c. An oscilloscope.

    d. Connectors and adapters.

    e. A calibrated detector.

10. An optical time-domain reflectometer is most valuable for

    a. precise characterization of connector loss.

    b. extending repeater spacing.

    c. measurement of fiber attenuation in the laboratory.

    d. location of cable faults in the field.

System Concepts

About This Chapter

The last several chapters have concentrated on specific elements of fiber-optic systems, from the fibers and light sources to measurement techniques. This chapter teaches you about the ways these components can be put together to make various types of fiber-optic systems. It concentrates specifically on the concepts used in fiber-optic systems. Chapter 16 will go into design details, and later chapters will cover specific applications.

Types of Transmission Systems

So far I've talked much more about fiber-optic components than about how they are put together to make systems. To understand system concepts, I will start by looking at the ways that communication systems are connected.

You can divide fiber systems into a few basic categories, according to the way signals are transmitted. Some simply carry signals back and forth between two points. Others carry signals to multiple terminals, either from a single source or from other terminals on the system. Many include switches that make temporary connections among different terminals attached to the system. Figure 15.1 shows each of these concepts. I will look briefly at how each is used in fiber optics.

FIGURE 15.1.

Types of transmission systems.

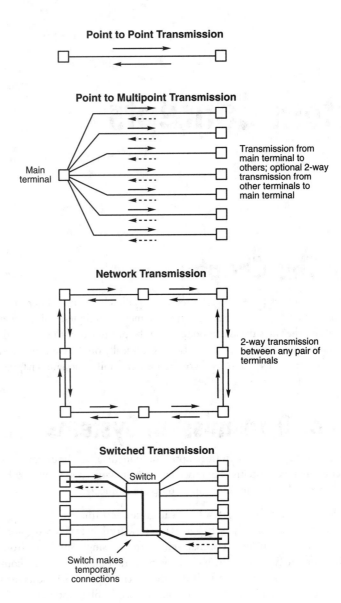

Point to Point Transmission

Point to Multipoint Transmission

Main terminal

Transmission from main terminal to others; optional 2-way transmission from other terminals to main terminal

Network Transmission

2-way transmission between any pair of terminals

Switched Transmission

Switch

Switch makes temporary connections

Point to Point

● Point-to-point transmission links pairs of terminals.

Point-to-point transmission is the simplest type of fiber-optic system. It provides two-way communication between a pair of terminals, each with a transmitter and receiver, that are permanently linked together.

Conceptually, the distance between terminals doesn't matter. The two could be on opposite sides of the room, or on opposite sides of the ocean. If the link is too long for the transmitter to send signals through the entire length of fiber, repeaters or optical amplifiers can be added to boost signal strength. Examples of point-to-point links range from a cable linking a personal computer and a dedicated printer to a transatlantic submarine cable.

If you look closely enough, you can break other types of fiber-optic systems into point-to-point links. That reflects the reality that any fiber system has a transmitter on one end and a receiver on the other, although couplers in between may split the signals among multiple fibers. It also reflects the fact that point-to-point fiber links generally are easy to build.

Point to Multipoint (Broadcast)

Another family of transmission systems sends the signal from one transmitter to many terminals. This is sometimes called broadcasting, because it is analogous to the way a radio or television transmitter sends signals to many home sets within its transmission range. In a fiber-optic system, the terminals may or may not return signals to the central transmitter. The return signal is often at a lower speed than the broadcast transmission.

Because they serve many terminals, point-to-multipoint transmitters generally send higher-power signals than those in point-to-point systems. The basic design can vary considerably, as shown in Figure 15.2. A tree or star coupler can split the signal from one transmitter to drive many terminals. Or the split signals can drive relay transmitters, which amplify and repeat the signal from the main transmitter and send it to terminals (or another stage of relay transmitters). Optical amplifiers may be used as relay transmitters or to boost the output of the main transmitter.

● *Point-to-multipoint transmission uses one transmitter to serve many terminals.*

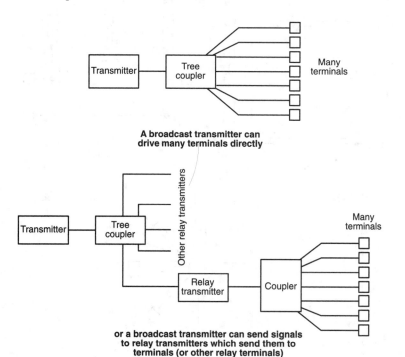

A broadcast transmitter can drive many terminals directly

or a broadcast transmitter can send signals to relay transmitters which send them to terminals (or other relay terminals)

FIGURE 15.2.

Point to multipoint transmission.

Like a point-to-point transmission system, a point-to-multipoint system is fixed, with permanent connections between transmitters and receivers. The most common examples of point-to-multipoint fiber systems are the trunking cables that carry signals from the head ends of cable television systems to regional distribution systems. Most current cable systems send few or no signals "upstream," from home subscriber terminals to the head ends where signals originate.

You should realize that the main transmitter in a point-to-multipoint system is more "important" than the terminals it serves. Even in a two-way system, it sends most information handled by the system, and is essentially an information provider (whatever you think of the offerings of your local cable system). Individual terminals provide little or no information, and that can go only to the main transmitter; they cannot communicate with each other. If the main transmitter dies, a point-to-multipoint system is off the air.

Networks

● Networks allow many terminals to talk with each other.

A network system differs fundamentally from a point-to-multipoint system because all terminals are treated (more or less) equally. All terminals can both send and receive signals, and all can send signals to any other terminal on the network. Figure 15.3 shows a few simple variations. Small networks are often called local area networks; larger ones may be called metropolitan area or wide area networks.

FIGURE 15.3.

Network transmission.

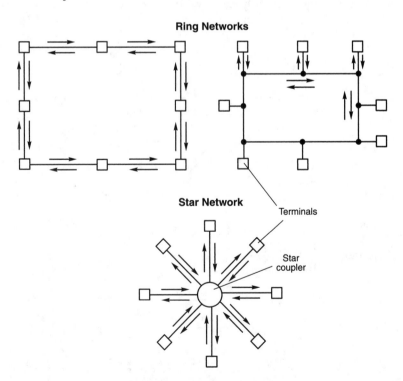

Ring Networks

Star Network

Terminals

Star coupler

Terminals can be connected to the network in various ways. One approach is to put them all around a ring or loop. Signals may be split or tapped from the loop to serve each terminal, or they may pass through each terminal, where they can be modified. Because fiber couplers have high attenuation, it is often easier to build networks as a collection of point-to-point links that are regenerated, with some modifications possible, at each terminal. That approach is used in some standard fiber networks, including FDDI. Another common design is the star network, shown in Figure 15.3, where signals to and from each terminal pass through a central point, either a passive coupler that divides input light, or an active coupler that receives and retransmits the signal. This approach is most often used with Ethernet and certain other networks where one node connects many terminals to the network backbone.

Like point-to-point and point-to-multipoint systems, networks have permanent connections to each terminal, and the overall architecture can be changed only by rewiring. This means that the same terminals always talk to each other unless they are rewired. Reconfiguration can be simplified if the network is designed with patch panels to attach and disconnect new terminals, an approach that works best with star networks. Local area networks for personal computers are a typical example.

Switched Transmission

Adding switches to a communication network makes it much more flexible. Switches allow any pair of terminals to send and receive signals directly to each other. The connections are inherently temporary, so each terminal can talk—at different times—to any other terminal. Depending on the system design, multiple terminals may be linked together at once. The basic idea is shown in Figure 15.1.

Switching adds a level of complexity to the system, because it requires switches, which are not easy to make with fiber-optic technology. However, it also adds tremendous power in permitting temporary connections between any pair of terminals that can be linked to the switch. The standard example of a switched network is the telephone system, which can connect any two phones anywhere on the globe.

Light can be switched in optical waveguides, as described in Chapter 12. However, in practice present switched fiber-optic systems work differently. Signals from the terminals go to optical receivers at the switch, where they are converted to electronic form. Then the signals are switched electronically to a transmitter attached to the switch that sends the signals to the other terminal. Thus present switched fiber systems are actually made of many point-to-point fiber-optic systems linking the terminals with the switches, and an electronic switch.

This still understates the complexity of the telephone network on two counts. One is that the actual connections to virtually all homes are made by copper wires, as you will see in Chapter 18, although fibers bring the signals to electronic transmitters that drive the wires. The other is that there are many levels of switching in the telephone network, some local,

⬤
Switching allows temporary connections between pairs of terminals.

some regional, and some at higher levels, as shown in Figure 3.4, and described in more detail in Chapter 17. A long-distance call is thus routed through a series of switching centers rather than a single switch.

Transmission Formats

Signals are transmitted through optical fibers by modulating the light from an optical transmitter with a signal. Different systems use different modulation techniques, depending on application requirements.

Digital and Analog Transmission

Digital transmission is used more widely than analog transmission.

In Chapter 3, you learned that signals could be transmitted as continuous variations in intensity (or some other parameter), or as digital pulses. Analog transmission was used for many years and remains common in cable television systems. It can pack more information into less bandwidth than digital signals, but it is much more vulnerable to noise and distortion and cannot be manipulated as easily as digital signals.

Digital transmission has grown much more popular for most telecommunications applications. It requires simpler electronics and can encode any form of information—making it possible to merge digital data streams from many different sources with little worry about interfaces. Digital signals are easy to process and manipulate. Although your home telephone remains an analog device, the telephone network converts its signals to digital form for easier switching and processing. Thanks to the low noise of digital telephony, a transatlantic telephone call sounds little different than one across the block.

Amplitude Modulation

Digital and analog signals are transmitted by amplitude modulation of a light beam.

Virtually all fiber-optic systems use amplitude modulation, in which the amount of light varies in proportion to the signal strength. You can see the basic idea if you go back and look at Figure 7.2. The signal varies much slower than the light waves (you couldn't see the light waves if they were drawn to true scale). The stronger the signal, the more light from the transmitter. With analog modulation, the light output varies continuously; for digital modulation, it is either off or on.

Series of digital bits can be encoded in several different ways.

Digital modulation is not simply off-on-off-on. There are several important digital modulation techniques, as shown in Figure 15.4. Each has its own distinct characteristics:

- NRZ (no return to zero) Coding—Signal level is low for a 0 bit and high for a 1 bit and does not return to zero between successive 1 bits.
- RZ (return to zero) Coding—Signal level during the first half of a bit interval is low for a 0 bit and high for a 1 bit. Then it returns to zero for either a 0 or 1 in the second half of the bit interval.

- Manchester Coding—Signal level always changes in the middle of a bit interval. For a 0 bit, the signal starts out low and changes to high. For a 1 bit, the signal starts out high and changes to low. This means that the signal level changes at the end of a bit interval only when two successive bits are identical (e.g., between two zeros).

- Miller Coding—For a 1 bit, the signal changes in the middle of a bit interval but not at the beginning or end. For a 0 bit, the signal level remains constant through a bit interval, changing at the end of it if followed by another 0 but not if it is followed by a 1.

- Biphase-M or Bifrequency Coding—For a 0 bit, the signal level changes at the start of an interval. For a 1 bit, the signal level changes at the start and at the middle of a bit interval.

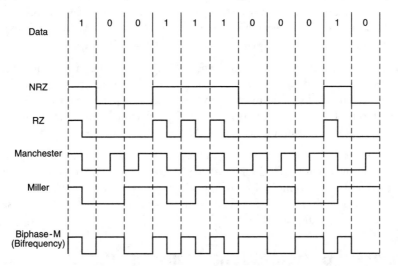

FIGURE 15.4.
Digital data codes.

NRZ coding is probably the most common in fiber systems, but each scheme has its advantages and disadvantages. Some, including RZ, Manchester, and bifrequency coding, can make two transitions during a bit interval. This requires more system bandwidth, but improves performance. For instance, the frequently switched Manchester code generates its own clock, whereas NRZ-coded signals can suffer loss of timing or signal-level drift during a long string of 0 or 1 bits.

All these amplitude-modulated signals are simple and easy to detect. However, amplitude modulation has inherent limits that have led developers to look at other options.

Soliton Transmission

Solitons are self-regenerating pulses that do not change shape along a fiber.

An important limitation at high speeds is pulse dispersion, which spreads out pulses as they travel along a fiber. Optical amplifiers can increase optical power, but they cannot compensate for dispersion. However, researchers have found that certain pulses with special shapes, called solitons, travel through fibers in an unusual way that regenerates them in their original shape at various points along the fiber.

Solitons are pulses that do a delicate balancing act between two competing effects that degrade the transmission of other pulses. One is chromatic dispersion, which causes pulses covering a range of wavelengths to stretch out in time as they travel along a fiber. The other is self-phase modulation, which causes the pulses to spread over a broader range of wavelengths as they travel. The details depend on some rather abstruse mathematics that you don't want to worry about, but if the right pulse is sent down fiber with the proper characteristics, the two types of stretching offset each other. First the pulse spreads out in time, and contracts in wavelength range, which causes it to contract in time and spread out in wavelength range. Eventually, the pulse reappears in its original shape.

Optical amplifiers can compensate for attenuation losses, letting solitons travel virtually unchanged.

Although the pulse shape survives unchanged, the light is attenuated as it travels through the fiber. However, optical amplifiers can make up for that loss, increasing signal strength to offset the attenuation. This makes it possible for soliton pulses to travel very long distances through fiber systems with optical amplifiers. Eventually enough pulse jitter accumulates to cause problems, but researchers have developed ways to avoid that.

Soliton pulses have proved surprisingly robust in optical fibers. The input pulses don't have to exactly match the ideal soliton shape (although they should not be dramatically different), because transmission through the fiber gives them the soliton shape. They can be modulated simply by an external modulator, which transmits 1s and blocks pulses corresponding to 0s in digital coding.

Coherent or Heterodyne Transmission

Coherent transmission has been demonstrated in the laboratory but has yet to find practical applications.

Another transmission alternative is coherent or heterodyne transmission, a concept similar to heterodyne radio transmission. Figure 15.5 shows the basic idea. The laser transmitter emits a frequency v_1, which is modulated by the signal. (The modulation can be in amplitude, phase, or frequency.) At the receiver, that light is mixed with light from a second laser at a nearby frequency v_0, giving an intermediate frequency signal at the difference frequency, v_1-v_0 (or, strictly speaking, the absolute value of that difference). That intermediate frequency signal can then be processed by the receiver electronics to give an output signal.

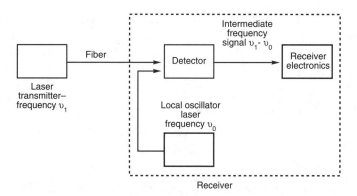

FIGURE 15.5.

Coherent or heterodyne transmission.

The major attraction of coherent detection is that it avoids noise encountered in direct detection, making the receiver sensitive to fainter signals, and allowing more loss between transmitter and receiver. It also allows the use of frequency-shift and phase-shift modulation, which won't work with direct detection. The greater sensitivity would let signals travel further between transmitter and receiver. Alternatively, that higher sensitivity could allow higher transmission speed, because the detector would need to see fewer photons to detect a bit. Another theoretical advantage of coherent transmission is that it might let system designers pack channels closer together in frequency than would otherwise be possible.

However, I should stress that the attractions of coherent transmission remain largely theoretical. It requires lasers that emit an extremely narrow range of wavelengths, which must be matched to equally narrow-line local oscillators. Matching those lasers and precisely tuning them are daunting tasks. So far, solitons look much easier.

⬤ Major advances are needed to make coherent detection practical.

Controlling Pulse Dispersion

Pulse dispersion is an important issue in system design because the spreading of pulses with distance limits data rate and bandwidth. There are two basic ways to control dispersion—by controlling the fiber or the light source. Each plays a part in practical system design.

⬤ Changing fiber type or light source can reduce pulse dispersion.

Single-Mode versus Multimode Fibers

As you saw in Chapter 4, single-mode fibers have much lower dispersion than multimode fibers because they avoid modal dispersion. Dispersion in single-mode fibers depends on the inherent dispersion of the fiber and on the wavelengths it transmits. Dispersion depends both on the particular wavelength and on the spectral bandwidth—the broader the range, the larger the dispersion.

You also saw in Chapter 4 that there are two standard types of single-mode fiber. The most widely used—step-index single-mode—has dispersion near zero at 1300 nm; dispersion-shifted fiber has lowest dispersion near 1550 nm. The choice between them is a basic one for system designers.

Single-Frequency Lasers

Another basic choice is the selection of light source. You saw in Chapter 6 that standard lasers have spectral bandwidth of about 3 nm, but single-frequency lasers have much narrower linewidth.

Pulse dispersion can be kept low with a standard laser source if the laser wavelength is matched to the lowest dispersion point of the fiber. That is, 1300 nm standard lasers can be used with step-index single-mode fiber, and 1550 nm standard lasers with dispersion-shifted fiber. Single-frequency lasers are needed at wavelengths where dispersion is high—1550 nm in step-index single-mode fibers, and 1300 nm in dispersion-shifted fibers. Single-frequency lasers are also needed to meet the most demanding transmission speeds in either type of fiber.

Dispersion Compensation

Developers are also working on a novel scheme to compensate for fiber dispersion. The idea is to splice together segments of fibers with different dispersion characteristics, so they cancel each other out, leaving a composite fiber with low overall dispersion.

This requires fibers with dispersion of opposite signs. If the speed of light in one fiber decreases as the wavelength increases, the speed must increase with wavelength in the other fiber. Ideally, the wavelengths that fall behind in one segment will catch up by the same amount in the other. Encouraging results have been achieved in the laboratory, but dispersion compensation has yet to find practical uses.

Multiplexing

Multiplexing is the simultaneous transmission of two or more signals through one fiber.

Another important system concept is multiplexing, which is the simultaneous transmission of two or more signals through one information channel. Multiplexing is found throughout communications, because it greatly increases transmission efficiency, reducing system costs. Telephone calls are multiplexed so one fiber or wire can carry multiple conversations. Many video channels are multiplexed on a cable television network—so a single fiber or coaxial cable carries all the video channels delivered to homes.

Two fundamentally different types of multiplexing are used in fiber-optic systems. One is the combination of multiple signals into a single combined signal that can modulate an optical transmitter. The other is the use of a single optical fiber to carry two or more optical signals (at different wavelengths).

Signal or Electronic Multiplexing

Electronic multiplexing began long before optical fibers were first used in telecommunications. Electronic equipment combines two or more separate input signals into a single output signal. That combined signal is transmitted through a communication system, then "demultiplexed" to break it into its original components. Multiplexing takes different forms in digital and analog systems.

Digital systems generally use time-division multiplexing, which combines several input signals into a single bit stream, as shown in Figure 15.6. In the example shown, four separate bit streams, each at 1.5 Mbit/s, feed into a multiplexer. The multiplexer combines the signals, selecting first one pulse from input 1, then a pulse from input 2, and so on in sequence. Essentially, the multiplexer shuffles the pulses together, and retimes them because the lower-speed pulses are too long to stuff into a faster stream of bits. At the other end of the system, a demultiplexer sorts the bits out, putting bit 1 into channel 1, bit 2 into channel 2, and so forth.

> Electronic multiplexers combine two or more signals to produce a signal that drives a fiber-optic transmitter.

> Digital multiplexers generate a single bit stream.

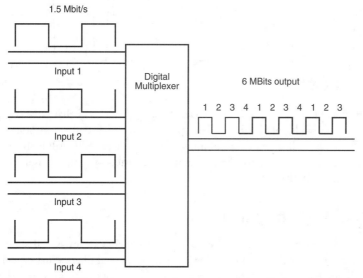

FIGURE 15.6.

Time-division digital multiplexing.

Analog multiplexers work differently, by shifting the frequencies of the input signals so each is at a separate frequency and does not interfere with the others. Then the signals are combined to generate a signal covering a wider range of frequencies. In Figure 15.7, each input signal covers a frequency range of 0-1 MHz. The first signal modulates a 10 MHz carrier, generating signals varying in frequency from 10 to 11 MHz. The second modulates a 12 MHz carrier, generating a signal from 12 to 13 MHz. The third and fourth signals modulate carriers at 14 and 16 MHz, respectively, to generate signals at 14-15 and 16-17 MHz. This is called frequency division multiplexing. Note that some dead space is

> Analog multiplexers shift inputs to different frequencies for transmission.

left between channels to avoid interference and that the carrier frequency is much higher than the signal frequency. The signals are sorted out by a demultiplexer at the other end of the system. Electronic bandpass filters separate the multiplexed signals that are mixed with the carrier frequencies to regenerate the original signals.

FIGURE 15.7.

Analog electronic frequency-division multiplexing.

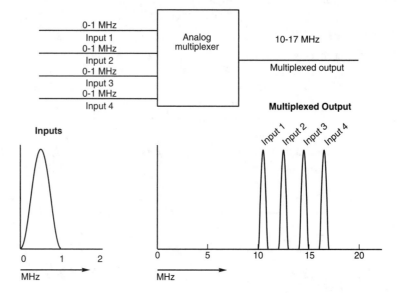

The outputs of both analog and digital electronic multiplexers behave like complete signals of that bandwidth or bit rate. A fiber-optic system can carry them without any special provisions because all the signal combining is done electronically. All the fiber-optic system has to do is carry the signal that drives its transmitter—whether or not the input is a composite of many different signals.

Optical Multiplexing

Optical wavelength-division multiplexing sends multiple signals down a fiber at different wavelengths.

Electronic multiplexing can combine many low-frequency or low-speed signals so they can effectively use the large transmission capacity of optical fibers. That capacity can be further expanded by sending optical signals at different wavelengths through the same fiber. This is possible because the signals do not interfere with each other and can be separated at the output end.

One simple form of optical or wavelength-division multiplexing (WDM) is bidirectional transmission. You saw earlier that most optical fibers carry only one signal in one direction. A second fiber is needed for two-way communications. However, a single fiber can carry signals in both directions if they are at different wavelengths. Beam splitters and combiners at each end merge or separate signals, as you saw earlier in Figure 12.5.

Wavelength-division multiplexing can also carry multiple signals at different wavelengths in the same direction through the same fiber. The simplest such systems may carry only two wavelengths, typically one in the 1300 nm window and the other at 1550 or 850 nm. This effectively doubles fiber capacity at the cost of extra transmitters, receivers, and optical components to combine and separate the two signals. More sophisticated optical systems can combine several wavelengths that are spaced much more closely. Developers are testing systems that would fit multiple signals, perhaps only 10 nm apart, into a single fiber window. So far they remain in the laboratory stage because of the problems in separating wavelengths spaced so closely.

Fiber Transmission Standards

For two people to communicate, they must speak the same language. Fiber-optic communication likewise works only if transmitters and receivers attached to the system speak the same language. That is, they must send and receive signals in the same format, and as you saw earlier, there are many possible formats. To see that this is done, engineers have developed standard formats for fiber transmission.

For simple point-to-point communications, all that is really needed is that the transmitters and receivers at the two terminals speak the same fiber-optic signal language. However, if they use a unique format, they cannot be connected to other equipment, and cannot benefit from the economies of scale that come with large-scale production of many terminals using the same format. The larger the network, and the more terminals attached to it, the more important standards become.

Several different standards have been developed for fiber-optic transmission. The differences come partly from the intended applications; high-speed local area networks have different operating requirements than long-distance telephone lines. Technological progress has also contributed to differences, because the standards were formulated at different times.

The evolution of standard-writing philosophy has also contributed to the proliferation of standards. Early standards used for fiber transmission were formulated for specific applications when telecommunications was more rigidly divided than today. The original hierarchy of digital telephone transmission rates was formulated in the 1970s, when telephone lines carried only telephone signals. Today, the high-speed lines operated by telephone companies are digital data highways that carry many other types of signals, from computer data and fax messages to compressed video signals. New standards have been devised to offer more flexibility.

Figure 15.8 shows how some of these standards are related. The top line shows various communication services, such as video, data, and voice communications, which are seen by end users. These may have their own standards, such as for digital and analog voice channels, and for television transmission. They are fed into the telecommunications network through adapters that convert those signals into a standardized common format for

●
Signal standards assure that different terminals can communicate with each other.

●
Different standards exist for end-user services, data interchange, and fiber transmission.

interchange, such as the 53-byte cells used in the Asynchronous Transfer Mode (ATM) standard. Strictly speaking, that standard format is designed to be insensitive to transmission format—it's just a way of repackaging the signals so they can all be carried by many different types of transmission systems.

FIGURE 15.8.

Relationships of communication standards for services, data interchange, and transmission.

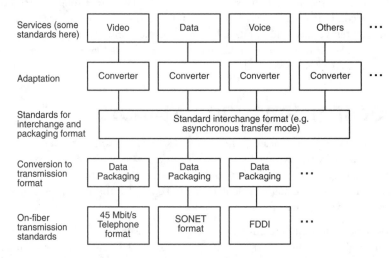

Once all the signals have been converted to the standard interchange format (through a device usually called a router or gateway), they still have to pass through another conversion layer, which packages them for transmission through optical fibers (or over other transmission media) in the standard formats used by those media. Figure 15.8 shows three possible transmission formats—the 45 Mbit/s level of the digital telephone hierarchy, the SONET telecommunication standard, and the FDDI standard developed for local area networks. The signals travel through the fibers in that format. When they reach the receiver at the other end, they must be converted back into the standard interchange format, and eventually back into the standard formats for the various services they carry.

All this may sound very complicated, but in fact it is a simplified version of the real complex array of standards and formats you may encounter. The communications world tends to organize things in ways that make sense to the engineers doing them but appear bewildering to others. Adding to the problem is a penchant for acronyms that generates charts full of confusing alphabet-soup labels, with definitions nowhere to be found. Some of the acronyms have other meanings in other fields, such as ATM, which in the communications world is Asynchronous Transfer Mode, not an automatic teller machine.

Trying to make sense of all these formats and protocols would take a book in itself, so I'll concentrate here on the standards of most importance for fiber optics. They will also be mentioned in later chapters on telephony and data communications.

Digital Telephone Hierarchy

Back in the days when AT&T was the only telephone company, it devised a set of standards for digital telephone transmission. Electronics convert a standard analog voice telephone signal into a digital signal at 56,000 bits per second. Other circuits combine 24 digital phone lines into a single bit stream at 1.5 megabits per second, called the T1 or DS1 rate. That, in turn, feeds into systems operating at successively higher rates. The original design went only to 274 Mbit/s, but much higher transmission speeds became possible with the advent of fiber optics. The fastest systems now operate at 2.5 gigabits per second, and 5 Gbit/s systems are being designed. A similar hierarchy is used in Europe, based on a somewhat different international standard.

I'll talk more about the telephone hierarchy in Chapter 17. What is important here is to recognize that it is a set of existing standards, although they become less formal at higher speeds. The hierarchy was designed for telephone transmission, and this imposes some constraints on its operation. One is that transmission capacity comes in sizes that are convenient for telephone transmission but may not meet the needs of other users. Another constraint is that you have to step back down the hierarchy to extract a lower-speed signal. That is, you have to break a high-speed signal into its component parts to extract one of the components. In addition, some features that would aid in operation of a modern network are missing because the hierarchy is based on old technology.

The digital telephone hierarchy is based on combining channels corresponding to individual phone lines.

SONET/SDH

To overcome these limitations, the telecommunications industry devised a new generation of standards based on a different type of multiplexing. The new standard organizes data into 810-byte "frames" that include data on signal routing and destination as well as the signal itself. These frames can be switched individually, without breaking the signal up into its component parts.

The new American standard is called the Synchronous Optical Network, or SONET. (A similar international standard is called the Synchronous Digital Hierarchy, or SDH.) The "synchronous" means that all network nodes ideally derive their timing signals from a single network master clock, but because that is not always practical, SONET and SDH can accommodate nodes with different master clocks. The details aren't worth worrying about.

SONET and SDH frames were designed to carry other signals as well as telephone traffic. The base (STS-1) rate is 51.84 Mbit/s. Three STS-1 channels can be merged into one 155.52 Mbit/s (STS-3) channel by merging the frames, producing frames that are three times larger. The standard provides transmission speeds to 2488.32 Mbit/s, the STS-48 rate, 48 times the SONET base rate.

SONET is a new standard that organizes data into "frames" for greater flexibility and transmits data synchronously.

Asynchronous Transfer Mode

Asynchronous Transfer Mode is a packet-switching standard that breaks signals into 53-byte cells for transmission.

Another inherent limitation of the digital telephone hierarchy is the way it packages transmission capacity—in discrete chunks with constant bit rates. If your needs fall in between levels, you have to buy more capacity than you need. If your transmission is bursty—at high speeds sometimes, but often low-speed or zero—you either cannot transmit at the peak rate, or you have to have to buy much more capacity than you need most of the time. The standard called Asynchronous Transfer Mode was devised to overcome this limitation.

ATM is one of a broad class of packet-switching technologies. The basic idea of packet switching is to collect bits of data into packets, then combine packets from many sources into a data stream. In ATM, the packets are cells of 53 bytes—a 5-byte header, which identifies the data and its destination, and a 48-byte information field, which contains the user data. ATM is an intermediate transmission format, in the sense shown in Figure 15.8. Services that users see, such as voice, video, and data communications, are converted into ATM cells in the second layer of the figure, and the ATM cells are then packaged for transmission through communication systems. Thus one of ATM's functions is to convert signals in various formats into a common format for switching and transmission.

Packet switching was first developed for data communications, but has since expanded to other applications. Figure 15.9 compares it with the time-division multiplexing used to allocate space in the digital telephone hierarchy. With time-division multiplexing, equal space is set aside for three terminals, although one is sending high-speed signals, one is sending only low-speed signals, and the third is sending nothing. With packet switching, blocks are made available to whichever terminal needs them. When one terminal is sending a lot of information, it produces most of the data packets—and no room is wasted on inactive terminals. Although time-division multiplexing is a synchronous technique (because all transmitters are working on the same clock), Asynchronous Transfer Mode does not depend on the transmitters sending signals in synchronous slots.

ATM converts signals from many sources into standard packets for transmission through various media.

The concept of packet switching is valuable because it gives system operators more flexibility in how they divide transmission capacity. High-speed video signals can be intermingled with low-speed voice signals, without having to worry about the details of splitting signal capacity into certain fractions. The system can assign priorities to different services, depending on how sensitive they are to delay. Thus video signals, which require a steady high bandwidth and can be disrupted by delays, might have a higher priority than computer data. Voice, which is sensitive to delays but requires little capacity, would have a different priority. Again, you don't want to worry about the details.

It is important that packet switching should not be apparent to the user (unless it does not disrupt transmission). You wouldn't notice packet switching of your telephone calls if it was fast and later signal processing smoothed out the fast and slow bits. But you would notice it if the result was that sometimes voices sounded like a tape recorder on fast forward, and other times rumbled too slow and low to understand.

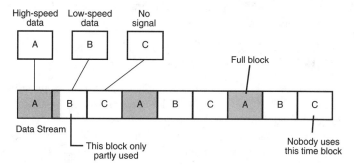

Time Division Multiplexed Transmission

High-speed data — A

Low-speed data — B

No signal — C

Full block

Data Stream

This block only partly used

Nobody uses this time block

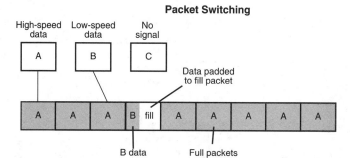

Packet Switching

High-speed data — A

Low-speed data — B

No signal — C

Data padded to fill packet

B data

Full packets from A

FIGURE 15.9.

Packet switching compared with time-division multiplexing.

ATM was developed as part of an overall effort to develop standards for a broadband integrated services digital network (B-ISDN) to carry an array of services in digital format. ATM is basically a standard for formatting and packaging data from many different sources. The standard deliberately avoids mentioning speeds or the transmission medium. It can format signals for transmission over either the standard digital telephone hierarchy or SONET. Hypothetically, it could even work with pigeons, each carrying one cell, but it would take lots of them.

Broadband ISDN

Standards for the Broadband Integrated Services Digital Network (B-ISDN) are really a framework for offering a variety of communications services to homes and businesses. It is an extension of the older concept of an Integrated Services Digital Network (ISDN), which would replace the present telephone network with one providing digital voice and data services, probably over the same wires. The broadband part of the name reflects the bandwidth added for higher-speed data transmission and to provide video capabilities. ISDN has a data rate of 140 kilobits per second; B-ISDN standards envision 155 Mbit/s transmission.

⬤ B-ISDN offers a family of services; the standard incorporates ATM coding and SONET transmission.

ATM coding and switching, and SONET transmission were intended to fit within the B-ISDN framework. Ironically, those two technologies seem to be gaining faster acceptance than B-ISDN itself.

Fiber Distributed Data Interface

FDDI is a 100 Mbit/s standard for network transmission.

Some fiber-optic transmission standards are designed for specific applications. An example is the Fiber Distributed Data Interface, FDDI, a networking standard for transmitting 100 Mbit/s. FDDI accommodates up to 500 nodes on a dual-ring network, with spacing up to 2 kilometers between adjacent nodes. The original standard called for 1300 nm transmission through graded-index multimode fibers, but transmission over much shorter distances is possible through specially conditioned copper wires. FDDI can also use single-mode fiber.

FDDI can handle ATM signals if they are delivered through a suitable interface. I will cover it in more detail in Chapter 20, along with other computer networks.

Fiber Channel

Fiber Channel is a standard "pipeline" for transmitting signals at four rates to 1062 Mbit/s.

Fiber Channel is an emerging standard that provides for transmission at 12.5, 25, 50, and 100 megabytes per second—corresponding to data rates of 133, 266, 531, and 1062 Mbit/s. The proposed standard allows use of single-mode or multimode graded-index fibers. It also allows shielded copper wires or coaxial cables, but those are not expected to be used except for very short spans inside equipment.

The standard is written for point-to-point transmission between pairs of terminals up to 10 kilometers apart over single-mode fiber, and for terminals up to 2 km apart over multimode fiber. The use of point-to-point transmission is intended to reduce overhead needed to direct signals, thus increasing efficiency. Point-to-point segments can be assembled into "fabrics," which include switches and function as networks. It is intended for use in networking applications similar to FDDI, but at higher speeds and possibly over a larger scale. Details are still being finalized.

Network Topology

Ring architectures are preferred because they offer backup in case of cable outages.

An important consideration in fiber-optic network design is how nodes are connected. Early systems had a branching design, as shown at the top of Figure 15.10. This seemed a logical approach, because it made the most direct possible connections between nodes and minimized the amount of cable used. In many cases, a single multifiber cable could carry all the traffic to and from a node.

Unfortunately, telecommunications companies found an inherent flaw in this design the hard way. These branched networks are very sensitive to single-point failures. If someone breaks the main cable going to a remote point, it can knock out all the service to that point.

This happened again and again in the late 1980s and early 1990s, forcing telecommunications companies to look at alternative topologies.

The design that has become the most favored is the ring topology shown at the bottom of Figure 15.10. Instead of single cable connections to each point, nodes are linked with a ring of fiber. If the cable fails at any one point, signals can be transmitted in the opposite direction, keeping all nodes on the network until the failure can be repaired. Rings can be nested, as shown, to connect local nodes served by a regional node, as well as the regional nodes themselves.

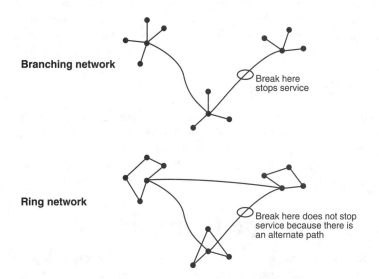

Branching network

Break here
stops service

Ring network

Break here does not stop
service because there is
an alternate path

FIGURE 15.10.

*Branch and ring
topologies for
connecting network
nodes.*

Optical Information Processing

System designers try to look far enough into the future to avoid dooming their products to instant obsolescence. A few concepts now in research and development deserve mention, because of their potential impact on fiber-optic communications.

Optical Switching

Modern fiber-optic networks still need electronics to switch signals. The optical switches I mentioned in Chapter 12 are not up to the high-speed, heavy-duty switching required to route telecommunications traffic. However, improved optical switches may someday permit optical signals to be switched directly in a communication network, without the need for converting them to electronic form.

●
Fiber optics may
find applications in
future systems for
information
processing.

Optical Interconnection

Optical communication techniques are being studied as a way to circumvent a major limitation on complex integrated circuits—the difficulty of interconnecting them. Electronic connections must be made to the sides of chips, but the perimeter is not growing as fast as the area or the number of components on chips. Getting information to and from integrated electronic circuits has become a major bottleneck.

One way to break this bottleneck might be to fabricate light emitters and detectors on the chip surface. They could send signals directly between chips, without having to pass through the electronic bottleneck at the edge. Optical signals can be sent rapidly, and they don't interfere with each other. Although most work centers on sending the optical signals through the air, another possibility might be making the connections with optical fibers.

Optical Computing

Optical devices can also be used to perform computing and logic operations. Optical computing is one of those perpetually promising technologies that has been lurking "just around the corner" for many years. The main problem has been that electronics had a long head start, and has also been moving forward rapidly. Optical computing has not caught up, but if breakthroughs arise, fiber-optic systems might have a role to play.

Optical devices can perform computing operations.

What Have You Learned?

1. Point-to-point transmission links pairs of terminals. Point-to-multipoint transmission uses one transmitter to serve many terminals.

2. Networks allow many terminals to talk with each other.

3. Switching makes temporary connections between pairs of terminals.

4. Digital and analog signals can be transmitted by amplitude modulation of a light beam. There are several different digital modulation patterns; the most common is NRZ (no return to zero).

5. Solitons are self-regenerating pulses that do not change shape along a fiber. Optical amplifiers can compensate for attenuation losses, letting solitons travel virtually unchanged.

6. Coherent transmission has been demonstrated in the laboratory, but major advances are needed to make it practical.

7. Pulse dispersion must be limited for high-speed transmission.

8. Multiplexing is the simultaneous transmission of two or more signals through one fiber.

9. Electronic multiplexers combine two or more signals to produce a signal that drives a fiber-optic transmitter. Digital multiplexers generate a single bit stream. Analog multiplexers shift inputs to different frequencies for transmission.

10. Optical wavelength-division multiplexing sends multiple signals down a fiber at different wavelengths.

11. Signal standards assure that different terminals can communicate with each other. Different standards exist for end-user services, data interchange, and fiber transmission.

12. The digital telephone hierarchy is based on combining channels corresponding to individual phone lines.

13. SONET is a new standard that organizes data into "frames" for greater flexibility and transmits data synchronously. The Synchronous Digital Hierarchy (SDH) is a similar international standard.

14. Asynchronous Transfer Mode is a packet-switching standard that breaks signals into 53-byte cells for transmission. ATM converts signals from many sources into standard packets for transmission through various media.

15. FDDI is a 100 Mbit/s standard for network transmission.

16. Fiber Channel is a standard "pipeline" for transmitting signals at four rates to 1062 Mbit/s.

17. Broadband-ISDN offers a family of services; the standard incorporates ATM coding and SONET transmission.

What's Next?

In Chapter 16, I will look at some simple examples showing how fiber-optic systems are designed.

Quiz for Chapter 15

1. What type of fiber-optic system is used to distribute cable television signals?
   a. Point to multipoint.
   b. Local area network.
   c. Switched.
   d. Point to point.

2. What type of system is used for telephone transmission?
   a. Point to multipoint.
   b. Local area network.
   c. Switched.
   d. Point to point.

3. How are digital signals modulated in present fiber-optic systems?

   a. Frequency modulation.

   b. Phase modulation.

   c. Amplitude modulation.

   d. Heterodyne modulation.

4. What unique characteristic makes solitons attractive for long-distance communications?

   a. Solitons are immune to attenuation.

   b. Soliton pulses can be amplified.

   c. Soliton pulses can carry digital signals.

   d. Soliton pulses recover their original shape after traveling through a fiber.

5. How are digital signals multiplexed together for transmission at a higher speed?

   a. They are transmitted at different frequencies.

   b. The bytes are interleaved.

   c. The signals are transmitted at different wavelengths.

   d. They must be sent through separate fibers.

6. Which of the following is not a type of multiplexing used in signals sent through optical fibers?

   a. Wavelength-division multiplexing, with signals at different wavelengths.

   b. Time-division multiplexing, with pulses at different times.

   c. Space-division multiplexing, with pulses at different positions in the fiber.

   d. Frequency-division multiplexing, with analog signals at different frequencies.

7. Which standards were developed as a family?

   a. B-ISDN, ATM, and SONET.

   b. FDDI, ATM, and SONET.

   c. Digital telephone hierarchy and FDDI.

   d. Fiber Channel, B-ISDN, and SONET.

   e. All standards were devised independently.

8. How many ATM channels can be sent through an FDDI network?

   a. 2.

   b. 20.

   c. 200.

   d. None, the formats are incompatible.

   e. Impossible to define, because ATM does not define signal rate.

9. What is the base-level (STS-1) speed of a SONET system?

   a. 56,000 bits per second.

   b. 1.5 Mbit/s.

   c. 51 Mbit/s.

   d. 155 Mbit/s.

   e. 2.5 Gbit/s.

10. What type of transmission can readily accommodate terminals sending signals at different rates?

   a. Packet switching.

   b. Time-division multiplexed.

   c. Wavelength-division multiplexed.

   d. Standard digital telephone hierarchy.

System Design

About This Chapter

Now that you've learned the general concepts of fiber-optic systems, it's time to see how systems are designed. This chapter covers general design concepts; Chapters 17 through 20 look at how fiber optics are used in various applications.

On top of any list of fiber-optic system design considerations is the loss budget, making sure that enough signal reaches the receiver for the system to operate properly. Other considerations close behind in importance are designing systems that are cost-effective and that offer the required transmission capacity—whether in the form of analog bandwidth or digital bit rate. This involves looking at trade-offs for which there are rarely hard-and-fast rules, such as the choices between fibers of various core diameters and light sources of various power levels.

This chapter won't prepare you for heavy-duty system design. However, it will prepare you to evaluate system designs and develop system concepts that should work. It concentrates on simple designs to give you a clear idea of how systems work.

Variables

Design of a fiber-optic system is a balancing act. You start with a set of performance requirements, such as sending 100 Mbit/s through a 5 km cable. You add some subsidiary goals, sometimes explicitly, sometimes implicitly. For example, you may demand cost as low as possible, less than another alternative, or no more than a given amount. Your system might need an error rate of no more than 10^{-9} and should operate without interruption for at least five years.

Design of fiber-optic systems requires balancing many cost and performance goals.

You must look at each goal carefully to decide how much it is worth. Suppose, for instance, you decide that your system absolutely must operate 100% of the time. You're willing to pay premium prices for transmitters, receivers, and super-duper heavily armored absolutely gopher-proof cable. But how far should you go? If that is an absolute must because of national security and you have unlimited quantities of money, you might buy up the entire right of way, install the cable in ducts embedded in a meter of concrete, and post guards armed with tanks and bazookas to make sure no one comes near the cable with a backhoe. If its purpose is just to keep two corporate computers linked together, you might be satisfied with the gopher-proof cable or with laying a redundant cable along a second route different enough from the first that no single accident would knock out both.

Many variables enter into system design.

That somewhat facetious example indicates how many variables can enter into system design. In this chapter, I will concentrate on the major goals of achieving specified transmission distance and data rate at reasonable cost. Many design variables can enter into the equation, directly or indirectly. Among them are the following:

- Light source output power (into fiber)
- Coupling losses
- Spectral linewidth of the light source
- Response time of the light source and transmitter
- Signal coding
- Splice and connector loss
- Type of fiber (single- or multimode)
- Fiber attenuation and dispersion
- Fiber core diameter
- Fiber NA
- Operating wavelength
- Receiver sensitivity
- Bit-error rate or signal-to-noise ratio
- Receiver bandwidth
- System configuration
- Number of splices, couplers, and connectors
- Type of couplers
- Costs

Many of these variables are interrelated. For example, fiber attenuation and dispersion depend on operating wavelength as well as the fiber type. Coupling losses depend on factors such as fiber NA and core diameter. Some interrelationships limit the choices available (e.g., the need to achieve low fiber loss may require operation at 1300 or 1550 nm).

Some variables may not give you as many degrees of freedom as you might wish. For example, you may need to interconnect several computer terminals. You have enough flexibility to pick any of the possible layouts in Figure 16.1, but you still have to connect all the terminals, and that requires enough optical power to drive them all.

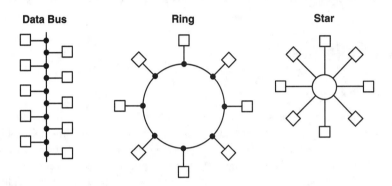

FIGURE 16.1.

Three ways to interconnect computer terminals.

Power Budgeting

Usually the first task in designing a fiber-optic system is power budgeting. That is, you need to make sure that subtracting all the system's optical losses from the power delivered by the transmitter leaves enough power to drive the receiver at the desired bit-error rate or signal-to-noise ratio. That design should leave some extra margin above the receiver's minimum requirements to allow for system degradation and fluctuations (e.g., degradation of a transmitter or splicing a broken cable).

> The first task in designing a fiber-optic system is budgeting optical power to deliver enough power to the receiver.

One note of warning before I get deeply into power budgeting: be sure you know what power you mean. Manufacturers tend to specify peak power, particularly for transmitters. For a normal digital fiber-optic system, peak power is about twice the average power. As long as transmitter output and receiver sensitivity are specified as the same type of power, you should have no problems, because power budgets are calculated in relative units of decibels. However, you might have 3 dB less power than you thought if transmitter output is specified as peak power and the receiver sensitivity is specified as average power.

In simplest form, the power budget is

TRANSMITTER OUTPUT - RECEIVER INPUT = Σ LOSSES + MARGIN

when arithmetic is done in decibels or related units such as dBm. The simplicity of these calculations is the main reason why units such as dBm and dBμ are used.

> The difference between transmitter output and receiver input equals the sum of system losses and margin.

All losses in the system must be considered. These include

1. Loss in transferring light from source into fiber

2. Connector loss

3. Splice loss

4. Coupler loss

5. Fiber loss

6. Fiber-to-receiver coupling loss

Some of these losses have been covered in detail earlier, but others deserve more explanation.

Light Collection

MATCHING LEDS TO FIBERS

●
Significant losses can occur in coupling light sources to fibers.

Typically, little light is lost in transferring from a fiber to a receiver, but large losses can occur in transferring light from the source into the fiber. The fundamental problem is matching the source emitting area to the fiber core. This is particularly true for LEDs with large emitting areas, as shown in Figure 16.2. If the emitting area is larger than the fiber core, some emitted light is lost in the cladding and dissipated from the fiber. In addition, some light rays are emitted at angles outside the fiber's acceptance angle. The smaller the fiber core and the numerical aperture, the more severe these losses. Losses are large for LEDs even when coupled to 62.5/125 multimode graded-index fibers. LED output in the 1 mW range can be reduced to about 50 μW—a 13 dB loss—by losses in getting the light into the fiber. Losses are even higher in coupling to single-mode fibers, so LEDs are almost never used with them.

FIGURE 16.2.

Light losses in transferring LED output into a fiber.

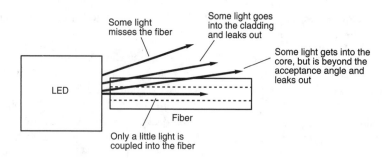

LASER SOURCES

●
A good laser can transfer a milliwatt into a single-mode fiber.

Semiconductor lasers deliver more power into optical fibers. Their advantages are a smaller emitting area, smaller beam spread, and higher output power. Where a good LED might couple 50 μW into a 62.5/125 multimode fiber, a good semiconductor laser could transfer a milliwatt or more into a single-mode fiber.

If this makes lasers seem like better light sources, that's largely because they are. Semiconductor lasers can be modulated faster and deliver more power into a fiber than an LED. However, cost counts in the real world, and LEDs are cheaper than lasers. They also last longer, and few LEDs require the cooling and stabilization equipment needed by many laser transmitters—additional factors that help make them cheaper than lasers.

FIBER CHOICE

Another factor in this light collection equation is the choice of fiber. Increasing core diameter and/or numerical aperture should increase the amount of light a fiber collects. The difference in transfer loss between two fibers—fiber 1 and fiber 2—is roughly

$$\text{LOSS DIFFERENCE (dB)} = 20 \log (D_1/D_2) + 20 \log (NA_1/NA_2)$$

where the Ds are core diameters and the NAs are numerical apertures of the two fibers. This relationship is valid as long as the emitting area is larger than the fiber core and no optics are used to change the effective size of the emitting area or the effective NA of the emitter.

For a numerical example, consider the differences in light collection by a fiber with 100 μm core and 0.3 NA and one with a 50 μm core and 0.2 NA. Plug in the numbers and you get

$$\text{LOSS DIFFERENCE} = 20 \log 2 + 20 \log 1.5 = 9.6 \text{ dB}$$

That factor-of-ten increase in coupling loss clearly indicates the need to consider the fiber type.

On the other hand, remember that larger core reduces transmission bandwidth as well as increases light-collection efficiency. The sacrifices are largest in moving from single-mode to multimode graded-index fiber and from graded-index multimode to step-index multimode.

> ●
> Increasing a fiber's core diameter and/or numerical aperture increases the amount of light it collects.

SINGLE-MODE FIBERS

The preceding example was for multimode fibers. Carry it a step further to a single-mode fiber with a nominal core diameter of 10 μm and NA of 0.11, and you will immediately see a big problem. A single-mode fiber collects 19.2 dB less light from an LED than 50/125 μm fiber.

If your first impulse is to say, "Forget it," you are in good company. However, there are ways to ease that coupling problem. Instead of just butting the fiber end against the LED or aiming LED output in the general direction of the fiber end, developers can focus the light onto the fiber end with tiny optics. Losses remain significant, but they are well below the exceedingly high levels predicted by this simple cookbook formula.

What about transferring diode laser output into fibers? The huge losses mentioned earlier go away because the light-emitting area of a laser is typically smaller than the core of a single-mode fiber. Laser light also does not spread out as fast as that from an LED. As shown in Figure 16.3, this makes light collection much simpler. In addition, lasers emit higher powers, so they can deliver a milliwatt or more into a fiber. Thus, a laser is the obvious choice if lots of power is needed because of high system losses (e.g., from transmission over long distances or because of the need to distribute signals to many terminals).

> ●
> Loss in transferring light from an LED into a single-mode fiber is about 19 dB higher than into a 50/125 fiber.

> ●
> Fibers collect laser output more efficiently because of the smaller emitting area and beam spread.

FIGURE 16.3.

Laser output is easier to transfer into a fiber.

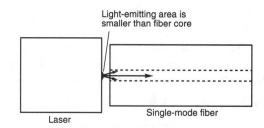

Light-emitting area is smaller than fiber core

Laser

Single-mode fiber

Fiber Loss

●

Fiber loss roughly equals attenuation multiplied by transmission distance, but transient losses can add to this for short fibers.

The simplest approximation to fiber loss is to multiply the attenuation (in decibels per kilometer) by the transmission distance. However, for multimode fibers and LED sources, this is only an approximation. An LED with large emitting area and numerical aperture excites high-order modes that leak out as they travel along the fiber. This transient loss, typically 1 to 1.5 dB, is concentrated in the first few hundred meters of fiber following the excitation source.

This means that for the first few hundred meters, multimode fiber loss is higher than calculated by multiplying the specified loss by distance traveled. Over a kilometer or so (depending on how the specified loss was derived), the actual loss becomes roughly equal to the calculated loss. If signals are sent longer distances through multimode fibers uninterrupted by connectors, the actual loss may be slightly lower than the calculated value, because higher order modes that contribute more to loss are not present further along the fiber.

Fiber-to-Receiver Coupling

●

Losses are normally small in transferring light from fibers to receivers.

One of the rare places where the fiber-optic engineer wins is in coupling light from a fiber to a detector or receiver. The light-sensitive areas of most detectors are larger than most fiber cores, and their acceptance angles are larger than those of multimode fibers. Of course, if you're determined to screw things up, you can find a detector with a light-collecting area smaller than the core of large-core multimode fibers. But that isn't likely.

Receiver Sensitivity

●

There are trade-offs among received power, speed, and bit-error rate or signal-to-noise ratio.

In much of the discussion that follows, receiver sensitivity is taken as a given. That is, assume that a receiver must have a minimum power input to work properly. Things aren't quite that simple because there are trade-offs between received power, speed, and bit-error rate or signal-to-noise ratio. These are indicated for a digital receiver in Figure 8.3. As data rate increases, a receiver needs more input power to operate with a specified bit-error rate. If the data rate is held fixed but the input power is decreased, the error rate can increase steeply.

These trade-offs are not always useful. Error rate can increase steeply as input power decreases. At the margin of receiver sensitivity, a 1 dB drop in power can increase error rate by a factor of 1000 or more! You gain more by lowering data rate, particularly near the receiver's maximum speed. However, many system designs do not allow much flexibility in transmission speed. As I describe later in this chapter, some gains in sensitivity are possible by switching bit encoding schemes, but the simplest course may be using a more sensitive receiver or going back to reduce loss or increase power.

Remember, too, that more power is not always a good thing. Too much power can overload the detector, increasing bit-error rate.

Other Losses

Splices, connectors, and couplers can contribute significant losses in a fiber-optic system. Fortunately, those losses are generally easy to measure and calculate. Connectors, couplers, and splices have characteristic losses that you can multiply by the number in a system to estimate total loss. However, there are two potential complications.

One is the variability of loss, particularly for connectors and splices. A given connector may be specified as having maximum loss of 1.0 dB and typical loss of 0.5 dB. The maximum is the specified upper limit for that type of connector; no higher losses should show up in your system (unless the connector was installed improperly or is dirty). The typical value is an average, meaning that average connector loss should be 0.5 dB but that individual connectors may be higher or lower.

You can calculate total loss in two ways for a system with four connectors. The worst-case approach is to multiply the highest possible loss (1 dB) by the number of connectors to get 4 dB. On the other hand, if the average connector loss is 0.5 dB, the most likely total loss is four times that: 2.0 dB. The prudent approach for so few connectors is to take the worst-case value, but it's much more realistic to take the average loss for systems with many connectors or splices. Remember, because detector overload can cause problems, you can run into trouble by seriously overestimating loss, as well as by underestimating it.

Transient losses following connectors can further complicate the picture for multimode fibers. Connectors near the transmitter may increase transient losses by effectively stripping away high-order modes, which otherwise would leak out further along the fiber. However, once light has traveled far enough to reach an equilibrium mode distribution (a kilometer or so), a connector can redistribute some light to higher-order modes, which tend to leak out of the fiber—a milder form of transient losses than experienced with light sources.

●
Total loss from connectors, couplers, and splices is their characteristic loss multiplied by the number in the system.

Margin

One quantity that always figures in the loss budget is system margin, a safety factor for system designers. This allows for the inevitable uncertainties in counting losses. You cannot rigidly specify component performance; real devices operate over ranges. The effects of concatenation, putting lengths of fiber end-to-end, are very hard to calculate and may vary with temperature and physical stresses applied to the fiber. Devices age, generally emitting less power and becoming less sensitive. Cables may be damaged or broken, so the fibers in them must be spliced, adding to overall attenuation.

To prevent these minor problems from disabling systems, designers add a margin of error. Depending on the application, the performance requirements, the cost, and the ease of repair, this may be 3-10 dB.

Examples of Loss Budgeting

To see how loss budgeting works, I'll step through the three simple examples shown in Figure 16.4, 16.5, and 16.6. Figure 16.4 (Example A) shows a short system transmitting 100 Mbit/s between two points in a building. Figure 16.5 (Example B) shows a telephone system carrying 565 Mbit/s between two switching offices 50 km apart. Figure 16.6 (Example C) shows an intrabuilding network delivering 100 Mbit/s signals to ten terminals. These examples are arbitrary and were picked to be illustrative, not representative, of actual design practice. Note that, in considering only loss budgeting, you don't directly consider the system data rate. Actual system design must consider transmission speed, as I describe later in this chapter.

EXAMPLE A

In Figure 16.4, designers need to transmit signals through 200 m of fiber already installed in a building. That means that they must route the signal through patch panels with connectors. In the example, they have six connector pairs, three on each floor: one linking the terminal device to the cable network for that floor, and one pair on each end of a short cable in the patch panel. (Connectors also attach the fiber to transmitter and receiver, but their losses are included under LED power transfer and receiver sensitivity.) The 50/125 graded-index multimode fiber has attenuation of 2.5 dB/km at 850 nm wavelength of the LED transmitter. The loss budget is as follows:

LED power into fiber	-16.0 dBm
Connector pairs (6 @ 0.7 dB)	-4.2 dB
Fiber loss (200 m @ 2.5 dB/km)	-0.5 dB
System margin	-10.0 dB
Required receiver sensitivity	-30.7 dBm

The calculation shows that the dominant loss is from the connectors. The fiber loss may underestimate transient loss, but the large system margin leaves plenty of room.

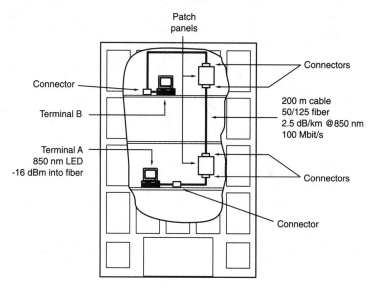

FIGURE 16.4.

Example A.

The calculated receiver sensitivity is a reasonable level, and system margin could be improved by picking a more sensitive receiver. This calculation started with a given loss, system margin, and input power, but you could start by specifying receiver sensitivity, system margin, and loss, to calculate the needed input power. Note that the LED transmitter provides a low input power.

EXAMPLE B

Loss sources in the telephone system in Figure 16.5 are quite different, as shown by the following sample calculation:

Laser power into fiber	0.0 dBm
Fiber loss (1300 nm) (50 km × 0.4 dB/km)	-20.0 dB
Splice loss (24 × 0.1 dB)	-2.4 dB
Connector pairs (2 @ 1.0 dB)	-2.0 dB
Power at receiver	-24.4 dB
Receiver sensitivity	-(-32.0) dBm
Excess power (system margin)	7.6 dB

Losses in a 50 km telephone system are dominated by fiber attenuation, even though it is only 16% of the level in the short-distance system.

The dominant loss is in the 50 km of fiber. Fiber attenuation is lower than in the first example because operation is at 1300 nm in single-mode fiber. Input power is higher because the long transmission distance justifies use of a more costly laser. Similarly, a more sensitive receiver can be cost-justified. The 24 splices join together segments of a buried or overhead cable. One connector (actually, a mated pair) is on each end, allowing connection of transmitter and receiver through patch panels. The use of cross-connecting cables in the patch panel, common in the telephone industry, would add an extra connector pair at each location, giving a total of four pairs for a 4 dB connector loss. Adding a more sensitive receiver or a more powerful transmitter could raise system margin.

FIGURE 16.5.

Example B.

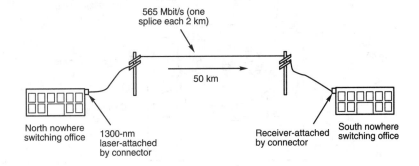

EXAMPLE C

Coupling losses are largest in a multiterminal network.

In Figure 16.6, there is a different complication—the need to divide the light signal among ten separate receivers. I try to meet this need with a 10×10 directional star coupler, which divides the input signal from any of ten input fibers among ten output fibers. I will assume it has excess loss of 3 dB and divides the rest of the input signal equally among ten output ports, for total input-to-output attenuation of 13 dB. (See Chapter 12 for more on coupler losses.) This one device dominates the loss budget:

LED transmitter (850 nm)	-16.0 dBm
Fiber loss (100 m @ 2.5 dB/km)	-0.25 dB
Connector pair loss (2 @ 0.8 dB)	-1.6 dB
Coupler loss (includes its own connectors)	-13.0 dB
Power at receiver	-30.85 dB
Receiver sensitivity	-(-30.0) dBm
System margin (insufficient)	-0.85 dB

The calculation of the system margin shows that I'm in deep trouble—the margin is negative! That means the receiver will not get enough signal to operate properly, even if the components continue to perform as well as I assumed. I need either a more powerful laser source or to redesign the system to eliminate the star coupler with its high loss.

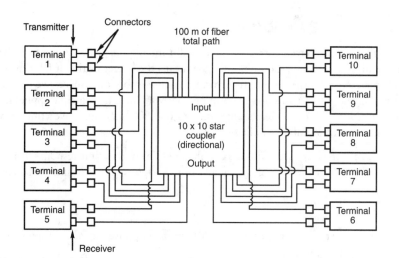

FIGURE 16.6.

Example C.

Bandwidth Budget

Calculating bandwidth budget is both simpler and more complex than calculating loss budget. The simplicity comes from being able to ignore components such as connectors and splices, which have no significant impact on system bandwidth. The complexity comes from the nature of the relationships that limit transmission speed, even after making some simplifying assumptions.

In my earlier description of system loss budget, there was no need to define systems as explicitly digital or analog. Both experience the same attenuation. Bandwidth and transmission speed considerations are somewhat different; it matters if the system is analog or digital, and even what digital coding is used. (For simplicity, here I take NRZ coding as standard.) Although details may differ slightly for other types of systems, the principles are the same.

One other initial assumption is needed to simplify calculations of bandwidth budget. I assume that I can calculate everything I need to know from time response to signal inputs without looking directly at frequency response. That is reasonable because there is a characteristic time per bit T at any transmission rate R:

$$T = 1/R$$

where R is the speed in bits per second. Thus, the bit time is 1 ns at 1,000 Mbit/s. (Roughly the same relationship exists for analog signals; there is a characteristic time that is the inverse of the frequency. Thus, a 1 ns time response corresponds to a 1 GHz analog signal bandwidth.) Although such simplifications do not give exact results, they do give a useful approximation.

●
Bandwidth budget calculations are not as straightforward as loss budgets, but some components, such as connectors and splices, can be ignored.

●
For a signal to be received correctly, the overall time response of a system must be less than the bit time.

Overall Time Response

For a signal to be received correctly, the overall time response of a system must be less than the bit time. Time response in this case means the longer of rise or fall time of the signal emerging from the system. If the time response is too long, successive pulses start overlapping and the system starts performing poorly. (The same principle applies to analog transmission.) Thus, a system that transmits 1,000,000 bit/s must have a time response faster than 1 μs (one-millionth of a second).

●

The overall time response of a system is the square root of the sum of the squares of response times of transmitter, receiver, and fiber.

The choice of time response simplifies calculations. The overall time response of a system is the square root of the sum of the squares of the response times of individual components:

$$t = (\Sigma t_i)^{1/2}$$

where t is the overall time response and t_i is the time response for each component.

Connectors, splices, and couplers do not affect time response significantly, but the transmitter, fiber, and receiver do. That means that the characteristic response time t of a fiber-optic system is

$$t = (t_{trans}^2 + t_{rec}^2 + t_{fiber}^2)^{1/2}$$

That is, the square root of the sum of the squares of the response times of the transmitter, receiver, and fiber.

Fiber Response Time

●

Fiber response time is the square root of the sum of the squares of modal and chromatic dispersion.

Rise and fall times of transmitters and detectors are given on data sheets and are ready to plug into the formula. Fiber response time must be calculated for the length of fiber from the specified value. It is composed of two components—modal dispersion and chromatic dispersion, which are combined by a sum-of-squares formula. Modal dispersion can be calculated from the specified value and the length of fiber, but calculating chromatic dispersion also requires knowing spectral width of the transmitter. Chromatic dispersion can be significant even for multimode fibers, so you should not ignore it.

The best way to see what is involved in response time calculations is to look again at Example A considered earlier in this chapter, transmission of 100 Mbit/s through 200 m of 50/125 fiber using an 850 nm LED. A typical commercial 50/125 fiber has modal bandwidth of 400 MHz-km, which is equivalent to a modal dispersion of 2.5 ns/km. For a 200 m length, that corresponds to a 0.5 ns time response, according to the formula

$$t_{modal} = DL$$

where t_{modal} is time response caused by modal dispersion, D is modal dispersion, and L is length.

To that must be added the effect of chromatic dispersion, $t_{chromatic}$, calculated from the formula

$$t_{chromatic} = D_c \times \Delta\lambda \times L$$

where D_c is chromatic dispersion, $\Delta\lambda$ is the range of wavelengths in the source output, and L is the length of the fiber. For a typical commercial 50/125 fiber, the chromatic dispersion is about 100 ps/nm-km at 850 nm, which combined with a 50 nm linewidth for a typical 850 nm LED gives a time response of 1.0 ns. Thus, chromatic dispersion is larger than modal dispersion.

The two terms add together by the sum-of-squares formula, giving a fiber time response of 1.1 ns. That leaves plenty of room for transmitting 100 Mbit/s, if a fast enough transmitter and receiver are used.

Receiver Response Time

The real limitations on short-distance transmission speed come from the transmitter and receiver. Suppose you pick an LED source with 7 ns response time and an inexpensive silicon detector with 10 ns response. Plug the numbers into the sum-of-squares law, and you get total time response of 12 ns, too slow to handle 100 Mbit/s signals. Getting a faster receiver, with 7 ns response time, would give you 10 ns overall response, an inadequate margin at 100 Mbit/s.

For short-distance transmission, the prime limitations on transmission speed are transmitter and receiver response time.

I chose a simple example to show how to calculate response time, but the same principles apply to more complex systems, like the 50 km, 565 Mbit/s system of Example B. In that case, much faster transmitters and detectors would be used, with typical response times of 0.5 ns or less. Fiber dispersion is very low at 1300 nm, but it cannot be neglected altogether because the laser emits a range of about 3 nm, which is unlikely to precisely match the zero-dispersion wavelength. Assuming a dispersion of 3.5 ps/nm-km at 1300 nm— large for standard single-mode fibers—I find that 50 km of fiber has a response time of about 0.5 ns. Combine this with a 0.5 ns source and detector, and the sum-of-squares rule gives total response time of 0.87 ns, adequate for 565 Mbit/s transmission.

Cost/Performance Trade-offs

So far, I have only mentioned in passing one of the most important considerations in real-world system design—cost. Minimizing cost is an implicit goal in all system design; a few guidelines for doing so follow. However, no book can give hard-and-fast rules for the tough job of making trade-offs between cost and performance. Ultimately, it is your judgment as a system user or designer whether pushing bit-error rate from 10^{-8} to 10^{-9} is worth an extra $1,000. What I can do is give you some ideas to apply in working situations.

Cost is vital in real-world system design. It is up to the user to make cost-performance trade-offs for specific applications.

In earlier chapters, I examined some trade-offs that can affect cost and performance but skimmed over others. Examples include the relative costs of components for 850 and 1300 nm systems, and the low marginal cost of adding extra fibers to heavy-duty cables. Many of these balances shift with pricing of commercial equipment and go beyond the scope of this introduction to the field. However, I can give you a set of rough-and-ready (but not necessarily complete) guidelines. '

Guidelines

I'll start the list of guidelines with a few common-sense rules:

- Your time is valuable. If you spend an entire day trying to save $5 on hardware, the result will be a net loss.

- Installation, assembly, operation, and support are not free. For a surprising number of fiber-optic systems, installation and maintenance costs more than the hardware. You may save money in the long term by paying extra for hardware that is easier to install and service.

- It can cost less to pay an expert to do it than to learn how yourself. Unless you need to practice installing connectors, it's much easier to buy connectorized cables or hire a fiber-optic contractor for your first fiber-optic system.

- You can save money by using standard mass-produced components, rather than designing special-purpose components optimized for a particular application.

You should also learn some basic cost trade-offs that people often face in designing fiber-optic systems.

- The performance of low-loss fiber, high-sensitivity detectors, and powerful transmitters must be balanced against price advantages of lower-performance devices.

- Low-loss, high-bandwidth fibers generally accept less light than higher-loss, lower-bandwidth fibers. Over short distances, you can save money and overall attenuation by using a higher-loss, more-costly cable that collects light more efficiently from lower-cost LEDs. (Because of the economics of production and material requirements, large-core multimode fibers are considerably more expensive than single-mode fibers.)

- The marginal costs of adding extra fibers to a cable are modest and much cheaper than installing a second parallel cable. However, if reliability is important, the extra cost of a second cable on a different route may be a worthwhile insurance premium.

- LEDs are much cheaper and require less environmental protection than lasers, but they produce much less power and are harder to couple to small-core fibers. Their broad range of wavelengths and their limited modulation speed limit system bandwidth.

Don't forget to apply common sense in system design. Labor is never free.

- Fiber attenuation contributes less to losses of short systems than losses in transferring light into and between fibers.

- Topology of multiterminal networks can have a large impact on system requirements and cost because of their differences in component requirements. Coupler losses may severely restrict options in some designs.

- Light sources and detectors for 1300 nm cost more than those for the 800-900 nm window, although fiber and cable for the longer wavelength may be less expensive.

- 1550 nm light sources cost more than 1300 nm sources. To provide reasonable bandwidth, they must have extremely low bandwidth or be used with dispersion-shifted fiber.

- Fiber and cable become a larger fraction of total cost—and have more impact on performance—the longer the system.

- Balance the advantages of eliminating extra components with the higher costs of the components needed to eliminate them. For example, it's hard to justify two-way transmission through a single fiber over short distances unless wavelength-division-multiplexing couplers are cheap, large installation savings are possible, or system requirements permit only a single fiber.

- In general, avoid repeaters because they require maintenance as well as capital expense. Remember that optical amplifiers are costly and currently available only at 1550 nm.

- Think of future upgrade possibilities. Fiber-optic technology is still developing, and transmission requirements are likely to increase. (To paraphrase one of Parkinson's Laws, "Communication requirements expand to fill the available bandwidth.") If a small extra expenditure now can open the way for much larger capacity in the future, it's probably worthwhile. A good example is installing cables with more fibers than needed at the moment, which can avoid the high cost of buying and installing additional cables later.

- Leave margin for repair and expansion. Although fiber-optic cables rarely fail by themselves, as you learned in Chapter 5, they can fail with human or other intervention. A careless foot can pull a fiber out of a connector, or an ice storm can take out overhead cables. It costs much less to allow for the possibility beforehand than to make up for it afterward. The same applies to leaving room for minor expansion (e.g., the addition of a few terminals to a network) without having to replace the system completely with a larger one.

- Remember light transfer between transmitter and fiber, and fiber and receiver. Some specification sheets list only transmitter output—not power actually transferred into an optical fiber. Coupling losses from fiber to receiver are smaller and, if properly designed, should be no more than those of a good connector pair.

● Remember to account for the costs of hardware such as splice enclosures and patch panels, and to make sure you have places to install them.

● Most fiber-optic failures are caused by human action rather than defective equipment. Careless contractors have broken so many cables they have become standard villains. However, one major failure occurred when technicians cut the wrong cable. Take the extra time and spend the extra money to make any important system less vulnerable to damage. This means labeling and document-ing the system carefully, as well as not leaving cables where people can trip over them. Failures are expensive both in disrupted service and in repair costs.

As you grow more familiar with fiber optics, you will develop some of your own guide-lines, based on your own experience. Indeed, many of the ideas listed are really only the application of common sense.

What Have You Learned?

1. Design of fiber-optic systems requires balancing sometimes-conflicting perfor-mance goals as well as costs.

2. The system loss budget is calculated by adding all system losses and subtracting them from the transmitter output power. The result equals the minimum power required by the receiver plus the system margin.

3. Significant losses can occur in transferring light from sources into fibers. LEDs in particular often emit light over a broader area and in a broader angle than many fiber cores can accept. Semiconductor lasers transfer light into fibers more efficiently. A good LED can couple 50 μW into a 62.5/125 multimode fiber, but a good laser can transfer a milliwatt into a single-mode fiber.

4. Coupling efficiency depends on the fiber type as well as the source. The larger the core diameter and numerical aperture, the more light the fiber will collect from a large-area LED source. Lasers, with small emitting area and small beam spread, can transfer light efficiently to smaller-core fibers.

5. Fiber loss roughly equals attenuation multiplied by transmission distance. However, transient loss of 1-1.5 dB occurs when large-area LEDs excite high-order modes that leak out in the first few hundred meters of multimode fibers.

6. There are trade-offs among received power, speed, and bit-error rate or signal-to-noise ratio for receivers, but these often are not useful.

7. Total loss from connectors, couplers, and splices is their characteristic loss multiplied by the number in the system. This can be either the worst case, calculated by multiplying maximum loss by number, or the most likely, calcu-lated by multiplying average loss by number of devices.

8. System margin is a safety factor to allow for aging of components and system modifications and repairs. Typical values are 3-10 dB.

9. Bandwidth budget calculations are not as straightforward as loss budgets, but some components, such as connectors and splices, can be ignored. These calculations are based on the system time response. For a signal to be received correctly, the overall time response of a system must be less than the bit time (i.e., less than the inverse of the bit rate or frequency).

10. Response time of a system is the square root of the sum of the squares of component response times. Calculations must include transmitter and receiver response times and both modal and chromatic dispersion of fibers.

11. Cost is a primary consideration in real-world system design. It is up to the user to make cost-performance trade-offs for specific applications.

12. Installation and operating costs must not be neglected. In some cases, installation may cost much more than the hardware. Small added expenses in hardware may mean major savings in installation.

13. Common sense is vital in making cost-performance trade-offs. Always try to anticipate future events, ranging from expansion to ways to repair a system in case of failure.

What's Next?

In Chapter 17, I will look at the use of fiber optics in long-distance telecommunications systems.

Quiz for Chapter 16

1. A large-area LED transfers 10 μW (10 dBμ) into an optical fiber with core diameter of 100 μm and numerical aperture of 0.30. What power should it couple into a fiber with 50 μm core and NA of 0.2?

   a. 10 dBμ.

   b. 9.5 dBμ.

   c. 3 dBμ.

   d. 1.0 dBμ.

   e. 0.4 dBμ.

2. A connector is specified as having loss of 0.6 dB ±0.2 dB. What is the maximum connector loss in a system containing five such connector pairs?

   a. 0.6 dB.

   b. 3.0 dB.

   c. 4.0 dB.

   d. 5.0 dB.

   e. None of the above.

3. A 10 Mbit/s signal must be sent through a 100 m length of fiber with eight connector pairs to a receiver with sensitivity of -30 dBm. The fiber loss is 4 dB/km, and the average connector loss is 1.0 dB. If system margin is 5 dB, what is the minimum power that the light source must couple into the fiber?

   a. -13.0 dBm.

   b. -13.4 dBm.

   c. -16.0 dBm.

   d. -16.6 dBm.

   e. -20.0 dBm.

4. A telephone system is designed to transmit 565 Mbit/s through 50 km of cable with attenuation of 0.4 dB/km. The system contains two connector pairs with 1.5 dB loss, a laser source that couples 0 dBm into the fiber, and a receiver with sensitivity of -34 dBm. How many splices with average loss of 0.15 dB can the system contain if the system margin must be at least 8 dB?

   a. None.

   b. 10.

   c. 20.

   d. 30.

   e. 40.

   f. None of the above.

5. A fiber-optic network uses a transmissive star coupler with excess loss of 1 dB to distribute signals to 30 terminals. The signals must travel through 200 m of fiber with 5 dB/km loss and through four connector pairs with 0.6 dB loss. If receiver sensitivity is -30 dBm and system margin 5 dB, how much power must the light source couple into the fiber?

   a. +9.4 dBm.

   b. 0.0 dBm.

   c. -5.8 dBm.

   d. -24.2 dBm.

   e. None of the above.

6. The time response of a system transmitting a 400 Mbit/s signal must be less than

   a. 1 ns.

   b. 2 ns.

   c. 2.5 ns.

   d. 4 ns.

   e. none of the above.

7. Response time of a fiber-optic system is

   a. the square root of the sum of the squares of the response times of components.

   b. the sum of the response times of components.

   c. the average of the response times of the components.

   d. not directly dependent on component response times.

8. What is the time response of a system using an 850 nm LED with 10 ns rise time to transmit though 2 km of 100/140 fiber with 50 ns/km modal dispersion to a receiver with 25 ns rise time? Neglect chromatic dispersion.

   a. 100 ns.

   b. 104 ns.

   c. 135 ns.

   d. 200 ns.

   e. None of the above.

9. Look again at the preceding example to calculate total chromatic dispersion along the system. Assume a value of chromatic dispersion of 110 ps/km-nm and an LED bandwidth of 40 nm. What is the chromatic dispersion?

   a. Under 1 ns.

   b. 4 ns.

   c. 5 ns.

   d. 8.8 ns.

   e. None of the above.

10. Your system design requires a transmitter that delivers -13 dBm into a fiber. You have a choice between two fibers—one that costs $0.50/m and requires a $100 transmitter and one that costs $0.60/m and can operate with a $30 transmitter. All other things being equal, using the more costly fiber will save money for distances shorter than

    a. 70 m.

    b. 200 m.

    c. 500 m.

    d. 700 m.

    e. 1000 m.

    f. none of the above.

Long-Distance Telecommunication Applications

About This Chapter

Now that you have learned about fiber-optic hardware, it's time to look at its major system applications. New technology and changes in regulations are breaking down traditional barriers, but fiber communications can still be loosely divided into four major realms: long-distance transmission, local telephone systems, video transmission, and data communications. In this chapter, I will talk about long-distance transmission of voice, video, data, and other signals, loosely defining "long distance" as distances of tens of kilometers or more—between major towns and cities, rather than between adjacent suburbs. These fiber systems are the backbones of national and international telecommunication networks.

Basic Telecommunications Concepts

Defining Telecommunications

Before looking at how optical fibers are used in long-haul telecommunications, I want to clarify what I'm talking about. The term "telecommunications" is deliberately broad. It dates back to the era when communication specialists were trying to lump telephones and telegraphs together under one heading. As the

● Telecommunications is a broad term, encompassing voice, data, facsimile, video, and other forms of communications.

● Many telecommunication networks exist; many (but not all) interconnect with the telephone system.

telegraph industry withered away, telephony became dominant, but the new word had caught on—and was useful because new types of communications were emerging. Telex became an accepted way to send messages around the globe. Facsimile systems began transmitting images of documents. Computer data communications grew rapidly. So did video transmission. They all fell under the broad heading of telecommunications.

Types of Telecommunications

Different types of telecommunications had different beginnings, and have evolved differently, but they are now converging. The reason is simple and compelling—it costs much less to build one common network than to build many separate ones. In practice, you can't always send many different signals over the same network; the phone wires that reach your home cannot carry cable television signals, for example. However, savings are possible wherever services can be combined. I'll look briefly at each major type of telecommunications.

THE TELEPHONE NETWORK

● The telephone network evolved into the backbone for the global telecommunications system. It offers worldwide connections.

The telephone network spread around the globe in the twentieth century and became the backbone for the international telecommunication system. Its purpose is to carry conversations between any two phones connected by the network. To do this, it has a network of connections extending to individual homes and offices. The signals from individual conversations are combined at telephone company facilities and switched to their destinations. Once mechanical switches connected electrical wires, but now electronic switches route signals through optical fibers and other media.

The telephone network offers both local and long-distance service; I will talk more about local service in Chapter 18. Telephone numbers are the keys to routing signals within the local network and over long distances. In North America, seven digits suffice to specify calls within an area code (phone companies have had to split some old area codes when they ran out of seven-digit numbers). Adding more numbers lets you call a larger area. In North America, we add three-digit area codes to specify other regions in the United States, Canada, and parts of the Caribbean; callers first dial 1 to specify long-distance routing. Overseas calls are prefixed by 011, a country code, a city code (equivalent to our area code), and the number. Thanks to this system, you can call most of the phones in the world from your home or office. (Although you may regret it when you get the bill.)

Each telephone line does not carry much information. A standard analog telephone line carries sound frequencies of only 300 to 4,000 hertz. That is enough for intelligible conversations, but far short of the 20 to 20,000 hertz range of the human ear. The analog signal is converted to digital format in modern phone systems, with one voice channel equal to 56,000 bits per second in North America.

There has been much talk about digital telephone service, and about a related service called ISDN, for Integrated Services Digital Network. Because these services are digital, they could handle data at higher speeds and with greater efficiency. They could also improve telephone quality. However, they would require new digital telephones, and improvements to the existing telephone network.

TELEPHONE-COMPATIBLE SERVICES

Although the telephone network was designed to carry voices only, its wide reach makes it attractive for carrying other kinds of communications that can fit over phone lines. The two most important examples are computer modems and facsimile machines.

Computer modems encode digital data as analog sounds, which whistle and warble over analog phone lines. Another modem at the other end converts those modem signals back to digital form. This way you can use your home computer and phone line to connect with a computer at work, without installing any new wires. However, the data rate is much smaller than the digital equivalent of a voice phone line. The fastest modems now transmit at 14,400 bits per second, but don't count on getting that much data through ordinary phone lines. Most phone lines can handle 2400 bits per second, and many can handle 9600 bits per second, but errors can occur, and noisy phone lines can make modem transmission difficult.

Fax machines work in a similar way. They digitally encode images, then transmit them as a series of bits. The standard rate is 9600 bits per second, but they can drop to lower speeds if the phone lines won't support that high speed. Like modems, they send the data as analog sounds, which may be converted to digital form for long-distance transmission.

Both fax machines and modems have standardized codes for data transmission, so different models can talk with each other. Fax modems, which attach to personal computers, speak both languages. Use of phone lines lets them reach anywhere the telephone network goes, but it limits their speeds to rates that the phone network can handle.

(In passing, I should note that telephone companies originally did not like the idea of attaching anything but their own telephones to the network. Only in the last few decades have phone companies opened their networks to other equipment—and found that it greatly increased their business.)

> ●
> Computer modems and fax machines send signals over standard analog phone lines.

OTHER DATA COMMUNICATION

Data communications via modem over phone lines is inherently inefficient. Organizations that do much data communications usually have dedicated transmission lines to carry digital data without the need for a modem. A simple example is a local area network (LAN) that ties together several personal computers, a file server, and one or more printers for the people in a work group.

> ●
> Local area networks and other dedicated digital systems carry data more efficiently than phone lines.

Higher levels of data communications are also common. A backbone network may link work groups to a corporate mainframe computer, carrying higher-speed signals. Special-purpose lines may link buildings on a corporate or university campus, or carry data between buildings on different sites. Large corporations may build their own national networks to carry their own data. Some universities and research organizations build their own high-speed fiber-optic systems to meet demanding transmission requirements.

However, in many cases the data may travel part or all of its route on public telephone/telecommunication lines. The digital signals may be directly input to telephone-company switching offices, for routing to other switching offices that serve other corporate facilities. Note the important difference that this data travels entirely in digital form, without modems, from one user to another. Users may send signals at many different rates, typically between 1.5 and 45 megabits per second. Many alternative carriers—other than traditional telephone companies—offer the same services.

VIDEO COMMUNICATIONS

●
Cable television receives signals from distant sources for local distribution.

The best-known video communications today are offered by cable television networks. However, cable television companies are not really in the long-distance business. Operators receive network feeds from long distances, but they deliver services only in local areas. Cable companies are gradually being integrated into the global telecommunication network, but their main role is as local carriers; I will cover them in Chapter 19.

Network television programs are transmitted over long distances from the point where they originate (a studio or news event) to local broadcasters and cable operators. Typically distribution is via satellite, but video signals are sometimes routed over fiber cables. Such signals are distributed one way only, unlike telephone signals, which are two-way communications. Cable operators with many franchises in the same area may link facilities in different towns via fibers.

●
Videoconferencing and video telephones are two-way systems.

There is a growing market for two-way video communications, for videoconferencing or video telephones. Extensive data compression can squeeze color pictures to fit on a single telephone voice channel, but the results suffer; the images are small, and you see a series of still pictures, rather than true motion. Higher-grade videoconferencing systems use multiple telephone channels and more sophisticated compression signals, giving better quality images and motion. However, they cost a lot more and are used only in business applications. (At this writing, home video telephones are also very expensive by telephone standards; you might consider them toys for rich grandparents.)

SPECIAL-PURPOSE COMMUNICATIONS

Some telecommunication systems don't fit neatly into the categories I described. Military organizations have separate communications networks designed to withstand hostile conditions at military bases. Utilities and railroads often have dedicated networks to meet their special needs, such as signaling for trains or monitoring operation of power lines. These special-purpose networks usually connect somewhere to the general telecommunication network. Some of them also offer commercial transmission services over surplus capacity—i.e., extra optical fibers laid along a utility right of way.

POINT-TO-POINT AND BROADCAST COMMUNICATIONS

I should stop to recall one fundamental division in telecommunication systems mentioned at the start of Chapter 15. Some systems transmit signals between pairs of points; others send signals from one point to many points. Satellites and radio transmitters are good at point-to-multipoint transmission, because they inherently spread their signals over broad areas. Cables are good at point-to-point transmission, because they deliver signals only where they are supposed to go to.

However, cables and radio can cross into each other's domains. Current cable television technology delivers the same signals to subscribers, in a sense "broadcasting" them through the cable, which delivers a stronger signal that is less noisy than one transmitted through the air. Conversely, cellular phones use radio signals to make point-to-point connections with moving vehicles and people.

● Point-to-point and broadcast communications are transmitted in different ways.

The Global Telecommunications Network

The pieces I have described so far are linked together, as shown in Figure 17.1, to form the global telecommunications network. Although many components were built separately, they have become increasingly interconnected because connectivity is vital to communications. That trend is sure to continue as all forms of communication spread.

● Diverse communications systems have been interconnected to form a global network.

The global telecommunications network operates on many levels and through many media. At the highest level, it connects national telecommunication networks. Submarine cables and satellite links cross oceans, linking continents with each other, with speeds of hundreds of megabits per second, or a few gigabits per second. Long-distance land cables connect adjacent countries. The nature of international links varies greatly, depending on geography, history, politics, economics, and a host of other factors. Closely allied countries, like the United States and Canada, are closely tied (to the extent that they share the same system of area codes). The countries of western Europe likewise are closely tied. On the other hand, only one submarine cable links the United States and Cuba, reflecting three decades of political hostility. Land and submarine cables carry traffic along common routes, such as between London and New York. Satellite links connect other places with lighter traffic, like isolated islands, as shown in Figure 17.2. (Satellites are also preferred to distribute signals to many points, as in broadcasting television signals.)

FIGURE 17.1.

Interconnection of telecommunication networks.

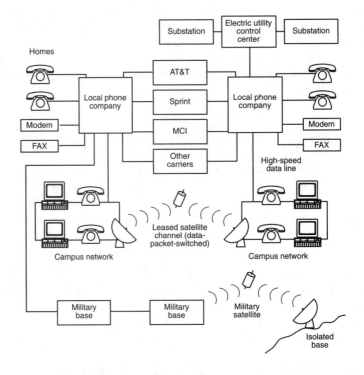

FIGURE 17.2.

Different media serve different long-haul transmission needs.

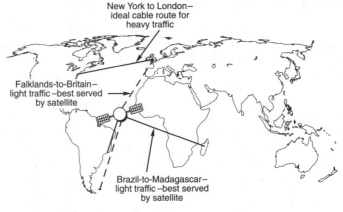

Fiber optics are the backbones of national telecommunication networks operating at speeds to gigabits per second.

National telecommunication networks differ in scale with the sizes of countries, but also usually operate at hundreds of megabits to gigabits per second. The United States network rivals that of western Europe in size, for example. Submarine cables play only minor roles in national networks (except for countries made up of many islands). The backbones of the switched telecommunication networks that carry telephone and other traffic are land-based fiber-optic cables. Satellites broadcast signals to multiple points (particularly video feeds to cable television operators), and carry limited other signals, particularly data or traffic to remote sites. Land-based microwave radio and coaxial cable systems are becoming rare, made obsolete by high-speed fiber-optic systems.

Regional networks are the next step down. In the United States, high-speed fiber-optic cables carry signals between regions at speeds from 400 Mbit/s to 2.4 Gbit/s. They often link major cities, then spread out to serve the surrounding area. Typically they branch out from points on the national backbone system, as you will see later.

Local telecommunication networks branch out from regional systems, and in practice can be complex. As you saw earlier, there are actually multiple local networks that are to some extent linked together. I won't talk much about local broadcasting of television and radio signals, or about cellular telephones, but in Chapters 18 and 19 I will cover the major cable-based services: telephone and cable television. As you will see later, the technology behind those two services is converging—thanks in part to fiber optics—and in coming years telephone and cable television networks will come to resemble and compete with each other.

There are also many other types of competition in the global telecommunications network. The United States has competing long-distance telephone carriers. Satellites and submarine cables compete for transatlantic traffic. Corporate structures are complex, but this is not a business book, and I'm not going to delve into that. Instead, I will look at the technical problems of merging many communication services into a single network.

Putting Networks Together

Putting the pieces together to make a single functional network requires combining many signals from diverse sources. There are two crucial concepts involved: multiplexing or combining many signals for transmission through a single carrier, and conversion of signals into a common format. These can be addressed in various ways, but the basic concepts and reasoning are the same. It's cheaper and easier to combine signals traveling the same path than to send them separately. Likewise, all the signals traveling through the system have to be converted to the same format.

You can compare the telecommunication network with the circulation system of your body. Blood flows from tiny capillaries to larger veins, which in turn feed larger veins that can carry more blood. After the blood passes through your heart and lungs, it is divided up into smaller and smaller arteries, and ultimately reaches the tiny capillaries. Voice telephone lines are the capillaries of the old telephone network, upon which the modern telecommunication system is based. Low-speed lines feed into higher-speed systems, which go longer distances. There is a standardized hierarchy of transmission rates.

All the information handled by any transmission system has to be translated into the same format. As you will see later, that format is changing as the network evolves, but it has maintained a strong compatibility with older hardware. With a pair of wire cutters and a screwdriver, you can attach a massive 1950-vintage dial phone in basic black to the same

Local telecommunication networks branch out from regional systems.

Signals must be combined and put into a common format for transmission through the telecommunication network.

standard analog telephone line as a high-technology combined fax-modem and answering machine.

The Digital Telephone Hierarchy

● Telephone-based systems are organized in a hierarchy of transmission rates.

The digital transmission hierarchy is itself evolving, with a new family of standards emerging. The old standards, still in wide use, are the North American digital telephone hierarchy of Figure 17.3, which was largely established in the days when AT&T was the only telephone company. The figure shows how low-speed signals (in units of voice channels) are multiplexed to higher speeds at each step of the hierarchy. In practice, all those slots in the higher-speed signals may not be filled. For example, rural areas may have too few telephone subscribers to use all the available slots. In addition, most phone companies now skip directly from the 1.5 Mbit/s DS1 rate to the 45 Mbit/s DS3 rate, not bothering with the intermediate DS2 rate.

FIGURE 17.3.

Multiplexing in the North American digital telephone transmission hierarchy.

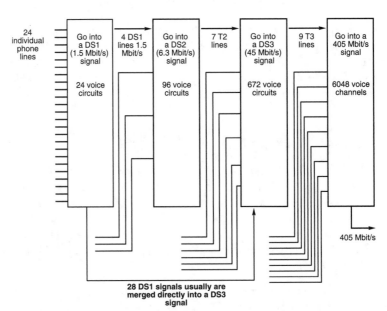

INTERNATIONAL STANDARDS

● Different digital telephone hierarchies are used around the world.

The North American digital hierarchy was never a global standard. The International Consultative Commission of Telephone and Telegraph, an arm of the International Telecommunications Union, known as CCITT from the French version of its name, set its own family of standards built around digitization of single voice circuits at 64,000 bits per second (rather than 56,000 in North America). It also multiplexed different numbers of channels together, getting different data rates. These standards, listed in Table 17.1 along with other standard rates, are used in Europe. (Japan set its own standards.)

Table 17.1. Transmission rates in North America and Europe.

Rate Name	Data Rate	Nominal Voice Circuits
North American Digital Hierarchy		
single circuit	56,000 bit/s	1
T1 or DS1	1.5 Mbit/s	24
T2 or DS2	6.3 Mbit/s	96
T3 or DS3	45 Mbit/s	672
T3C or DS3C	90 Mbit/s	1344
T4 or DS4	274 Mbit/s	4032
400 Mbit/s	405 or 417 Mbit/s[*]	6048
565 Mbit/s	565 Mbit/s[*]	8064 (56 kbit/s equiv.)
810 Mbit/s	810 Mbit/s[*]	12,098
1700 Mbit/s	1700 Mbit/s[*]	24,192
2400 Mbit/s	2400 Mbit/s[*]	36,290
European (CCITT standard)		
single circuit	64,000 bit/s	1
Level 1	2.048 Mbit/s	30
Level 2	8.448 Mbit/s	120
Level 3	34.304 Mbit/s	480
Level 4	139.264 Mbit/s	1920
Level 5	565.148 Mbit/s	7680
SONET/Synchronous Digital Hierarchy		
STS-1/OC-1	51.84 Mbit/s	672 (28 DS1s or 1 DS3)
STS-3/OC-3(STM-1)	155.52 Mbit/s	2016
STS-12/OC-12(STM-4)	622.08 Mbit/s	8064
STS-48/OC-48(STM-16)	2488.32 Mbit/s	32,256
STS-96/OC-96(STM-32)	4976.64 Mbit/s	64,512
STS-192/OC-192(STM-64)	9953.28 Mbit/s	129,024

[*]Actual line rate depends on design and overhead bits and is not standardized.

HIGHER-SPEED SYSTEMS

When AT&T established the North American Digital Hierarchy, no one envisioned sending signals faster than the 274 Mbit/s DS4 rate. Fiber optics quickly passed that mark in the 1980s, but the arrival of competition on the long-distance scene meant that no one entity could set transmission rates in North America. Manufacturers soon began offering systems at higher speeds, first 400 Mbit/s, then 565 Mbit/s and up. As of 1993, the highest speeds in commercial use are 2.4 Gbit/s and 3.4 Gbit/s (the latter achieved by simultaneously transmitting 1.7 Gbit/s at 1310 and 1550 nm). Developers are working on 5 and 10 Gbit/s systems for long-distance transmission.

In the absence of industry-wide standards above the 565 Mbit/s CCITT format, developers have defined their own proprietary formats. This means that actual data rates differ from the nominal values, depending on things like differences in overhead bits added to aid in system operation.

SYNCHRONOUS DIGITAL HIERARCHY/SONET

Unhappy with the chaotic lack of standards, the telecommunications industry wrote a new generation of standards, which operate at higher speeds and can accommodate many kinds of signals besides telephone traffic. This is called SONET, the Synchronous Optical Network, in North America, and SDH, the Synchronous Digital Hierarchy, by CCITT. (I mentioned these earlier, in Chapter 15.) The two standards are functionally virtually identical, although they assigned different number levels to transmission rates, as shown in Table 17.1. They can be extended to higher transmission rates; I show up to 10 Gbit/s, the highest proposed rate.

If you look carefully, you will note that the basic STS-1 SONET signal carries as many phone lines as a DS3 signal, but transmits more bits per second. The basic SONET rate was designed to have the same capacity as a DS3 carrier. The bit rates differ because more overhead information is added to the SONET signal to allow the system to monitor its transmission and direct signals properly. Although that may make SONET transmission seem less efficient, the high capacity of fiber optics makes it less important to pack as much signal as possible onto the line.

FITTING INTO THE HIERARCHY

Although it was designed to transmit only telephone signals, the digital telephone hierarchy can handle other signals—as long as they're converted into a compatible format. That's what fax machines and modems do automatically—convert digital signals into analog tones compatible with the phone system. Digital signals can also be transmitted at higher speeds, including the 56 kbit/s equivalent of a voice channel, and the 1.5 or 45 Mbit/s rates used for multiplexed phone signals.

The problem—from the user standpoint—is that signals must fit into one of those slots. You can lease a 1.5 Mbit/s line, but not a 2 or 3 Mbit/s line. Nor is the system designed to accommodate "bursty" data like computer file transfers, which ideally should be sent rapidly, but require line capacity only rarely. It's like a shoe store that offers only three sizes; they may meet the most common requirements but won't fit everybody. It became clear that telecommunication networks needed to become more flexible.

New Multilevel Standards

Telecommunications organizations are developing a new family of standards that offers much more flexible services. The SONET transmission speeds described previously are only one part of those standards. I won't bore you with the details (standards documents can combat insomnia), but I will stop to review the basic concepts, many of which were covered in Chapter 15.

● Emerging tele-communication standards are designed to accommodate many services.

The new standards are written in "layers," as shown in Figure 17.4. The upper layers are the services that users "see"—like voice, video, or data transmission. The middle layers are interfaces and switching formats used within the network. The lowest layer (called the physical layer) specifies the format that is actually transmitted.

● New standards are written in layers.

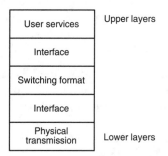

FIGURE 17.4.

Layers in a telecom-munication standard.

A signal enters the system through user services, then is converted into the switching format (Asynchronous Transfer Mode in the B-ISDN standard), and then is converted again into the physical transmission format (SONET in B-ISDN) at the transmitter. The process is reversed at the other end. Ideally, the user need not worry about the details but sees only the upper layer.

The goal of these standards is to let a single network handle many different communication services, much more flexibly than the digital telephone hierarchy. I covered how in Chapter 15, but I should highlight the differences between the new system and the old digital hierarchy.

Multiplexing Differences

B-ISDN systems and the digital telephone hierarchy combine signals in different ways.

The digital telephone hierarchy can be adapted to carry non-telephone signals, but it's a force-fit, not an easy one. The new standards were designed for general-purpose telecommunications, so they can handle many different formats. Two of the most important differences between the new network standards and the old digital telephone hierarchy are in the way they package data for transmission. Both give the newer networks more flexibility.

Asynchronous Transfer Mode (ATM) is basically a way to package signals in small chunks for transmission. As you saw in Chapter 15, the chunks are small, but their small size makes it easier to even the signal flow through the system. Different interfaces exist that convert different signals into ATM format, and they produce data packets at different rates. The network packages them further for transmission. This lets ATM systems accept data at essentially any rate—not just the fixed levels of the digital telephone hierarchy. It also lets them adjust loads, to even out bursts of data.

SONET also packages data, but in somewhat larger chunks. (You can think of ATM packets as standard containers being packaged into boxcars on a SONET train.) As you saw in Chapter 15, the digital telephone hierarchy simply interleaves lower-speed signals to produce higher-speed signals. To extract one signal, you have to demultiplex down to the level at which it originated. SONET organizes the signals into frames, so one low-speed signal can be switched out of a higher-speed without going back all the way back down. This makes switching much easier.

Although SONET and ATM were developed as part of the same B-ISDN standard, they do not have to be used together. ATM packages can be loaded onto other transmission trains, including the digital telephone hierarchy. SONET can also accept other inputs.

Submarine Cables

Submarine fiber-optic cables offer long-distance transmission at high speeds with few repeaters.

The best way to understand the global telecommunications network is to look at some examples. I'll start with the longest links in the system, submarine cables, then move to national networks in this chapter. Chapters 18, 19, and 20 will describe smaller-scale systems.

Submarine cables come in many types. Some cross the few kilometers of seawater separating an island from the mainland; one of the first to use fiber was an 8-kilometer cable from Portsmouth, England to the Isle of Man off the English Coast. Many cross tens or hundreds of kilometers of sea; the Mediterranean and Caribbean seas are crisscrossed with submarine cables. Some span thousands of kilometers of ocean; the longest run across the bottom of the Pacific. Figure 17.5 shows the biggest planned and installed submarine fiber cables.

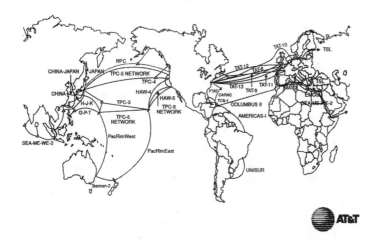

FIGURE 17.5.

Major submarine fiber cables. (Courtesy of AT&T)

Fiber optics became the standard choice for submarine cables in the 1980s because they neatly fit system requirements. Transmission capacity should be as high as possible because cables are costly to make, lay, and operate. The distance between repeaters should be as long as possible, because submerged repeaters are costly and are potential trouble points that are extremely hard to fix under the ocean. The cable should transmit digital signals cleanly to be compatible with modern equipment. Those specifications veritably call out "fiber optics."

The History of Submarine Cables

TELEGRAPH CABLES

The need for electrical cables first arose with the electric telegraph in the early 19th century. It wasn't long before engineers laid waterproof cables underwater to carry telegraph signals. In 1850, the first submarine cable was laid in the open ocean, between Britain and France, but it carried only a few messages before a fisherman caught it and hauled it to the surface. He thought it was a rare type of seaweed!

The first transatlantic cable carried telegraph signals more than a century ago.

That experience taught submarine cable engineers an important lesson—waterproof isn't enough. The biggest dangers to cables in shallow waters are fishing trawlers and anchors. The next telegraph cable laid between Britain and France in 1851 was armored with ten 7 mm wires of galvanized iron. That protected it from fishermen well enough that it worked for many years.

The next bold step was to lay a cable across the entire Atlantic Ocean, which is far deeper and wider than the English Channel. Two efforts failed in 1857, but the following year a third worked for about three weeks. However, the principles of long-distance telegraphy were not well-understood at the time, and an operator fried the cable by putting 2000 V across it. The Civil War and other problems slowed progress, and it was not until 1866 that the first transatlantic telegraph cable was working regularly.

TELEPHONE CABLES

● Submarine telephone cables were not made until the mid-20th century because they needed submerged electronic repeaters.

The telephone came along soon afterward, but submarine telephone cables were much harder to build. One key problem was the need for a repeater to boost telephone signals so the cable could carry them long distances. Mechanical relay amplifiers could handle the dots and dashes of the telegraph, but electronic amplifiers are necessary to handle voices. The required electronic technology wasn't available until well into the 20th century. Instead, the first transatlantic phone calls were made by short-wave radio signals, or "wireless," which could travel long distances because it bounced off the atmosphere, but was known for its crackle, static, and fade-out.

Britain laid the first submarine telephone cable with an underwater repeater in 1943 between Holyhead and Port Erin. The first such cable in North America—and the longest-operating—was laid between Key West, Florida, and Havana, Cuba, in 1950; it operated until 1989. The first transatlantic telephone cable, the TAT-1 (TransATlantic-1) cable between Britain and Canada, began operation in 1956.

● Transmission capacity and repeater spacing were limited in submarine coaxial cables.

Those early telephone cables were made of coaxial cable, which offers the highest bandwidth of any metal cable. However, coaxial cable has limitations. Its attenuation increases with the square root of transmission frequency v and decreases as the inside diameter D of its outer conductor increases:

$$\text{ATTENUATION} = C \times v^{1/2}/D$$

where C is a constant depending on cable characteristics.

Increasing transmission frequency raises cable capacity but requires either a smaller repeater spacing (to compensate for higher losses) or a larger cable diameter (to reduce transmission losses). Either one is a problem. Repeaters are costly, can fail, and require electrical power. Thicker cables are hard to lay in the ocean.

The TAT-1 cable, which carried 36 telephone circuits, was 1.6 cm in diameter and had repeaters 70.5 km apart. The last transatlantic coaxial cable, TAT-7, which was put into service in 1983, is 5.3 cm in diameter and requires a repeater each 9.5 km. Speech interpolation techniques can crowd 4200 analog phone circuits onto the cable, but that is the practical limit of coax technology.

SATELLITES VERSUS CABLES

Communications satellites arrived in the 1960s. The first transatlantic communications satellite, Intelsat I ("Early Bird"), was launched April 6, 1965. It could carry 240 phone calls, nearly twice as many as any extant transatlantic cable. By 1970, satellites offered more transatlantic voice circuits than cables, and cables were in trouble.

Cable developers were not about to concede defeat to satellites, but coax had reached its limit. Satellite channels have some disadvantages. Radio waves take a quarter of a second to make the round-trip to and from geosynchronous orbit 37,000 km (22,000 mi) above the earth. That delay is barely perceptible, but it can be annoying if a telephone call makes two bounces. Microwave transmission to and from satellites can be intercepted by electronic eavesdropping or spy satellites. Phone calls sent over analog satellite channels can suffer echoes that sound louder than the person at the other end.

Perhaps even more important was the fact that satellites and cables were traditionally owned and operated by different organizations. Most submarine cables are owned and operated by international consortia of telecommunication companies and government communication authorities (e.g., AT&T, British Telecom, and the German Post Office, which is responsible for that nation's telecommunications). Most satellites are owned and operated by the International Telecommunications Satellite Organization (Intelsat) or by private companies. Cable operators couldn't simply shift to satellites; they would be forced out of the international telecommunications business if they gave up altogether on cables. To avoid that fate, they started developing submarine fiber cables. They were impressively successful, and the technology continues to evolve rapidly.

> ● Satellites offered many advantages over coaxial cables, forcing cable developers to seek a different technology.

Types of Undersea Cables

Before I start looking at submarine cable designs, you should recognize that not all undersea cables are alike. They can be divided into three broad categories: short systems that are typically part of a national telecommunications network, moderate-distance systems connecting different countries but not crossing entire oceans, and transoceanic systems spanning thousands of miles (or thousands of kilometers). Each has somewhat different characteristics.

Short systems are typically 100 km or less, and link islands with nearby continents or other islands. Examples include cables running between islands in the Japanese archipelago. Although some early fiber systems used repeaters, designers generally prefer to avoid repeaters. From the network standpoint, these cables are merely links in a national network that happen to run under water. Some countries have linked coastal cities with submarine cables because they are cheaper than land cables.

Moderate-distance cables run to over 1000 km and include repeaters or optical amplifiers. Many cross the Caribbean, Mediterranean, and North Seas; others link islands in

> ● There are three type of submarine cables: short unrepeatered, moderate-distance repeatered, and long-distance repeatered transoceanic.

countries like Indonesia and Malaysia. These are significant links in the global telecommunications network, with the amount of traffic depending on the areas served. Typically designers include extra capacity for future expansion.

Transoceanic cables run thousands of kilometers between continents and contain many repeaters or optical amplifiers. These are among the most critical links in the global telecommunications network. System requirements have consistently pushed fiber technology; innovations such as optical amplifiers and solitons have been developed in large part to meet their needs.

Design of Submarine Cable Systems

Submarine fiber system design has evolved rapidly.

The design of submarine fiber-optic cable systems has evolved very rapidly. Early work on TAT-8, the first ocean-spanning fiber-optic cable, helped launch the rapid spread of single-mode fiber systems operating at 1300 nanometers. It was quickly followed by longer, higher-speed systems, and shifts to the 1550 nm wavelength, where lower attenuation allows longer repeater spacing. A new generation of systems in development will be built around erbium-doped fiber amplifiers and carry signals at speeds much higher than TAT-8.

Table 17.2 summarizes the key design parameters of three important current and planned transoceanic cables. Looking at the dates alone may exaggerate the progress somewhat. AT&T Bell Laboratories, the primary developers of TAT-8, were very cautious in their design, and subjected their equipment to extensive testing. This largely reflects the nature of submarine cables. The initial fiber-optic ones represented huge investments in what at the time was still new technology. Fixing submarine cable is very costly and time-consuming, so the hardware was tested extensively before the cable was laid. Designers were particularly cautious about the electro-optic repeaters that had to be installed on the ocean bottom, largely because early semiconductor lasers had had limited lifetimes. However, much of the technology that had been developed and tested for TAT-8 could be adapted for later systems—with much shorter lead times.

Table 17.2. Submarine cable designs.

System	TAT-8	TAT-10	TAT-12/13
Date Operational	1988	1992	1995
Spans	Atlantic	Atlantic	Atlantic
Data rate	278 Mbit/s	565 Mbit/s	5 Gbit/s
Working Pairs	2	2	1
Repeater spacing	50 km	over 100 km	no repeaters
Wavelength	1300 nm	1550 nm	1550 nm
Technology	E-O repeater	E-O repeater	Optical amps

TAT-8 operates at 1300 nm, with repeater spacing of 50 km and raw data rate of 278 Mbit/s in each of two fiber pairs (a third pair is kept as a spare). That corresponds to capacity of 4000 voice circuits per pair, but speech compression packs five voice channels into each digital phone circuit, giving total capacity of 40,000 voice channels.

The next steps were to higher data rates through each fiber, 565 Mbit/s, and to the 1550 nm, where lower attenuation allows longer repeater spacing, if dispersion-shifted fibers or narrow-line lasers are used to prevent dispersion from limiting bandwidth. The first regular uses of 1550 nm transmission were in repeaterless cables, where transmitters and receivers could be kept above water at the terminal points. This relieved the understandable concerns about putting developmental hardware in submerged repeaters. It also extended repeaterless transmission distances, which can exceed 200 kilometers—much longer than possible at 1300 nm—depending on transmitter design. Some countries are laying short submarine cables along their shores to save the costs of paying for right of way on land. Figure 17.6 shows how Italy is laying submarine cables to connect coastal cities. They transmit at 1550 nm through dispersion-shifted fiber, avoiding repeaters.

⬤ 1550 nm systems carry 565 Mbit/s, with repeater spacing beyond 100 km.

FIGURE 17.6.
Submarine fiber cables link coastal cities in Italy; land lines are also shown. (Courtesy of Corning Inc.)

Rapid improvements in 1550 nm technology soon relieved fears about installing 1550 nm DFB lasers in submerged repeaters. TAT-8 began service in 1988; by 1992, submerged 1550 nm repeaters were carrying 565 Mbit/s signals across the Atlantic in the

TAT-9 and -10 cables, which use standard single-mode fibers. Their repeater spacings are well over 100 km.

• Optical amplifier cables will operate at 2.5 or 5 Gbit/s.

Contracts have already been signed for the next generation of submarine cables, in which erbium-doped fiber amplifiers will replace submerged repeaters. The compelling attraction of optical amplifiers, as you saw in Chapter 9, is that they can operate at essentially any data rate, whereas the data rates of submerged repeaters are fixed by their internal electronics. This will allow upgrading of optical-amplifier systems simply by replacing terminal electronics. The new systems are designed to operate at data rates of 2.5 or 5 Gbit/s. Current contracts call for them to be installed across the Atlantic and the Pacific, and in the Caribbean.

Long-Haul Fiber Systems on Land

• Long-distance telecommunication networks on land interconnect major population centers.

Long-distance telecommunication networks on land, like submarine cables, carry high-speed signals and serve as the "backbone" of the global telecommunication network. However, there are some important differences that affect their function. Most submarine cables provide point-to-point transmission along busy routes (although a few have branch points, and some new systems form rings). Long-haul land networks are indeed networks, linking together many population centers. You saw the Italian network in Figure 17.6. Figure 17.7 shows another such network, the backbone system connecting regions of the United States, separated by several different long-distance carriers.

FIGURE 17.7.

Long-haul fiber systems in the United States. (Courtesy of Kessler Marketing Intelligence, Newport, RI)

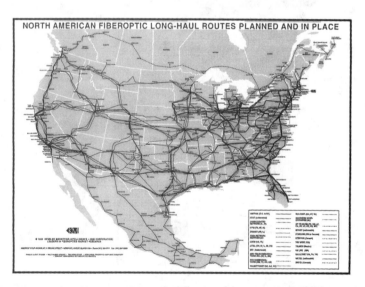

Note also that long-haul land networks do not have the vast uninterrupted cable runs of transoceanic submarine cables. A cable that crosses the Atlantic has no logical stopping points between Europe and North America. However, a fiber cable crossing North America might stop in Pittsburgh, Cleveland, Chicago, Omaha, Denver, and Salt Lake City to distribute signals at those points.

National telecommunication networks also tend to carry more traffic than intercontinental submarine cables. Most business and personal calls are made within the same country, and high costs discourage international calls. A one-minute call from Boston to California costs 24 cents; one from Boston to England costs $1.43. One result is that backbone land cables operate at higher speeds than submarine cable installed through 1993, although that may change with the advent of optical amplifiers.

A critical functional difference is that land systems are much easier to access than submarine cables. Most land cables run along railroad rights of way, highways, or utility lines. A service crew can reach and repair a defective repeater or broken cable much easier than in submarine systems. (Conversely, land cables are also more vulnerable to damage because they can be more easily reached.) The much greater cost and difficulty of repairs mean that submarine cables may be designed more cautiously, to give longer times between failures. A standard submarine cable design calls for no more than two failures that require hauling the cable up from the ocean floor during the 25-year lifetime of the cable.

⬤ Land systems are easier to access than submarine cables.

National backbone systems like the ones shown in Figures 17.6 and 17.7 link with the global network of submarine cables shown in Figure 17.5. The backbone network also distributes signals regionally by connecting to regional networks at each major population center. These regional networks spread out over smaller areas, and may operate at lower speeds, but in other ways resemble national systems.

Backbone Systems

Backbone systems on land generally operate at high speeds. The first American systems transmitted 400 Mbit/s at 1300 nm over standard single-mode fiber, with typical repeater spacings 40 km. During the 1980s, maximum data rates steadily increased, first to 565 Mbit/s, then to 800 Mbit/s and 1.7 Gbit/s. The highest direct modulation rate in use in 1993 is 2.4 Gbit/s, but 10 Gbit/s systems have been demonstrated in the laboratory. Wavelength-division multiplexing of two 1.7 Gbit/s signals gives a total data rate of 3.4 Gbit/s, without requiring electronics operating at that speed.

Most older systems were built with standard single-mode fiber having minimum dispersion at 1300 nm, and this fiber has stayed in place. Indeed, long-distance companies quickly realized the economic advantages of installing cables with several spare fibers, which initially were not connected to any terminal electronics. These "dark fibers" provide spare capacity for future expansion, avoiding the need for costly installations of extra cables at the low initial cost of putting extra fibers into the same cable sheath.

⬤ Standard single-mode fibers were used in older long-haul systems; many cables included spare fibers for future expansion.

Thanks to this planning, long-distance companies have been able to upgrade transmission capacity by replacing terminal electronics or adding electronics to unused fibers as needed. Capacity can be increased with faster transmitters and receivers, by adding

● **Existing cables can be upgraded to higher capacity with new transmitters and receivers.**

another transmission wavelength, or both. Some fibers are still carrying 400 Mbit/s signals, where that capacity is adequate, but new systems are not being installed at that rate. With suitable transmitters, the same pair of fibers that carried 400 Mbit/s at 1300 nm in 1985 can carry 1.7 Gbit/s at both 1300 and 1550 nm—a total of 3.4 Gbit/s. Repeater spacing is left unchanged at 40 to 50 km as long as the 1300 nm wavelength is used; shifting entirely to 1550 nm with narrow-line lasers might allow elimination of every other repeater.

New systems continue to be installed, usually at higher speeds. Some new systems transmit at 1550 nm through dispersion-shifted fiber, which allows repeater spacings of 90-100 km, about double that of 1300 nm systems. The exact distance depends on the loss budget allowed by the transmitter-receiver pair. Standard single-mode fiber has slightly lower loss at 1550 nm than dispersion-shifted fiber. However, dispersion-shifted fibers reduce demands on 1550 nm transmitters, which must emit only a very narrow range of wavelengths for high-speed transmission through standard single-mode fiber.

● **Wavelength-division multiplexing and optical amplifiers can enhance long-distance systems.**

So far, most wavelength-division multiplexing has been with one wavelength in each of the two windows, 1300 and 1550 nm. However, wavelengths can be more closely spaced to expand capacity further. Optical amplifiers can amplify several signals in the 1550 nm window.

Optical amplifiers allow other enhancements to long-haul systems. At the transmitter and receiver ends, they can increase output power and raise receiver sensitivity. They can also replace electro-optical regenerators on the transmission path, avoiding the need for costly and complex repeaters in remote areas, although obviously these concerns are not as serious on land as in submarine cables.

Although the general tendency has been to higher and higher transmission speeds, there is a down side to concentrating traffic on a few fibers. As more than one telephone company has learned the hard way, it takes just one mistake to break a cable—and if that cable is carrying all the traffic to an important location, this can cause big problems. This is a major reason why telecommunications carriers are moving to ring networks, which are better able to survive cable failures without service interruptions.

Future Technology

● **Demands for transmission capacity seem certain to continue to increase.**

Laboratory developers keep pushing transmission speeds and distances higher and higher. At the 1993 Conference on Optical Fiber Communications (OFC), AT&T Bell Laboratories reported sending 10 Gbit/s signals 20,000 km (over 12,000 miles) using optical amplifiers and the soliton transmission scheme described in Chapter 15. That might seem adequate for all practical purposes; the distance is halfway around the Earth, and it's hard to envision any need for cables longer than that. Yet at the same conference, Nippon Telegraph and Telephone reported a somewhat less practical experiment that sent 10 Gbit/s repeatedly around a loop of fiber until it had traveled 100 million kilometers—far enough to reach Mars!

Are things getting out of hand? In part, such experiments are driven by the sheer quest for knowledge and technical expertise. It's a kind of race, and the laboratory tests are called "hero experiments" because it takes a heroic effort to get them to work. Many yield no practical results, but some do. I can remember hero experiments that used technology that's now in practical use.

The real answer is that demands for transmission capacity are rising rapidly, and probably will soar even higher as more new technology is brought into the telecommunications network. Anyone who owns a personal computer with a hard disk knows that information expands to fill the space available—like junk expands to fill closet space. Likewise, you can say, "Communication requirements expand to fill the available bandwidth." New video and data services require more bandwidth. Image transmission is bandwidth-hungry. Someday those trends, like any other, may change, but we don't seem to be near that point yet.

What Have You Learned?

1. Telecommunications is a broad term, encompassing voice, data, facsimile, video, and other forms of electronic (and optical) communications. The global telecommunications network that interconnects with many other systems is an outgrowth of the telephone network.

2. Many separate telecommunication networks exist; many (but not all) interconnect with the telephone system to form a global network that can route signals around the world.

3. Computer modems and fax machines encode digital signals and send them over standard analog phone lines. Their data rates are lower than the 56,000 bit/s digitized equivalent of a phone line. Dedicated digital systems and local area networks carry digital data more efficiently.

4. Cable television receives video signals from distant sources for local distribution. Video conferencing and video telephones are two-way services; most require more bandwidth than a standard analog voice phone line.

5. Terrestrial fiber-optic cables are the backbones of national telecommunications networks operating at hundreds of megabits to a few gigabits per second. Submarine fiber cables and satellite links carry intercontinental traffic at similar speeds. Regional networks branch out from national ones.

6. Many signals are merged together (multiplexed) for transmission over long distances and separated at the other end. The telecommunications industry has a hierarchy of transmission rates, with lower-speed systems feeding into faster ones. The higher-speed systems generally run for longer distances.

7. Different digital telecommunication rates are used around the world, and the standards are evolving. The old North American telephone hierarchy went up to

only 274 Mbit/s, but many systems were built at rates to 1.7 Gbit/s. The CCITT standards used in Europe are somewhat different. Both are being replaced by SONET/SDH standards that go into the gigabits per second range.

8. The old telephone hierarchy requires signals to have data rates matching one level of the standard. SONET/SDH systems are more flexible and can handle many different data rates.

9. B-ISDN (Broadband Integrated Services Digital Network) standards are written in layers. SONET covers physical data transmission. The related Asynchronous Transfer Mode (ATM) is a common format used for switching data signals from many different sources.

10. Submarine fiber-optic cables can send signals long distances at high speeds with few repeaters. The first transatlantic fiber-optic cable operated at 276 Mbit/s with 50 km repeater spacing. Systems now being planned will span the Atlantic entirely with 1550 nm optical amplifiers.

11. Terrestrial fiber-optic cables in national telecommunication systems operate at data rates of 400 Mbit/s to 2.4 Gbit/s. Some carry two 1.7 Gbit/s channels at separate wavelengths, for total speed of 3.4 Gbit/s.

12. Telephone networks are designed for future expansion. Fiber systems allow for upgrades by including spare fibers in cables and by upgrading transmission speed on operating fiber pairs. Many have been upgraded from 400 or 565 Mbit/s to 1.7 Gbit/s by replacing transmitters and receivers. Wavelength-division multiplexing and optical amplifiers can enhance long-distance systems.

What's Next?

In Chapters 18 and 19, I'll look at local telephone and cable television networks, which connect individual subscribers to telecommunication services. Once quite distinct in function and technology, the two are converging, and may ultimately be hard to tell apart.

Quiz for Chapter 17

1. Which types of signals travel onthe global telecommunications network?

    a. Voice telephore.

    b. Digital data.

    c. Facsimile.

    d. Video.

    e. All of the above.

2. How are signals carried on the global telecommunications network?

    a. They are digitized and multiplexed to produce high-speed combined signals that can be routed long distances.

b. Analog and digital signals are carried on separate networks.

c. Local networks feed signals to regional networks, which route them to national backbone systems and international lines.

d. a and c.

e. All of the above.

3. What is the highest rate in the North American Digital Hierarchy?

a. 45 Mbit/s.

b. 274 Mbit/s.

c. 400 Mbit/s.

d. 1.7 Gbit/s.

e. 10 Gbit/s.

4. Which of the following is not a SONET/SDH transmission rate?

a. 45 Mbit/s.

b. 51.84 Mbit/s.

c. 155.5 Mbit/s.

d. 622 Mbit/s.

e. 2.4 Gbit/s.

5. How does signal combination differ in SONET systems and the North American digital telephone hierarchy?

a. The only difference is in data rates.

b. SONET can accept signals that do not exactly match rates in its hierarchy; the digital hierarchy requires signals at lower rates in the hierarchy.

c. SONET is completely incompatible with the digital hierarchy.

d. SONET cannot be used at low speeds.

6. How are B-ISDN, SONET, and ATM related?

a. They are three completely different standards.

b. B-ISDN is an international standard; SONET and ATM are North American equivalents.

c. SONET and ATM are different layers in the B-ISDN standard.

d. ATM and SONET are incompatible.

7. Which best describes Asynchronous Transfer Mode?

a. A way to package digital signals in small chunks for switching and transmission.

b. A standard detailing requirements for optical transmission at 1550 nm.

c. An analog-to-digital conversion technique.

d. Data transfers that have gotten out of synch.

8. What advantages do optical fibers have over coaxial cables for submarine transmission?

a. Longer repeater spacing.

b. Smaller cable size and greater flexibility.

c. Higher transmission capacity.

d. All of the above.

9. A 400 km submarine fiber cable
   has repeater spacing of 50 km.
   How many undersea repeaters
   must it include?

   a. 7.

   b. 8.

   c. 9.

   d. 10.

   e. None of the above.

10. What must be done to upgrade a
    400 Mbit/s fiber-optic system to
    transmit 1.7 Gbit/s?

    a. The fiber must be replaced
       and transmission wave-
       length changed to 1550
       nm.

    b. Optical power must be
       increased from the trans-
       mitter.

    c. The transmitter and
       receiver electronics must be
       replaced with ones operat-
       ing at the higher speed.

    d. The fiber must be cleaned
       and the cable recondi-
       tioned.

    e. Upgrades are impractical.

Regional and Local Telephone Systems

About This Chapter

You saw in Chapter 17 that the global telecommunications network grew largely from the telephone network but now carries many other kinds of signals. Regional and local telephone systems also carry signals other than traditional telephone conversations, but they retain much of their traditional telephone identity, and in this chapter I will look at them separately. Optical fibers are used throughout regional and local telephone service, although so far very few reach all the way to individual homes. I'll explore those uses of optical fibers and also show how the local telephone network is growing more similar to the cable television networks described in Chapter 19.

Basic Concepts

A few decades ago, it was simple to define the telephone system because all it did was connect telephones. The modern telephone system is considerably more complicated, although it maintains much of the same structure and features (and can still use the heavy black dial phone your grandmother has used for the last 40 years). You've already learned generally about the global telecommunications network; now I'll focus on the local telephone system and related concepts.

UNDERSTANDING

Telephone Networks

The telephone
network connects
individual phones
through cables and
switching equip-
ment.

The telephone network is built of local and regional components. The telephone wires in your home connect to a local switching office or automatic switching equipment, probably only a few miles away (except in rural areas). Switching centers, or "central offices" in telephone jargon, typically serve 25,000 to 50,000 phone lines, the equivalent of a large town or small city. If you call the local pizza shop, the switching office routes your call to that phone line without sending the signal beyond your local area, as shown in Figure 18.1. If you're calling someone in the next town, it passes the call on to that town's central office, which routes it to the proper person. If you're calling further away, it routes the signals to a long-distance carrier, which connects the call through the distant switching office, also shown in Figure 18.1.

FIGURE 18.1.

Routing telephone calls.

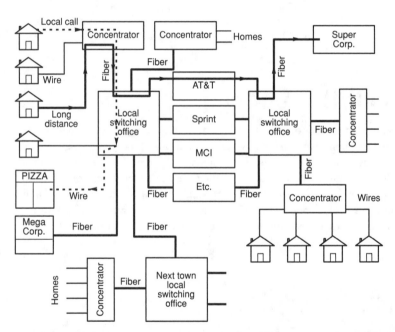

The calls pass from your home on wires to a concentrator in your neighborhood, which typically serves a few hundred phone lines. The concentrator multiplexes your calls with phone signals from your neighbors and combines them into a digital signal transmitted through optical fibers to the switching center. (The details may differ somewhat in older installations, but phone companies are moving toward this model in new systems.) The switching center routes local calls in its area and multiplexes other signals together for transmission to other switching centers or to long-distance carriers.

The telephone industry has its own terminology for this distribution system. Each phone line is a voice circuit. Cables from the central office to the subscriber are part of the local loop or subscriber loop. Those carrying many voice circuits between telephone switching facilities are trunks. Virtually all trunk cables are fiber-optic and operate at 45 Mbit/s and above.

Originally, the telephone network carried entirely analog signals, which were routed by banks of mechanical switches. Today, the analog signals from your phone line are converted to digital format at the concentrator or switching office. All signals from home phone lines are analog; even modems and fax machines convert their digital signals to analog format to go over the first part of the phone network. (The word "modem" originated as modulator-demodulator.) Switches and other digital phone equipment process and route the digital signals, which travel through the entire system in digital format before being converted back to analog form for the receiving telephone.

●
Analog phone signals are converted to digital form for switching.

Private Lines and Bypass

The telephone network described earlier is geared to serving individual telephone lines. Those services are fine for homes or small businesses, but larger businesses need more services. These might include the following:

●
Larger businesses may require special services such as private and high-speed transmission lines.

● A transmission line directly from the central office to a corporate telephone-switching system that provides voice mail and perhaps data services. The company may lease one or more 1.5 Mbit/s T1 lines to carry multiplexed signals directly to and from the switching office. Larger companies may need 45 Mbit/s T3 lines. The company's phone system serves essentially the same role as the phone-company concentrators shown in Figure 18.1.

● A cable to carry signals directly from company offices somewhere else, without using the local telephone system (e.g., between two large plants separated by a few kilometers, from a plant to a satellite dish across town, or from a plant directly to a long-distance phone network).

● A transmission line capable of carrying digital data directly between company computers and telephone-switching facilities, without the degradation inevitable with modems and digital-to-analog conversion. Typically, signals are carried at the 1.5 Mbit/s T1 rate, the 45 Mbit/s T3 rate, or the 100 Mbit/s rate of the FDDI network standard.

A transmission system leased from the phone company is called a private line. If somebody else owns and operates it (either the company or a third party), it's called a bypass because it bypasses phone company lines. Some companies may contract with independent companies that operate their own urban telecommunication networks, sometimes called metropolitan area networks, which typically carry digital data as well as (digitized) phone calls. The terminology can get complicated, and it is still evolving.

Why bother with private lines or alternative carriers? They can be valuable either to save money or to get better transmission—or sometimes both. By going directly to digital signals, the company can transmit more data faster than otherwise possible, with fewer errors; it can also improve the quality of voice phone service. Digital phone lines are also compatible with advanced voice-messaging systems.

Integrated Services Digital Network (ISDN)

ISDN is a digitized switched service that can operate on phone lines.

The Integrated Services Digital Network, or ISDN, is a concept that has been "just around the corner" for an embarrassing number of years. The idea is to digitize voice and data transmission signals right at the subscriber's home or office, combining them into a single stream of bits that could be switched by the telephone network just as digital phone signals are now. The standard calls for transmission of 144 kbit/s, including two 64 kbit/s channels (equivalent to digitized voice signals), and one 16 kbit/s channel (intended for data signals). The twisted wires that carry present analog phone signals to homes should be capable of carrying ISDN, although some may require conditioning.

One big attraction of ISDN is that it digitizes signals right at the start, so they can be switched efficiently through the entire telecommunication network. Another is that it gives customers a digital pipeline that can carry any type of digitized signal at that rate. What has delayed implementation of ISDN is the need for special equipment both in the telephone switching office and at the customer end. Only recently have telephone companies installed the equipment they need to offer ISDN services to many customers, but that's the easy part. To take advantage of ISDN services, people have to buy equipment that costs several hundred dollars and offers few dramatic advantages for ordinary phone users. (Bigger benefits come with the addition of other services, such as high-speed data communications or video telephones.)

ISDN has its passionate advocates, but it has yet to catch on in North America. It is in danger of being preempted by a family of more advanced services that both telephone and cable television companies hope to offer in the near future.

Advanced Services

Telephone and cable television companies both want to offer advanced voice and video services.

Federal regulations have long kept telephone companies out of the cable television business, except in rural areas or in places where they do not provide phone service. However, technological advances, particularly in fiber optics, are pushing both phone and cable companies toward offering a family of "advanced services." These new services include choices of many more video programs than now available (as described in Chapter 19), data communications, and wireless communications (either cellular phones or personal communication services that serve much the same purpose), as well as services now offered over the wired telephone network.

I'll talk a little more about these advanced services later in this chapter. For now, it is important to realize that they could greatly enhance the demand for telecommunication services in general and lead to a dramatic expansion of telephone network capabilities—if they take off.

Hardware Realities

One miracle of modern merchandising is the ability to put the same product into several different packages and sell it to different customers looking for different things. If you've been reading carefully, you may begin to suspect that this is happening in fiber-optic communications. Do telephone companies and the companies offering bypass services buy the same type of hardware to carry signals across town? Of course they do. Is there really a difference between a trunk line laid in telephone company ducts and one operating at the same speed along an electric utility's high-voltage power lines? Yes, but the difference is mostly in the packaging. The same type of fiber might be used in different cables; the fiber is immune to electromagnetic interference caused by the power lines, but the transmitters and receivers would require shielding.

In this book, my concern is the realities of hardware, and that's what I'm going to talk about for the rest of this chapter. I'll look first at the trunk lines that carry multiplexed signals between switching offices, then at the fiber systems that are extending closer and closer to homes. I will close by looking toward the future, and the likely convergence of phone and cable television networks. Except for the convergence of phone and cable systems, I won't worry much about who offers the services or operates the equipment.

● Many different telecommunication networks can use the same fiber-optic hardware.

Trunk Systems

The basic requirements for trunk transmission are to carry multiplexed signals at 45 Mbit/s and up between switching offices, with the higher speeds growing more common. However, there can be considerable variation in what "between switching offices" means. In urban and suburban areas, switching offices are typically several kilometers apart, whereas rural switching offices are smaller and may be tens of kilometers apart. In areas with reasonable population densities, fiber optics can transmit signals between switching offices without repeaters.

● Trunk systems operate at 45 Mbit/s and up over tens of kilometers.

Several factors pushed the early spread of fiber optics for trunk cables. Early graded-index fibers were the first technology well-suited for operation at the 45 Mbit/s T3 rate, which was convenient for connecting switching offices. Fibers offered higher speeds than other cables and did not require repeaters over the distances typical between switching offices. Fiber-optic systems were much less vulnerable to power surges, water damage, and electromagnetic interference than wire cables. (Fibers themselves are immune to power surges and interference, but transmitters and receivers can be affected.)

Fiber cables quickly gained favor in metropolitan areas because of their small size and high capacity. Most cables are routed under city streets through hollow ducts such as that shown in Figure 18.2. Ducts protect the cables and offer room for future expansion, but they fill up as transmission needs increase. Once no more duct space is left, the phone company must dig up the streets again, and that can cost much more than new transmission hardware.

● Four fiber cables can fit in an underground duct that can hold only one metal cable.

FIGURE 18.2.

Hollow plastic ducts house cables under city streets.

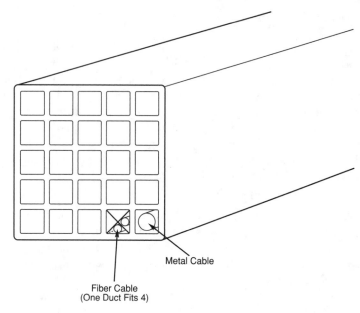

Metal Cable

Fiber Cable
(One Duct Fits 4)

Phone companies found that the small size of fiber cables let them recover duct space. Four standard fiber-optic cables can fit in a duct that would hold a single metal cable with much lower capacity. Extra fibers in the cable leave room for future expansion. By the late 1980s, urban phone companies were pulling working copper cables out of ducts to make room for new fiber cables.

Rural cables are normally installed overhead on utility poles or underground by direct burial. (A tractor with a plow opens a trench into which the cable is lowered, then the trench is filled.) Because cable size is not as crucial a concern as with cable pulled through ducts, it took a little longer for fibers to become accepted for such applications. However, fibers are now the standard medium for trunk cables.

Trunk System Architecture

Fiber trunk technology has evolved rapidly.

The first fiber-optic trunk systems used 850 nm GaAlAs lasers to transmit several kilometers through graded-index multimode fibers. By about 1980, phone companies began installing graded-index multimode fiber systems with 1300 nm InGaAsP lasers. Step-index single-mode fibers operating at 1300 nm soon followed, offering much faster transmission speeds and longer repeater spacing. A fourth generation of systems, operating at 1550 nm, is emerging, but their major advantage—longer repeater spacing—is not as important for trunk systems as for backbone systems, because few switching offices are separated by very long distances.

Table 18.1 summarizes some important features of these systems. The figures are representative of what each generation can offer, but not all systems require the maximum data rate or repeater spacing. You might think that phone companies would pick the simplest technology that meets their needs, but the economics don't work that way. Graded-index fibers are more expensive than single-mode fibers, and—more important—telecommunications companies prefer to leave their options open for future expansion. As you saw earlier, cable-installation economics make it much less costly to install cables with extra fibers, then add transmitters and receivers when more capacity is required. Thus single-mode fibers are standard for trunk installations and for virtually all other applications in the telephone network.

Table 18.1. Typical performance of fiber-optic trunk systems with laser sources.

Generation	First	Second	Third	Fourth
Wavelength	850 nm	1300 nm	1300 nm	1550 nm
Fiber	graded-index multimode	graded-index multimode	step-index single mode	single mode, step-index or dispersion-shifted
Data rates	45 Mbit/s	90 Mbit/s	400 Mbit/s to 2.5 Gbit/s	to 2.5 Gbit/s now; 10 Gbit/s in lab
Fiber loss	3-4 dB/km	1 dB/km	0.35 dB/km	0.25 dB/km
Repeater spacing	8 km	20 km	40-50 km	80-100 km (thousands of km with optical amps)
Limits	Loss, fiber bandwidth	Loss, fiber bandwidth	Loss, source bandwidth	Loss, source bandwidth

Telephone companies normally install aerial, direct-burial, or ducted cables, which were described in Chapter 5. The degree of armor and other protection used depends on system requirements. In lightning-prone areas, for instance, totally nonmetallic aerial cable is preferred. Armored cable is needed for direct-burial applications to protect against gnawing rodents (particularly gophers) in areas where they live.

Special cables are made for electric utilities, as shown in Figure 18.3. These serve two functions: as ground wires for high-voltage power lines and as fiber-optic cables for telecommunication. The outer metal part of the cable serves as a ground wire. Because optical signals are not affected by electric currents, power transmission does not affect the optical

Phone companies install aerial, ducted, and direct-burial cables.

signals carried by fibers in the cable core. Here, too, the major advantage is savings in installation cost, in this case by including optical fibers for signal transmission (at modest extra cost) in a power line being installed to carry electricity. It is even possible to combine electric power and fiber-optic signal transmission in a underwater cable.

FIGURE 18.3.

Cable transmits both electrical power and optical signals.

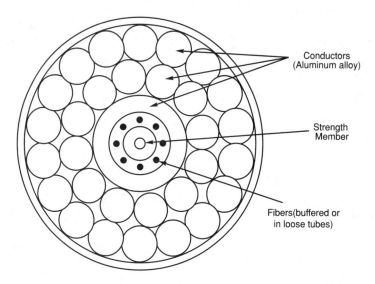

Conductors
(Aluminum alloy)

Strength
Member

Fibers(buffered or
in loose tubes)

Older generations of technology remain in use.

The rapid progress of fiber-optic trunk technology does have some important practical consequences in working with existing rather than new systems. Although multimode trunk systems are no longer being installed, some remain in the field, operating at 850 or 1300 nm. Those systems may no longer be cost-effective to install, but—unlike old copper wire cables—they are not being pulled out and thrown away. The old cables are likely to stay in use as long as they work and provide adequate capacity. Thus you should not blithely assume that all telephone systems use single-mode fiber.

Third-generation systems at 1300 nm remain the most common and are still the type most often installed. However, the push to higher transmission capacity has led to interest in 1550 nm systems. In some cases, the 1550 nm wavelength may be added to an existing 1300 nm system, for wavelength-division multiplexing. In other cases, new 1550 nm systems may be installed for their higher data rates. Eventually, optical amplifiers may be needed for 1550 nm systems, but few interoffice trunks require transmission distances longer than the 80-100 km repeater spacing at 1550 nm.

Fiber in the Subscriber Loop

In the 1980s, the subscriber loop was the backwater of the telephone network. Phone companies installed sophisticated new fiber-optic systems in regional networks and trunk cables; long-distance carriers invested heavily in even more sophisticated fiber systems that spanned continents and oceans. But the part of the phone network between you and the switching office, the subscriber loop, stayed just about the same. Old-fashioned copper wires still carried analog phone signals, just as they had in the days when all phones were heavy black things with dials, and mechanical switches routed signals with an audible clickety-clack.

That picture has changed rapidly in the last few years. Phone companies have turned to fiber-optic systems to carry signals part-way between switching office and individual phones. Look back at Figure 18.1, and you can see a box called a "concentrator" on the path between some phones and the switching office. All the phone lines in a neighborhood feed into the concentrator, where they are digitized and multiplexed. Then the digitized signal is transmitted through wires or optical fibers to the switching office, where it can be routed to its destination. Figure 18.4 shows this in more detail, but still in simplified form.

● Fiber optics are spreading rapidly in the subscriber loop but don't yet reach all the way to homes.

● Fibers are used to connect subscriber-loop concentrators to switching offices.

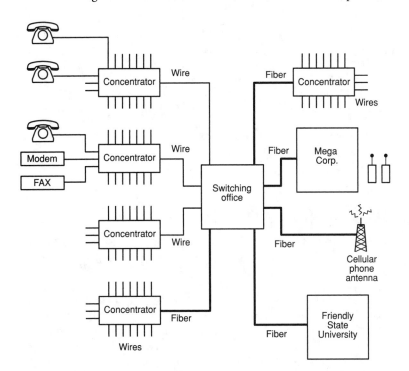

FIGURE 18.4.

The telephone subscriber loop.

Large organizations such as corporations and universities typically have their own dedicated fiber-optic cables that connect their internal phone (and data) networks with the telephone switching office. Other fiber cables connect with the antennas that serve cells in cellular phone systems. I've simplified the drawing for clarity; the details are typically more complex, and the network far more extensive.

In practice, the connection from your home phone to the switching office may go through two levels of concentration, as shown in Figure 18.5. The wires from your home phone go to a serving terminal, probably within a block of your door. Wires from several serving terminals go to a serving area interface, somewhere in the neighborhood. Wires or fibers from that concentrator carry multiplexed signals to the switching office. You can think of the different levels as your home, your block, your neighborhood, and your town.

FIGURE 18.5.

Path of phone signals between home and central office.

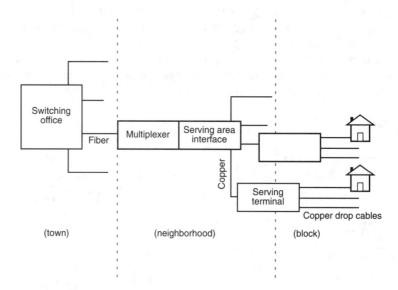

Plain Old Telephone Service

●

The main business of the local phone system remains POTS, Plain Old Telephone Service.

Before I go into the structure of the subscriber loop, I should stress that the basic business of the local phone system is still providing what the industry calls POTS—Plain Old Telephone Service.

Plain old telephone service requires analog bandwidth of 4 kHz to give what telephone companies consider intelligible speech. Intelligible is not high fidelity. Telephones transmit sounds at about 300-4,000 Hz; a good stereo can reproduce sound to the limits of human hearing, 20-20,000 Hz. The 4 kHz bandwidth also includes control signals and the sounds of rotary and touch-tone (push-button) dialing. (By the way, push-button dialing is analog, based on unique pairs of audio-frequency tones transmitted when each button is pushed. Old-fashioned rotary dials send signals as a series of clicks—ironically a more digital approach than tone signals.)

Engineers have found that POTS lines can carry other things besides telephone conversations. Modems and fax machines rely on phone lines to carry digital signals converted into an analog format then reconverted to digital form inside the phone system. This is inherently inefficient; the best modems can squeeze only 14,400 bits per second down a phone line, only a quarter of the nominal 56,000 bits per second of a digitized phone call. And even that doesn't work very well on poor phone lines, which may suffer data errors when modems operate above 1200 bits per second. Noisy phone lines can also interfere with fax transmission.

Nonetheless, good wire lines are perfectly adequate for voice, modem, and fax POTS. In fact, with proper conditioning, twisted wire pairs can transmit 1.5 Mbit/s a kilometer or so without repeaters. That's usually enough to reach a concentrator in urban or suburban areas.

Why, then, do you need fiber optics for the subscriber loop? The answer is that you don't for traditional POTS to existing residential areas with one phone line per house. Copper cables connect lots of concentrators to switching offices and give adequate surface. So far few fibers go beyond the concentrator closest to the switching office, the serving area interface shown in Figure 18.5. However, the idea of stringing fiber further out in the network becomes more attractive as fiber costs decline, phone companies offer more services, and subscribers use more phone lines.

⬤ POTS lines carry modem and fax signals.

Fiber-Optic Applications

Fiber applications in the subscriber loop have benefited tremendously from the decline in prices of fiber-optic equipment. Fiber still costs more than metal cables for low-speed transmission, but the difference is much less than it used to be. Moreover, phone companies have found that labor accounts for much of the cost of cable installation. If a new cable is needed, it doesn't cost much more to install fiber—but the fiber gives extra capacity that leaves plenty of room for future expansion. Thus, fiber is widely used in places like cables between concentrators and switching centers, where the capacity is likely to be beneficial. (As you will see later, it is the cost of terminal equipment that has kept fiber from going all the way to homes.) Fibers and concentrators are often used to serve new developments, where new construction is required. Phone companies are not pulling out working wires in the local loop to install fiber—but they may replace defective metal cables with fibers.

⬤ Fibers carry high-speed multiplexed signals to remote distribution points, or concentrators.

Fibers are also used to connect large organizations directly to switching offices. Typically, the cables connect directly to the organization's phone system and computer networks. Some organizations specifically request high-speed fiber connections for digital transmission of data, voice, and other signals. They often buy transmission capacity over fibers in units of 1.5 Mbit/s T1 carriers or 45 Mbit/s T3 carriers.

● Single-mode fiber is now preferred, but some multimode fiber has been used in the subscriber loop.

The technical requirements for subscriber-loop transmission to concentrators or large organizations are not demanding, so multimode fibers could be used. However, multimode fibers have fallen out of favor because they leave little room for future expansion—one of the most important selling points of fiber connections. Virtually all installations today use single-mode fiber, although some multimode fiber may remain in use.

Emerging Services and Spreading Fibers

● Telephone and cable television networks are evolving to offer similar services.

Optical fibers are not needed to carry POTS to homes. Installing the vast bandwidth of single-mode fibers for that task is like building a freeway for oxcarts. However, technological advances and the rapid integration of other services in the global telecommunication system open other possibilities. A single fiber-optic system could integrate voice and data communications now carried over phone lines with video services now carried by cable television—and with other services only now emerging, such as videophones and high-speed data for people working at home.

Telephone companies are talking about new services such as "video dial tone." In essence, this is a transmission channel for video signals that other organizations would provide. For example, you might use the phone system to request a program from an online video library, and it would be transmitted to your home on demand (i.e., when you want it). Or broadcasters might lease video channels from the phone company.

Meanwhile, cable television companies are talking about adding telephone and data services to their offerings. I'll talk more about those plans in the next chapter, but the important point is that the two systems are evolving toward a common goal—providing a broad array of telecommunication services to homes and businesses.

● Current plans would bring fibers to the curb, not to homes, and offer many new services.

Current plans would not bring fibers all the way to homes, but the fibers would come close. Initially, fiber cable might serve a cluster of 500 homes, but developers envision later bringing fiber "to the curb"—where one fiber would serve a cluster of perhaps a dozen homes. Wires would go from there to the home. Although wires can't carry high-speed signals very far, they don't have to if the fiber is just down the block. Developers have shown that short twisted wire pairs or coaxial cables have plenty of bandwidth for many services if they simply drop the signals at homes. Combined with digital video compression technology, that would greatly enhance transmission capacity, even before fiber reaches all the way to homes.

Future Services

Optical fibers do not have unlimited transmission capacity, but they could bring many new services to homes. Table 18.2 lists some that have been suggested, with rough indications of the capacity they would require. You may have your own ideas for adding more services. The requirements listed as "Low" could probably operate with less than a single 56 kbit/s digitized voice channel. Many services are listed as requiring the equivalent of one digital voice channel.

Table 18.2. Possible services on fiber-optic networks.

Service	Bandwidth Requirement
Wire telephone	56 kbit/s
Cellular telephone	56 kbit/s
Computer data	56 kbit/s or up
Compressed video (conventional)	1-2 Mbit/s per channel
High-definition TV (compressed)	20 Mbit/s per channel
Switched video on demand	20 Mbit/s per channel
Interactive video	56 kbit/s - 1 Mbit/s
Video telephone/videoconferencing	56 kbit/s - 1 Mbit/s
Telecommuting	56 kbit/s - 1 Mbit/s and up
Simultaneous voice/data transmission	Circa 100 kbit/s
Electronic mail	Low
Automatic meter reading	Low
Fire/burglar alarms	Low
Home monitoring systems	Low
Home remote-control systems	Low
Computer conferencing	56 kbit/s or up
Videotex	56 kbit/s or up
Information services	56 kbit/s or up
Home banking	Low
Home shopping (teleshopping)	56 kbit/s - 1 Mbit/s
Educational services	56 kbit/s - 1 Mbit/s
Advanced voice capabilities	56 kbit/s or up
Facsimile transmission	56 kbit/s
High-quality digital audio	100 kbit/s
Energy management	Low
Video surveillance for home security	56 kbit/s - 1 Mbit/s

●
Advanced electronics can compress digital video by a factor of 60.

Tremendous progress has been made recently in video compression. Hardware for 10-to-1 compression without degrading transmission is on the market. High-definition television (HDTV) developers hope to achieve 60-to-1 reduction in data rates—from about 1 Gbit/s for a raw digitized signal to around 20 Mbit/s. Cable television operators likewise believe they can squeeze several compressed digital video channels into the 6 MHz bandwidth that has been used for each standard analog television channel. How much compression can be achieved depends on the type of signal being transmitted; the more changes between successive frames, the higher the data rate required to transmit the images properly. Thus sports events would require more capacity than "talking heads" news programs. I'll talk more about entertainment video in the next chapter, but as you will see, the distinctions between the telephone and cable television networks are blurring.

●
High-capacity fiber networks may offer videophones and videoconferencing.

Some of these advanced services are available to some extent today. Science-fiction writers have daydreamed about video telephones since the 1920s, as shown in Figure 18.6. Telephone companies followed; AT&T made a famous public demonstration of its "Picturephone" at the 1964 World's Fair in New York. It seemed a logical extension of communications via wire—as well as a way for phone companies to make more money. However, video images have always cost much more to generate, transmit, and receive than sound. Videophones and videoconferencing systems exist today, but they are far more expensive than ordinary telephones. One that sends small color images over ordinary phone lines costs about $1,000—and sends only a series of still images, not moving video. More costly videoconferencing systems send better-quality images, but they require the equivalent of two or more phone lines.

The biggest problems for videophones and videoconferencing may not be costs but who wants the services. Videoconferencing systems are used by businesses to avoid costly and time-consuming travel. They are predictable and usually scheduled in advance. They seem to be effective in some organizations—especially among people who have met each other in person—but they remain in limited use. The general public has shown little interest in buying videophones for regular use, at least partly because videophones would open their lives to random callers. (Your bleary-eyed and unshaven face might discourage brokers trying to sell you stocks, but that isn't the image you'd want to present to your boss or paramour.) Rather than worrying about the etiquette of when to turn on the video, most people seem not inclined to buy the things.

●
Telecommuting systems would link home offices to workplaces.

There may be more potential for telecommuting systems that would open a broad voice, data, fax, and video link between home offices and workplaces. The basic idea is to keep employees who work at home in close contact with their co-workers and supervisors. It is not likely to totally supplant the daily commute, but some people already use voice, data, and fax communications to work from home for part of the time, saving the time and costs of daily commuting. The higher the transmission capacity, the easier it is to transfer images and other information, and to remain in close contact with colleagues at distant sites.

What is important to stress for all the new services proposed in Table 18.2 is that they are proposed services. No one knows if people are willing to pay enough for them to be worthwhile. For example, how much would you pay for a system that could remotely monitor your home? Are you a person who always worries that you might have left the stove on or the water running somewhere? Would your electric power and gas companies be willing to reduce your rates if you installed automatic metering systems that relayed billing information directly to them? The answers will depend on the costs, competing technologies, and people's interest in the concepts.

●
Demand for new service remains to be seen.

Advanced System Concerns

A host of practical concerns remain that will affect how and when advanced services are delivered—if at all. They range from technical and marketing concerns to social and regulatory issues. I will look briefly at these practical concerns before moving on to consider the technical issues.

●
Any new network will have to be compatible with the existing one.

- Any new network will have to be compatible with the old. For example, old and new telephones must be able to talk with each other because the entire network will not be changed at once. Likewise, the new system should be compatible (at least for a transition period) with existing programs and equipment. Consider your reaction as a consumer if the electric company installed new wires and said you had to buy all new appliances to use them.

- Will telephones work if the electric power is off? Phone companies have been agonizing over this for years. Customers have come to expect it, and it's easy with copper wires. But it's much harder to do with a fiber-optic system.

- The new services will have to justify the extra cost to consumers.

- The new services will have to be packaged in ways that give customers choices. People without computers won't want to pay for access to data lines.

- Controversies over access to certain services are sure to arise. Phone companies have been forced to let customers block access to pay services, like telephone sex lines or costly 900 numbers. There are sure to be battles over video pornography.

New services will require changes in regulations.

- Existing telecommunications regulations must be altered if one network is to carry both video and phone services.

- Corporations will be competing with each other to offer services; they may get further competition from other technologies, such as direct satellite broadcasting.

- Somebody has to supply the money needed to build the network.

- Regulators and society must deal with questions of equitable access to services. Residential telephone service is now considered a necessity, especially for the elderly and house-bound, so phone companies are required to offer "lifeline" services at low cost to elderly people of limited means. Likewise, residential phone rates are set at standard levels that don't depend on how much it costs to service each residence. What happens if phone companies install new services only in affluent areas? What happens if poor people are shut off from educational opportunities or jobs because they cannot afford or access networks offering new services?

These issues range from the seemingly mundane to major. Some need to be resolved before an advanced network can be built; others may not arise until after it is partly completed. This is not the place to explore them in great depth, but you should be aware of them.

Designs for the Future Network

The key technical problem in designing the future network is providing the high transmission capacity needed to carry advanced services at a reasonable cost. Putting fiber in the subscriber loop (fiber in the loop systems, or FITL) is only part of the solution. Advances in other technologies, such as digital compression of video images and improved transmission over copper wires, can greatly enhance performance. The goal is to get the maximum performance at minimum cost.

Visions of the future fiber-optic network have evolved over the years. At first, developers wanted to bring fiber all the way to homes. In recent years, they have shifted their attention to bringing fiber close to homes but making the final connections with copper wires. Advocates of that approach believe it offers many of the advantages of all-fiber connections at a fraction of the cost. I'll look more closely at the two ideas, then move on to implementation plans.

● An advanced network must provide high transmission capacity at reasonable cost.

Fiber to the Home (FTTH)

Telecommunication visionaries have been looking at ways to bring optical fibers all the way to homes since 1972, when John Fulenwider first suggested the idea at the International Wire and Cable Symposium. The first experimental system was installed in Higashi-Ikoma, Japan, in 1978. Others have been tested since then, but all fiber to the home (FTTH) systems share a common limitation—cost.

● Fiber to the home systems are expensive.

Costs can be broken into three components: hardware, installation labor, and maintenance. Fiber-optic hardware is more expensive than wires, but unless you're installing sophisticated broadband laser transmitters in each home, the difference is not dramatic. Cost of terminal equipment—such as special interfaces for fiber-optic phones—is more of a problem. Designers are working on ways to reduce those costs, but the problem has not been solved.

Labor now accounts for a large share of installation costs. The costs of fiber and copper installation are comparable for providing services to new developments—but most people live in existing houses. It costs a lot more to install a new fiber-optic connection to an existing house than to leave the old phone wires in place.

Maintenance is another concern. Fiber optic systems generally require less maintenance than copper wires, at least in the trunk and long-haul network. However, phone companies worry that more problems could arise in fiber links to homes—especially if do-it-yourself homeowners tried to fix their own fiber systems.

Phone companies remain interested in fiber to the home systems, but they would prefer cheaper and simpler alternatives. The most attractive one is Fiber to the Curb (FTTC).

Fiber to the Curb (FTTC)

I've stressed in earlier chapters that fibers are much better than wires for sending high-speed signals a long distance. The reason is that transmission losses in wires increase rapidly with signal frequency. However, suitable wires can carry high-speed signals short distances. Twisted wire pairs or coaxial cables have decent loss and bandwidth if you are sending signals only a few hundred feet. This is the basic idea of fiber to the curb technology.

Figure 18.5 shows how the phone network branches out from the switching office to individual homes. Fiber goes from the switching office to the serving area interface—your neighborhood box. In the drawing, copper goes from that box to the serving terminal on your block. Replacing that length of copper with fiber would take fiber bandwidth to within a few hundred feet of your home. Your telephone and cable television "drops" could carry signals from that point to your home—with no necessary changes to anything in your house.

The fiber part of the network covers most of the distance between the switching office and the home, giving high bandwidth over a distance of a few kilometers. Each fiber cable in that network serves many subscribers, so the cost of installing it can be divided among many customers. The only wires that are left are the ones that cost the most to replace but do comparatively little damage to signal transmission.

Both telephone and (as you will see in Chapter 19) cable television companies are talking about building fiber to the curb systems. The details depend on the type of system. Figure 18.7 is adapted from a system being considered by NYNEX. It uses a ring of fiber rather than a branching network, to reduce its vulnerability to cable breaks. Fiber branches out from add/drop multiplexers on the ring, carrying signals to optical network units that each serve a few subscribers.

System Architecture

Fiber to the curb architecture is still evolving, and its final shape will depend strongly on the services it carries. If, as seems likely, regulatory barriers are removed between the telephone and cable television industries, the fiber network will probably carry a mixture of signals switched to individual homes, like present phone signals, and signals broadcast to the entire network, like present cable television networks.

Signals brought by fibers to the local optical interface (called the optical network unit in Figure 18.7) will be divided and transmitted to individual homes through copper. Some signals, like cable television broadcasts, will be distributed to all subscribers, although the household decoder may translate only some of them into usable form. Other signals, like telephone calls and data transmission, will be switched in some way so only certain households will receive them. Figure 18.8 shows one possible way to distribute signals through wires.

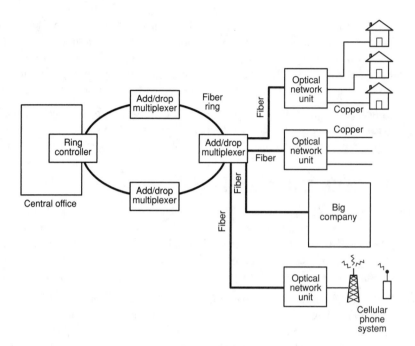

FIGURE 18.7.

Fiber ring in a fiber to the curb system.

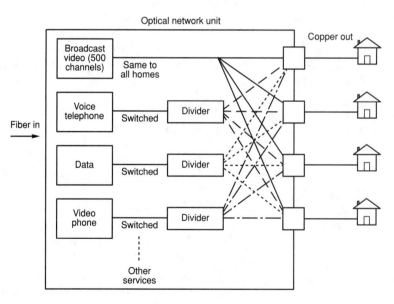

FIGURE 18.8.

Signals will be divided at the end of the fiber system for transmission over wires to individual homes.

Early fiber to the home systems being developed commercially are leaning toward an approach called passive optical networks because they divide and switch signals in an essentially passive way. Instead of using active switches, a passive optical network uses couplers and filters. As a simple example, you could imagine signals for each home being at a different wavelength, with optical filters transmitting only the desired wavelength.

Passive optical networks are attractive because they reduce the numbers of costly and complex active or switching components required, and concentrate the complex components at a limited number of sites. The overall effect is to reduce costs and improve performance.

Plans for Subscriber Networks

Long-term efforts
continue around the
world to bring fibers
to homes.

Advanced fiber-optic plans have been widely publicized in the United States, but other countries are also working on similar ideas for bringing advanced services to homes over optical fibers. Japan's Nippon Telegraph and Telephone Corp. is particularly active and has talked of switching its network completely to fiber optics by 2015. European countries are also planning fiber-optic systems.

It seems clear that fiber optics will be coming closer to homes in the next decade, but how far they come and what services they deliver remains to be seen. The tremendous advances of the technology justify optimism, but the history of consumer response justifies caution. Neither the telephone nor cable television was an overnight success, and some new services have failed. One memorable example was videotex, which sought to put information from remote computers on home television screens, only to die a quiet death from market apathy—unable to deliver enough information to satisfy users, and outdone technologically by the spread of personal computers. We are coming into truly interesting times for home communication technology.

What Have You Learned?

1. Most telephone cable is in the subscriber loop, the part of the network linking subscribers to switching offices. Fibers reach from switching offices to concentrators serving a few hundred phone lines.

2. Analog signals from standard phones are converted to digital form for switching and transmission through the telephone network. Larger businesses may have high-speed digital lines, supplied by the local telephone company or other carriers.

3. ISDN is switched digitized service that can operate over standard phone lines. It carries two 64 kbit/s channels and one 16 kbit/s channel, but it is not in wide use.

4. Telephone and cable television companies both want to offer advanced voice and video services over networks to homes, including hundreds of video channels, video on demand, and cellular phones. They would use similar hardware, but their competition is limited by federal regulations.

5. Trunk systems transmit 45 Mbit/s and up between central offices; most operate at 1300 nm over single-mode fiber. They are installed in underground ducts, on overhead poles, or by direct burial.

6. Fibers usually do not extend beyond concentrators today, but they are beginning to extend to feeder boxes that serve clusters of homes. This concept is called fiber to the curb.

7. The main business of local phone companies is POTS, Plain Old Telephone Service, including fax service. Telephone companies are looking to offer new voice, data, and video services, probably with fiber to the curb.

8. Video compression is a key technology enabling development of advanced systems to deliver new video services, including hundreds of cable channels, video on demand, video conferencing, high-definition television, and video telephones.

9. The emerging new network will have to be compatible with the existing phone network.

10. Fiber to the home systems might offer more benefits than fiber to the curb, but the high cost of installing them is hard to justify.

11. Demand for advanced services has not yet been demonstrated.

What's Next?

In Chapter 19, you will see how fiber optics are used for video transmission and why the cable television industry has made little use of fiber.

Quiz for Chapter 18

1. What area is served by a typical switching office?
   a. A major city of over 1 million people.
   b. An entire area code.
   c. 25,000 to 50,000 phone lines.
   d. 500 phone lines.
   e. A large corporation.

2. What is the bandwidth and format of the signal generated by a standard telephone?
   a. 300-4,000 Hz, analog.
   b. 20-20,000 Hz, analog.
   c. 56,000 bits/s digital.
   d. 64,000 bits/s digital.

3. What is the Integrated Services Digital Network?

    a. A meaningless marketing buzzword.

    b. The existing telephone network is an integrated digital network.

    c. A future all-fiber network.

    d. Digital services offered over existing phone lines at 144 kbit/s.

4. Typical distances of trunk cables in urban areas are

    a. 100-1000 m.

    b. 1-10 km.

    c. 10-100 km.

    d. over 100 km.

5. What technology is used in a third-generation fiber-optic system?

    a. Multimode fibers transmitting 850 nm several kilometers between repeaters at 45 Mbit/s.

    b. Multimode fibers transmitting 1300 nm up to 20 km between repeaters at 90 Mbit/s.

    c. Single-mode fibers transmitting 1300 nm for 40 km between repeaters at 45 Mbit/s to 2.5 Gbit/s.

    d. Single-mode fibers transmitting 1550 nm for 90-100 km between repeaters at 45 Mbit/s to 2.5 Gbit/s.

    e. Single-mode fibers and optical amplifiers operating at 1550 nm and 10 Gbit/s.

6. Optical fibers are used in the subscriber loop to

    a. multiplex many home phone lines at a concentrator for transmission to the central office.

    b. transmit signals on trunk lines between central offices at 45 Mbit/s to 2.5 Gbit/s.

    c. carry individual phone circuits to home subscribers.

    d. replace damaged wires in existing cables.

7. What is the main service offered over home telephone lines?

    a. High-speed digital data.

    b. POTS, Plain Old Telephone Service.

    c. Video telephone.

    d. Cable television.

    e. Teleshopping

8. What size area would be served by each fiber drop in a fiber to the curb system?

    a. A small town.

    b. 500 phone lines.

    c. About a block.

    d. Individual homes.

    e. Only large business.

9. Digital video compression makes it easier to offer which of the following services?

    a. Video conferencing.

    b. Hundreds of home cable television channels.

    c. Video on demand.

    d. Video telephones.

    e. All of the above.

10. Which of the following is not important in offering new services?

    a. Compatibility with the existing phone network.

    b. Affordability.

    c. Changes in telecommunication regulations.

    d. Bringing fiber optics to every home in the country.

Video Transmission

About This Chapter

Many types of video signals are transmitted for many purposes, and the field is evolving rapidly. You saw in the last chapter that telephone and cable television networks are evolving in similar directions and adapting similar fiber-optic architectures. In this chapter, I will look from the cable television standpoint, to see how optical fibers are spreading through the cable system. I will also look at other types of video transmission. First, however, I will look at the basics of video transmission.

Video Basics

Video transmission is rather more complicated than sending voice or data signals. Old-fashioned analog voice telephones simply convert the continuous variations in sound intensity of your voice into continuous variations in an electrical signal. Data signals are strings of 1s and 0s that correspond to the binary data. Video signals have to encode continually changing pictures and sound, and that gets complicated.

- Video signals encode continually changing pictures and sound.

- Video requires much more transmission capacity than sound or equivalent digital data.

Television pictures are a series of images, displayed one after another on the screen. Look closely, and you will see that the pictures are made up of many parallel lines, drawn one after another on the screen. The video signals carry the information needed to draw these lines, saying what parts should be light and what parts should be dark and how the colors should be displayed. They also encode the accompanying sound.

The result requires much more transmission capacity than telephone sound or equivalent digital data. It's often said that one picture equals a thousand words, but Table 19.1 shows that the picture requires considerably more transmission capacity than 1,000 words of written text. Sophisticated digital compression can reduce data rates by factors of 10 to 60 for video transmission, but they depend on transmitting a series of pictures, so single images cannot be compressed as much. Note that the spoken word is much less efficient when converted into digitized sound.

Table 19.1. Comparison of video, voice, and text transmission requirements.*

Transmission	Analog Equivalent	Digital Equivalent (uncompressed)	Digital (compressed)
Standard Television channel (NTSC)	6.3 MHz	100 Mbit/s	2-10 Mbit/s
Telephone circuit	4 kHz	56 kbit/s	about 10 kbit/s (rarely used; assumes silent intervals)
1,000 words spoken on phone (6 minutes)	—	20 Mbits	?
TV-quality video frame	—	3.3 Mbits	1 Mbits*
1000 words of text	—	60 kbits	30 kbits (2-to-1 compression)
HDTV (US proposal)	—	1.2 Gbit/s	20 Mbit/s (proposed standard)

*Approximations, based on 3-to-1 compression for single video frames; much higher compression is possible for a series of frames.

You can appreciate these differences if you use a personal computer with sound and graphics programs. For example, a screen full of plain text takes up about 2,000 bytes of memory (word processors with fancy formatting add more), whereas a black-and-white shot of the same screen of text takes up about 20,000 bytes. Complex graphic images can occupy a megabyte or more, and a digitized voice of a single "hello" can occupy 40,000 bytes. In short, get into high-quality audio and video and your transmission requirements go up fast.

Video signals are transmitted in standardized formats so that transmitters can talk to receivers. These formats have evolved for historical reasons and sometimes are not ideal. Europe and North America use separate standards, based on the different frequencies used for electric current (60 Hz in North America and 50 Hz in Europe). The North American standard for broadcast color television was designed to be compatible with older black-and-white sets, at some sacrifice in performance. Newer standards are being developed, notably the much-publicized high-definition television (HDTV). Like earlier standards, the new ones will not be perfect or universal; Europe, Japan, and North America may wind up with different HDTV standards. However, developers hope to avoid serious compromises and may be able to make next-generation television compatible with computer displays.

● Video is transmitted in standardized formats.

Standard Broadcast Video

The present standard for broadcast television programs in North America is the NTSC format, from the National Television System Committee. The analog signals carry information representing the lines that compose the screen images. Pictures are displayed as 525-line frames (although a few lines do not actually show up on the screen). Nominally, NTSC shows 30 frames a second, but to keep the image from flickering to the eye, NTSC uses an interlaced scanning technique. First it scans odd lines on the display, then the even lines, then the odd lines again, as shown in Figure 19.1. Technically, this interlaced scan displays 60 half-images (called fields) a second, with only 267 lines of resolution each, but this fools the eye, giving the appearance of high resolution while avoiding the flicker of slower scanning speeds.

● NTSC video displays 30 analog 525-line frames a second.

Nominally, the NTSC video bandwidth is 4.2 MHz, with the sound carrier at a higher frequency. However, for broadcast the video signal is used to modulate a radio-frequency signal, a process that increases overall bandwidth to 6 MHz, which is the amount of radio-frequency spectrum allocated to each broadcast television channel in the United States. Figure 19.2 shows the structure of this signal, which extends from 1.25 MHz below the carrier frequency to 4.75 MHz above the carrier. The NTSC format is used in North America, Japan, Korea, the Philippines, and much of South America.

FIGURE 19.1.

Interlaced and progressive scanning.

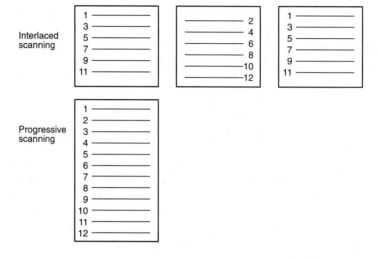

FIGURE 19.2.

Structure of NTSC broadcast video signal.

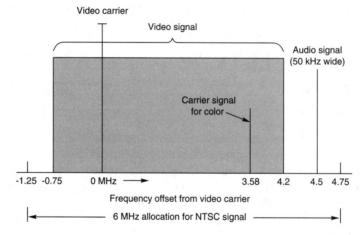

●

PAL and SECAM are interlaced scanning systems showing 25 frames of 625 lines each per second.

Two other broadcast television standards are in wide use: PAL and SECAM. Both are interlaced scanning systems that show 25 frames (50 fields) per second, each frame with 625 lines. These have nominal video bandwidth of 5 to 6 MHz and broadcast channel bandwidth of about 8 MHz. PAL and SECAM systems are used in Europe, mainland Asia, Africa, and parts of South America.

These standards were set for television broadcasting from ground-based transmitters. National and international standards set aside specific frequencies for television broadcasting, with each channel allocated the required bandwidth (6 MHz for NTSC channels). These standards have come into wide use for other types of video because NTSC, PAL, and SECAM equipment is readily available.

You should also realize that broadcast video standards were established decades ago, when color television came on the market. (The NTSC standard was a modified version of the original North American standard for black-and-white television, which goes back to 1948.) This means that these standards were developed to deal with the limitations of the vacuum-tube technology available in the electronic stone age.

● Standard video formats are decades old.

Computer Displays and Video Formats

It might seem logical to use standard television displays for computers, but the two technologies are not readily compatible. Television sets are adequate displays for computer-based video games, and sufficed for some early personal computers. However, text displayed on a screen does not show up well with interlaced scanning, because the interlacing effectively mixes information from successive frames for part of the time. The best displays for computers use progressive scanning, in which all lines are scanned one at a time, then the entire screen is rescanned, shown in Figure 19.1.

● Computer displays require progressive scanning to show text clearly, not NTSC format.

Progressive scanning demands more bandwidth and faster electronics than interlaced scanning, because it transmits 60 complete frames a second to avoid flicker, compared to the 60 fields (or half-frames) for NTSC video. The benefit is somewhat higher resolution than interlaced scanning. This was one of the technical trade-offs accepted by the broadcast industry when it developed its standards.

Multimedia or interactive video displays are hybrids, based on a combination of computer and television technologies. At this writing, they don't have their own special display standards; those that play on computers use computer formats, those played on television sets use television formats.

Digitized Video and Compression

Like voice signals, video can be digitized. The raw data rate is considerably higher than the corresponding analog bandwidth, as shown in Table 19.1. However, raw image data contains much redundant information, and suitable software can compress it to occupy less space.

● Emerging video systems will use compressed digital video.

Some compression is possible with all files, by software that recognizes long strings of identical bits or bytes and replaces them with shorter codes containing the same information (e.g., a code indicating that the next 20 digits are 1s). However, only limited compression is possible for a single image or data file. Video signals that contain a series of images can be compressed much more by using other techniques that transmit the changes between images rather than an entire new image. Most images do not change completely between frames, so it takes much less information to convey the changes than to send a whole new image.

Impressive progress has been made in video compression recently; 10-to-1 compression is readily available, and developers believe 60-to-1 compression should be no problem soon. Compression works best for images that change little between frames, such as video-conferences or news broadcasts showing talking heads. Sports events are more difficult because they have rapid motion and change. Emerging video technology, such as high-definition television and high-capacity cable television, will use digital video compression.

High-Definition Television (HDTV)

HDTV will double resolution for larger screen televisions.

The electronics industry has been pressing for a new generation of high-definition television (HDTV) technology for several years. The goal is to offer larger images of much better quality (and not incidentally to make more money for the electronics industry by selling a new and expensive generation of television sets). The number of lines per screen will be roughly doubled, and the screens will be made wider, with a width-to-height ratio of 16 to 9 rather than the 4 to 3 ratio of NTSC video. The improvement is evident on large screens, but not as obvious on smaller screens, and smaller sets may not take advantage of all the possible resolution. (Remember that video resolution is measured in lines per screen, not lines per inch.)

Japan and Europe developed analog HDTV systems, and the Japanese system has been marketed to television studios in the United States. However, the U.S. has decided to develop a compressed digital transmission system that would squeeze an HDTV signal to 20 to 22 Mbit/s (down from a raw data rate of around 1.2 Gbit/s). Details have yet to be worked out, but plans call for progressive scanning of 780 to 1100 lines per screen, in a format compatible with computer displays yet also at least somewhat compatible with interlaced displays.

Transmission Media

So far I haven't said much about how video signals are distributed. Although our main interest is fiber optics, I will mention other technologies to point out important differences.

TERRESTRIAL BROADCAST

Television began with broadcasts from local stations on the ground.

Television began as a broadcast medium, with radio-frequency signals transmitted from tall antennas. NTSC, PAL, and SECAM standards were all based on the assumption that the signals would be broadcast from terrestrial towers, which reach tens of miles, depending on frequency, power, antenna height, and local obstacles. Stations are allocated specific frequency channels and must broadcast signals at those frequencies (which can be allocated to other stations outside of interference range). Broadcasting remains important, although cable transmission has become common in the United States.

CABLE TELEVISION

Cable television began as community antenna television (CATV, an acronym still used by the industry) to serve areas not normally reached by broadcast television signals. The idea was to build one big antenna to pick up remote broadcasts, then distribute the signals via coaxial cables to local homes. Eventually the concept spread to urban and suburban areas where broadcast quality was better, but the choices were limited. Economics and interference limit the number of broadcast channels in a metropolitan area, but cable systems can pick up many more channels (from satellites and distant stations) and distribute them to homes. Cable systems can also offer extra-cost "premium" services to customers who rent special decoders. (The signals are scrambled and sent to all subscribers, but only those who pay the premium for the decoders can unscramble them.) Existing cable television systems serve essentially the same function as broadcast television—they distribute the same signals to everyone. This makes them function like one-way pipes (although some have a limited two-way capability).

Present cable systems carry analog video signals in NTSC format. Each channel is assigned a frequency slot and signals are multiplexed together for transmission over the cable. Typical cable systems carry dozens of 6 MHz channels, but some new ones can carry up to 100. The signals are carried through coaxial cables at a broad range of radio frequencies to about 550 MHz. Set-top cable boxes demultiplex the signals, picking out the one selected by the viewer.

Early cable systems were built entirely from coaxial cables, but fiber optics have come into increasing use. Emerging technology will let cable systems carry much more two-way traffic than at present and transmit several hundred video channels at once. You will see how later in this chapter.

DIRECT BROADCAST SATELLITES

A direct broadcast satellite is essentially a television transmitter in geosynchronous orbit that stays in place above the equator broadcasting signals over a wide area. The signals are transmitted at microwave frequencies (above 1 gigahertz), and received by a dish-shaped antenna on the ground. The signals must be decoded to be viewed on a home television set.

Satellite broadcasts serve a similar function to cable systems. They are particularly attractive for rural areas that lack cable service, because they don't require a large investment in cable to serve few homes. They can also carry more channels than many present cable systems, but they provide only one-way service because they broadcast signals to large areas. (Two-way satellite communication systems use separate satellites designed for that purpose.)

●
Cable television carries dozens of analog NTSC video channels in a radio-frequency signal.

●
Direct broadcast satellites can serve large areas with microwave transmission.

Other Video-Transmission Applications

Not all video transmission is broadcast through the air or sent over cable television. There are a variety of other video transmission applications, ranging from studio production to security monitoring over closed-circuit television. I summarize a sampling of these other uses in Table 19.2.

Table 19.2. Video transmission applications outside of broadcasting and cable television.

Application	Requirements	Special Notes
Broadcast-signal delivery to transmitter or cable head-end	High signal quality	
Closed-circuit television (e.g., security monitoring)	Vary	Need not be NTSC
Electronic news gathering	Link mobile camera to fixed equipment	Camcorders often preferred
Network Feeds	Deliver programs to remote stations	Usually satellite
Studio Transmission	High quality	Inside studio
Remote Pickup	From fixed remote site to studio	
Videoconferencing and video telephones	Handle minimal motion, work over phone network	Two-way communication
Multimedia and interactive video	Computer-generated or stored images	Must be computer-compatible

Transmission requirements differ widely. In some cases (e.g., electronic news gathering), a single channel is sent between two points, and the transmission medium must be portable and not subject to interference. In others (e.g., remote pickup), point-to-point transmission can go over fixed commercial telecommunication lines, including satellite channels and fiber optic cables. Network feeds require simultaneous transmission to many points. Although many applications demand high-quality images, a few—notably closed-circuit surveillance cameras—do not. Videoconferencing and video telephones are unlike other video applications because they are switched two-way services like telephones; they also accept trade-offs in image quality to allow transmission over telephone lines.

This chapter is devoted primarily to cable television, but you should be aware of these other video applications. Many of them use some fiber optic transmission, over dedicated cables or over the public telecommunications network.

The Present Cable Network

The cable television network is evolving rapidly, and to cover cable properly you need to look at both the current system and the technology that will appear over coming years. As you saw in Chapter 18, the cable television network is evolving to resemble the telephone network in many aspects. For now, however, the two are quite distinct. Telephone signals are low-bandwidth and switched; cable signals are high-bandwidth and distributed to all subscribers. The telephone network makes extensive use of digital electronics; the present cable network carries analog NTSC video signals.

To understand the role of fiber optics, I will first look at the present cable network and its architecture, and the places where fibers are used today. Then I will step back and look at emerging cable technology and see how it will make more extensive use of fibers.

Traditional Cable Architecture

Figure 19.3 shows the traditional architecture for cable television systems. Signals from satellites and television stations are collected at a control center or head end, which functions as the heart of the system. The head end may connect with other cable networks in the area, reflecting the growing integration of cable companies. The head end combines video signals from various sources and sends them to regional hubs, which distribute them to distribution nodes and then to homes. The number of levels depends on the size of the community served.

The links between cable head ends and hubs are called supertrunks. They must carry 30 to 100 video channels from a cable system head end to remote points many kilometers away, from which signals are distributed to subscribers. A key requirement is that signal-to-noise ratio at the output be at least 53 dB to ensure that signal quality is adequate when it reaches subscribers. Early cable systems used coaxial cables or terrestrial microwave towers to send signals across such distances. However, those systems were comparatively noisy, a problem for cable television systems trying to market better transmission quality than broadcast stations. Coaxial cable systems also required many amplifiers or repeaters—one about every 0.5 to 0.6 km (0.3 to 0.4 miles)—and failure of any one of those repeaters could knock out the whole system. Optical fibers offered ways around those problems, and by the late 1980s, fibers had become standard for supertrunks. Typically the signals are divided among multiple fibers in a single cable sheath.

● Present cable systems transmit analog NTSC video signals.

● Fiber-optic links are used as supertrunks between head ends and distribution hubs.

FIGURE 19.3.

A traditional cable television system.

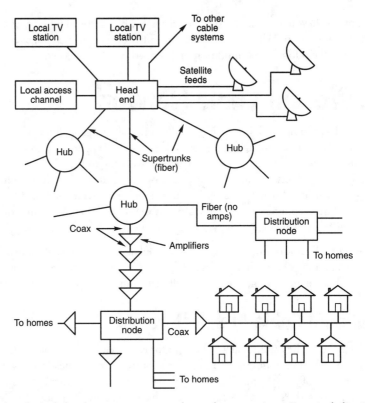

A critical technical development came at about the same time. Practical distributed-feedback lasers reached the market, and their prices began dropping. This was vital because cable television requires analog transmission, which is inherently sensitive to laser noise. The more controlled emission from distributed-feedback lasers reduced noise and made system response more linear, greatly improving analog video transmission quality, without a huge price premium.

> Distributed-feedback lasers give analog fiber systems a linear response.

High-quality analog fiber systems soon began spreading further into the cable distribution network shown in Figure 19.3, replacing coaxial cables between hubs and distribution nodes. For cable companies, the big improvement was eliminating long chains of amplifiers between the hub and the distribution node, which caused noise and made the system prone to failure. Cable system planners now envision fibers running out to each node, serving between 500 and 2000 households.

Cable Evolution

Present analog cable systems carry signals at frequencies between 400 and 550 MHz. As cable companies expand their capacity, they are considering adding digital channels at higher frequencies, up to about 750 MHz, to take advantage of new capacity added by the expansion of fiber links.

Existing cable systems do not carry signals at the low end of the spectrum, at frequencies below about 30 MHz. Developers envision using those frequencies for return signals, originated in the subscriber's home, to offer additional services. Eventually, these changes will bring about a new generation of cable systems.

● Digital signals will be added to existing cable systems at higher frequencies.

Emerging Cable Technology

Bringing fiber to the distribution node sets the stage for the next step in cable television evolution. Cable companies hope to add new services and push fiber out to serve smaller clusters of households, as shown in Figure 19.4. Fiber need not reach all the way to homes to greatly enhance transmission capacity. Short lengths of coaxial cable have a large bandwidth and do not require amplifiers; a 100-meter (109-yard) length of coax has bandwidth well over 1 GHz. Extending fiber to the curb could let cable companies tap that bandwidth.

● Cable companies want to bring fiber close to homes, using coax to carry signals the last short distance.

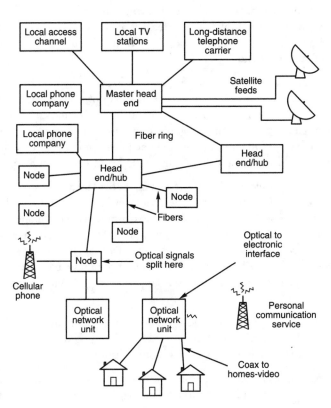

FIGURE 19.4.

Future cable television system with fiber to the curb.

If this sounds suspiciously like what telephone companies are thinking of doing, you've noticed the common evolution of cable and telephone systems. Both are thinking of "fiber to the curb" systems, which would use wires to carry signals into individual homes. Both want to expand the services they offer. Cable companies want to offer two-way services and mobile phones; phone companies want to offer video. Competition may at last come to local home telecommunications.

Proposed new cable systems resemble fiber to the curb systems for telephone systems.

The fiber technology is somewhat similar to that of the new telephone systems I described in the last chapter. Fibers will carry signals from the head end to remote nodes, where they will be passively split (using 1-to-n couplers) among several output fibers serving different optical network units. Each ONU will serve a small number of homes. It's easy to envision the ONU as the service provider for the block, while the nodes serve neighborhoods.

Passive optical networks are an important trend because they address technological limitations that had slowed the spread of optical fibers in cable systems. The high loss of optical couplers had discouraged their use in branching networks like traditional cable television systems. This has been offset by the emergence of optical amplifiers and high-power transmitters, which provide much stronger optical signals, carrying enough power to be split among many more branches of the network. Telephone companies are taking advantage of the same trend.

Cable companies plan to change the services more dramatically than their network plan. Digital video compression will let them pack hundreds of video channels onto the same network that now carries no more than about 100. Those channels may not all be different; some might carry the same programming, but phased in time. For example, a premium movie service might transmit the same movie on four different channels—one starting at 8:00 pm, another at 8:15, a third at 8:30, and a fourth at 8:45.

Digital compression technology will let cable systems carry hundreds of channels and video on demand.

Digital technology would also let cable companies offer video on demand—access to an online video library. A computer at the head end would store compressed digital versions of popular movies, which you could ask to have transmitted to your home at any time you wanted. You could pause the program or replay portions, just as you can a videotape, but you wouldn't have to run out to the rental store. Hardware developers believe the computer memory could store hundreds or thousands of compressed digitized films, depending on design details. (For comparison, one large video chain claims its stores stock 8,000 movies.)

The switched cable network could also handle mobile phones or personal communication services. The network would distribute calls to the appropriate local nodes, which would make the connection via radio waves broadcast from its towers.

Cable companies are also looking at some of the other services I talked about in Chapter 18. The details are far from clear, but the technology is moving fast, and by the end of the century it may be hard to tell the cable companies from the phone companies.

Other Video Applications

Cable television is the most important video application of fiber optics, but Table 19.2 lists several other types of video transmission where fibers are used. Many of those jobs are simple and straightforward point-to-point transmission. Examples include electronic news gathering, closed-circuit television, signal distribution in studios, and low-noise video transmission to and from cable head ends. Although metal cables can do most of these jobs, fibers may offer benefits of lighter weight, smaller size, longer transmission distance, immunity to electromagnetic interference, and avoidance of ground loops and potential differences.

●
Small, lightweight fiber cables are valuable for portable systems and to avoid EMI and ground loops.

Fiber transmission also offers more subtle advantages, notably avoiding the need to adjust transmission equipment to account for differences in cable length. Television studio amplifiers are designed to drive coaxial cables with nominal impedance of 75 ohms. However, actual impedance of coaxial cables is a function of length. As cable length increases, so does its capacitance, degrading high-frequency response if the cable is longer than 15-30 m (50-100 ft). Boosting the high-frequency signal, a process called equalization, can compensate for this degradation, but proper equalization requires knowing the cable's length and attenuation characteristics. Compensation also becomes harder with cable lengths over 300 m (1,000 ft) and is impractical for cables longer than about 900 m (3,000 ft). There is no analogous effect in optical fibers, so operators need not worry about cable length.

Videoconferencing and videophones are important potential video applications that differ in important ways from other video systems. Both are two-way signals that must be switched and routed over limited-bandwidth telephone lines—not one-way signals that are delivered to all subscribers over standard cable networks. Both are in limited use today.

What Have You Learned?

1. Video signals encode continually changing pictures and sound. They are transmitted in standard formats and require considerably more capacity than voice or digital data.

2. Analog NTSC video displays 30 analog 525-line frames a second with interlaced scanning. Each NTSC channel requires 6 MHz of broadcast spectrum. PAL and SECAM are interlaced scanning systems that each second show 25 analog frames of 625 lines each. These formats are decades old.

3. Computer displays need progressive scanning to show text clearly, not the interlaced scanning of NTSC, PAL, or SECAM. Progressive scanning demands more bandwidth and faster electronics.

4. Emerging video systems will use digital compression, which can compress files and transmission requirements by 10-to-1 to 60-to-1.

5. High-definition television will double resolution for larger screens. The U.S. has decided to use compressed digital video for transmission and storage; Japan earlier picked an analog approach.

6. Video signals can be broadcast from a ground station to serve a local area. Microwave transmission from direct broadcast satellites can serve large areas; customers require satellite dishes and converters.

7. Cable television systems carry dozens of analog NTSC video channels in a radio-frequency signal; customers require set-top converters to access premium services.

8. Cable television systems use fiber-optic supertrunks to carry video signals from the control center or "head end" to distribution hubs. Coaxial cables carry signals from the hubs to subscribers in most present systems. Coax is inexpensive but is limited by the need for closely spaced repeaters.

9. Digital signals will be added to existing cable systems at frequencies above the 400 to 550 MHz limit of present analog signals.

10. Cable television operators want to bring fibers closer to homes, with coaxial drops from fiber distribution points. Each fiber node would serve a small area, about a block, with an architecture similar to that of fiber to the curb systems for telephone systems. Digital compression technology will let cable systems carry hundreds of channels and video on demand.

11. Small, lightweight fiber cables are valuable for portable systems and to avoid EMI and ground loops in applications such as electronic news gathering and closed-circuit television.

What's Next?

In Chapter 20, you will learn how optical fibers are used in the digital world of data communications.

Quiz for Chapter 19

1. What is analog bandwidth of one standard NTSC television channel?

   a. 56 kHz.

   b. 1 MHz.

   c. 6.3 MHz.

   d. 25 MHz.

2. How many lines per frame do standard European television stations show, and how many full frames are shown per second?

   a. 525 lines, 25 frames per second.

   b. 625 lines, 25 frames per second.

c. 625 lines, 30 frames per second.

d. 1125 lines, 25 frames per second.

3. The proposed HDTV standard in the U.S. will require 20 Mbit/s after digital compression. How much digital compression is planned, and what would the data rate be without it?

    a. 3-to-1 compression, 90 MHz.

    b. 10-to-1 compression, 200 Mbit/s.

    c. 60-to-1 compression, 1200 Mbit/s.

    d. None of the above.

4. What key development made the quality of analog fiber-optic transmission adequate for cable television trunks?

    a. Highly linear distributed-feedback lasers.

    b. Inexpensive single-mode fiber.

    c. Dispersion-shifted fiber.

    d. Digital video compression.

    e. Optical amplifiers for 1550 nm systems.

5. What is the most important advantage of fiber optics over coax for distributing cable television signals from head ends to hubs?

    a. Fiber optics are hard to tap, so they reduce signal piracy.

    b. Fiber repeater spacing is much longer, avoiding noise and reliability problems with coax amplifiers.

c. Fiber can be extended all the way to subscribers.

d. Fiber cables are less likely to break.

6. How are video signals distributed to subscribers on present cable television systems?

    a. All subscribers receive the same signals, which require set-top decoders to show premium services.

    b. Signals from set-top controls are used to switch designed signals to the home.

    c. Equipment at the head end switches selected services to each subscriber.

    d. One pair of optical fibers runs directly from head end to home.

7. What signal format is used by present cable television systems?

    a. Each system has a proprietary format.

    b. Analog NTSC signals, with 6 MHz bandwidth, assigned to radio frequencies of 400 or 550 MHz.

    c. Digitized compressed video at 20 Mbit/s.

    d. Analog PAL format in North America.

8. What new video services are cable television companies considering offering with compressed digital video?

    a. High-definition television only.

    b. Videoconferencing and telecommuting.

    c. Hundreds of video channels.

    d. Video on demand and hundreds of video channels.

    e. High-quality digitized audio programs.

9. How do cable television planners envision connecting homes to future networks?

    a. With purely fiber-optic systems.

    b. They expect to lose the market to direct broadcast satellites.

    c. With twisted wire pairs.

    d. With coaxial cables running from the home to a fiber interface serving several homes.

10. What is happening to the difference between telephone and cable television networks?

    a. Nothing major is changing.

    b. Most technical distinctions are disappearing, as both move to a fiber to the curb architecture.

    c. The two are using different types of fibers.

    d. Regulations are keeping the two industries entirely separate.

Computers and Local Area Networks

UNDERSTANDING

About This Chapter

Fiber-optic data communication has been considered promising since the late 1970s, but only in recent years have fiber optics become important in local area networks and other computer communications. The major reason has been that few computer applications needed the speeds and long transmission distances fibers offer. Fiber optics have spread as local area networks have moved to higher speeds and spread over larger areas. Some fiber-based networks are already in commercial use; others are in development for the higher transmission speeds that come with more powerful computers. To understand the use of fiber optics for data communications, I will first review computer communication basics and relevant fiber capabilities. Then I will look at how fibers meet system requirements.

Basic Concepts

Computers think and communicate in binary bits—1s and 0s. Identicall computers—and circuit boards within the same computer—decode binary information in the same way. Different computers must convert information to a common format before they can interchange it. This is the job of data communications software and hardware.

Computer data rates are normally measured in baud, which means signal-level transitions per second. In coding schemes where there is one transition per bit, baud equals bit rate. In schemes such as Manchester coding, where there are two transitions per bit, the baud rate is twice the bit rate. Most computer data transfer speeds are slow by the standards of fiber-optic telecommunications. The

Computers require a common digital format to interchange information.

fastest personal computer data interchange over phone lines is 14,400 baud, whereas high-speed transfers to and from hard disks are measured in megabaud. Electrical wires can handle those data rates as long as distances are modest or there are no special problems such as electromagnetic interference.

Point-to-Point Transmission

Much data is exchanged between pairs of devices.

Much data transmission simply moves information between pairs of devices (e.g., from a computer to a printer or external hard disk), as shown in Figure 20.1. Even if data goes from the external disk to the printer, it must pass through the computer, as shown in Figure 20.1.

FIGURE 20.1.

Point-to-point connections between a personal computer and individual external devices.

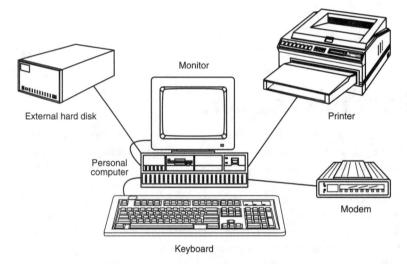

External hard disk

Monitor

Printer

Personal computer

Modem

Keyboard

Point-to-point transmission is the simplest of tasks for fiber optics because it requires only a transmitter, a receiver, and some fiber. However, in most cases, wires are even simpler because they can carry the electrical output from one device directly to the other without a special transmitter or receiver. Optical fibers enter the picture when data rates become too high, distances too long, or the environment too noisy, or when other factors make it hard for wires to work. Such cases are becoming more common.

Local Area Networks

A local area network interconnects many nodes.

Networks link many terminals, nodes, or devices. Many personal computers are linked to local area networks (LANs), which serve many nodes or devices in the area, as shown in Figure 20.2. (You might think that the array of devices attached to a single personal computer forms a network, but in practice local area networks connect multiple computers and other terminal devices.) Typically, a local area network serves a work group, department, or small business, whereas a wide area network links many local area networks to serve an entire large company. Local area networks typically have interfaces with other communication services; the example in Figure 20.2 has links to a wide area network and (through a fax modem) to the telephone network.

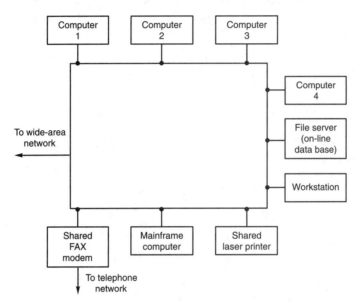

FIGURE 20.2.

A LAN interconnects many nodes that can send messages to any other node.

Details vary widely among local area networks, but the key idea is that all nodes can interchange data with each other. A single medium—wires, optical fibers, or radio waves—carries signals to all nodes on the system. Data packets carry header information to route them to particular nodes. Many terminals can use the network at the same time. For example, the users of computers 1 and 2 could each retrieve data from the file server at the same time, computer 3 could print a report on the laser printer, and computer 4 could access the corporate wide area network to communicate with another department.

Local area networks can take a variety of forms. Three basic approaches to LANs are shown in Figure 20.3. In the star topology, all signals pass through a central node, which may be active or passive. (An active star, which switches signals to particular nodes, functions like a telephone switch.) In a ring network, the transmission medium passes through all nodes, and signals can be passed in one or both directions (sometimes over two parallel paths for redundancy). In the data-bus topology, a common transmission medium connects all the nodes but is not closed to form a loop (i.e., the signal does not pass through all nodes in a series). Variations on these approaches make classification more complex than it might sound, but I will ignore those here.

The major media used for LANs are twisted wire pairs (shielded or unshielded), coaxial cables, and optical fibers. (Wireless LANs are not in wide use and have inherent limitations.) The difficulty of tapping into optical fibers at nodes makes some designs awkward

● Common LAN types are the ring, the star, and the bus.

or uneconomical. To overcome this problem, fiber LANs may be made with fiber-optic cable connecting nodes, at which signals are converted back into electrical format; those electrical signals can be sent to the terminal equipment and used to drive a transmitter sending optical signals to the next node. This approach avoids excess losses, but the use of extra transmitters and receivers drives up costs. One result is that completely fiber networks are usually used only at high speeds, although fibers are often used as segments of slower networks.

FIGURE 20.3.

Star, bus, and ring LAN architectures.

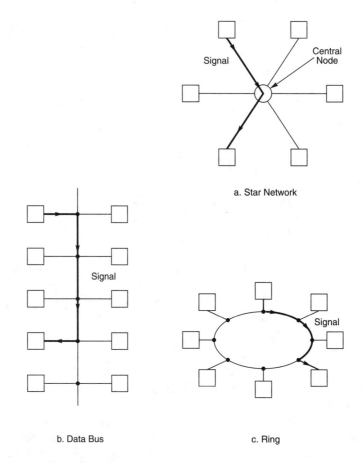

a. Star Network

b. Data Bus

c. Ring

Wide Area Networks

Wide area networks link local area networks.

The next step up from the local area network is the wide area network or WAN (sometimes called a metropolitan area network or MAN). A wide area network links local area networks, and often other terminals, over a large area. For example, a wide area network in a large company may connect departmental LANs and have a direct link to a corporate mainframe computer. This arrangement clusters groups that work together and need to communicate more intensely in local area networks, while still allowing communication among different groups. It avoids performance degradation that comes from adding too many terminals to a local area network.

On a simple level, a wide area network is a network linking networks instead of linking terminals. As a result, it has higher transmission speeds than local area networks.

Wide area networks are not inherently limited to data communications. In fact, the term is often used for networks that carry voice and data signals over a large area—for example, to all company plants in a large metropolitan area. This covers metropolitan networks operated by some telecommunications carriers that are really alternative telephone companies—another indication of how telephone networks are evolving into other forms.

⬤

WANs are not inherently limited to data communications.

I covered the networks oriented toward local telephone communications in Chapter 18. Here I will concentrate on corporate or campus networks that nominally carry mainly data communications. Remember, however, that the technological boundaries are hazy.

Optical Interconnection

All data transfer is not between separate boxes. Data must be moved between circuit boards in computers, between chips on circuit boards, and even within chips. In some cases, such as supercomputers, data must be transferred at very high speeds, and wires and other circuit components must be very tightly packed. Fibers can help both by allowing faster data transmission and by avoiding interference between adjacent wires. Gigabit data rates may require fiber optics.

⬤

Optical signals can make interconnections within computers and even within chips.

Optical links may also carry information directly to and from integrated-circuit chips. Such connections grow harder as the scale of integration increases, because the number of circuit elements on the chip surface can increase much faster than the room, to make electrical connections along the edges of the chip. Optical interconnection offers a way around this problem.

One possibility is to multiplex several digital signals together, forming a single high-speed signal that could be transmitted through one fiber rather than several wires. Other alternatives involve optical devices that emit and receive light from the chip surface, with or without fibers. Certain developmental semiconductor lasers have mirrors above and below the active layer, and emit light from their surfaces. This light could be collected by external optics or optical fibers, then focused onto a detector on another chip, which would convert the signal into electrical form.

Some current high-performance computers already use high-speed fiber-optic connections between circuit boards. More such links are likely as computer speeds increase, and developers are working on the high-speed Fiber Channel standard partly for this purpose. Direct connections from chip surfaces have been demonstrated in the laboratory but are further in the future and may not use optical fibers at all.

Why Use Fiber Optics?

Copper wires are adequate for the vast bulk of computer data transmission over point-to-point links and local area networks. As you saw earlier, copper wires can transmit high-speed signals a short distance, and slower signals considerably further. Optical fibers make more sense in wide area networks and are used in many more of them. However, wires remain the usual choice for data transmission.

The main advantage wires offer is lower cost. Wires can directly carry the electrical signals generated by computers and peripheral devices over short to moderate distances. Optical transmitters and receivers are needed to convert those signals to and from optical form for transmission through an optical fiber. You don't want to pay that premium where you don't have to. The inherent limitations of signal splitting in fiber-optic couplers can also drive up costs, especially for networks that connect many devices.

A more subtle advantage of wires is their compatibility with existing computer and data-transmission equipment. The back of your personal computer has wire connections but no fiber-optic sockets. Likewise, many high-speed computer ports are designed for parallel transmission through multiwire cables, not for serial transmission of bits through a fiber-optic cable. You can always add adapters, but they cost extra and can be hard to find.

Despite these limitations, fiber optics are gaining popularity for data transmission. Costs are coming down, and data rates and network sizes are both going up—trends that favor fibers. I'll look next at the important advantages fibers offer that can offset higher costs.

High-Speed or Long-Distance Transmission

The steady rise in computer data transmission speeds, coupled with growth in network sizes and the need to connect remote devices, pushes the spread of fiber optics. For many wide area networks, fiber may be the only technology capable of providing enough data capacity over long enough distances. However, this is less common for current point-to-point links and local area networks; even 100 Mbit/s signals can go short distances through wires.

As with residential telephones, the fiber need not go all the way to the end user. High-capacity fiber cables may run from the basement to each floor, then branch out on each floor to serve several local nodes. Copper wires can branch out from the local nodes to carry the signals the short final distance to desktops.

Upgrade Capability

Even if the high capacity of fiber is not absolutely necessary now, it may be useful in the future. Few people are going to rip out a functioning wire system to make room for expansion sometime in the future. However, the prospects for expansion make fiber more attractive where new cabling must be installed, particularly in new buildings. Labor

The vast bulk of computer data communications is over wires.

Wire links are cheaper than fibers.

Some applications require fibers to meet speed and/or distance requirements.

Fiber-optic cables make future upgrades easier.

accounts for much of the cost of cable installation, and it takes much less labor to install cable in a new building than to string new cable through existing walls.

Suppose, for example, you were moving into a new building with an expected life of 20 years and were given the cost estimates shown in Table 20.1. The fiber hardware would cost 40% more, but installation would be only slightly more expensive, so overall cost would be 18% higher. If you're on a tight budget, that might seem to tip the scales toward copper. But then look at the high labor costs of retrofitting fiber after the building is completed. With data-transmission demands expected to continue increasing in coming years, spending a little extra on fiber initially to allow for upgrading could save big money in the long term. (This example shows that the penny-wise saving of $3,000 by picking copper in the first place could cost $33,000 extra in the long run.)

Table 20.1. Hypothetical costs of fiber and copper installation in new building.

Costs	New Copper	New Fiber	Retrofit Fiber Later
Hardware	$5,000	$7,000	$7,000
Labor	$12,000	$13,000	$26,000
Total Cost	$17,000	$20,000	$33,000

Immunity to Electromagnetic Interference

One advantage of optical transmission is its immunity to electromagnetic interference (EMI), a common source of noise. EMI arises from the basic properties of electromagnetism. Changing currents generate magnetic fields, and magnetic field lines can generate (or induce) an electrical current when they cut across a conductor. This can generate noise, as shown in Figure 20.4, when two wires come close together. The strong current in the upper wire induces a current in the lower wire, which added to the input signal becomes noise. If the noise power is high enough, it can overwhelm the input signal. A noise spike, caused by a sudden surge of current, can add a spurious data bit to a digital signal.

You don't need two wires next to each other to get EMI. All you need is passage of magnetic flux through a conductor. That's how a radio or television antenna picks up broadcast signals. Stray magnetic fields can induce a current in any wire exposed to them. Often those currents are simple noise (e.g., the crackle on AM radio when a nearby light is switched on). Sometimes the result is crosstalk (e.g., a radio program in the background on your telephone line).

⬤
Fibers are immune to electromagnetic interference, which can cause noisy data transmission in wires.

FIGURE 20.4.

Induction of noise currents in a wire carrying an electrical signal.

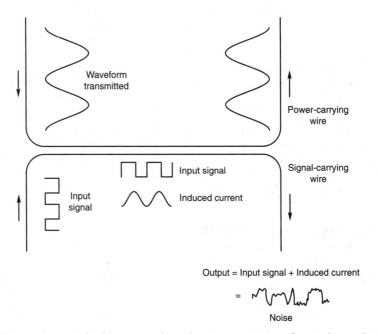

Shielding, as in coaxial cables, can reduce electromagnetic interference by weakening the magnetic field that reaches the inner conductor, but even coax isn't immune to EMI. Optical fibers are immune because they transmit signals as light rather than current. Thus, they can carry signals through places where EMI would otherwise block data transmission—from power substations to hospitals with radio paging systems.

(I should note, however, that fiber-optic transmitters and receivers include electronic circuits that can pick up EMI, so fiber systems are not perfectly immune to EMI. Careful design is required to shield transmitters and receivers exposed to EMI.)

Data Security

● Fibers do not emit electromagnetic fields that can be tapped by eaves-droppers.

Magnetic fields and current induction work two ways. They don't just generate noise in a signal-carrying conductor; they also let information leak out. Changes in signal current cause fluctuations in the induced magnetic field outside a conductor, which carry the same information as the current passing through the conductor. That's good news for potential spies because they can eavesdrop on these magnetic fields without cutting the cable. Shielding the wire, as in coaxial cable, can alleviate the problem, but bends and imperfections can let some signals leak out.

There are no radiated magnetic fields around an optical fiber—the electromagnetic fields are confined within the fiber. That makes it impossible to tap the signal being transmitted through a fiber without bending the fiber (to induce bend losses) or cutting into the fiber.

That would increase fiber loss sharply, in a way easy for users of the communication channel to detect, making fibers a much more secure transmission medium. This has led to fiber-optic links in some secure government data networks.

Nonconductive Cables

Subtle variations in electrical potential between buildings can cause problems in transmitting electrical signals. Electronic designers assume that ground is a uniform potential. That is reasonable if ground is a single metal chassis, and it's not too bad if ground is a good conductor that extends through a small building (e.g., copper plumbing or a third wire in electrical wiring). However, the nominal ground potential can differ by several volts if cables run between different buildings or sometimes even different parts of the same building.

That doesn't sound like much—and in the days of vacuum tube electronics, it wasn't worth worrying about. However, signal levels in semiconductor circuits are just a few volts, creating a problem called a ground loop, which isn't mentioned in many engineering textbooks. When the difference in ground potential at two ends of a wire is comparable to the signal level, stray currents start to cause noise. If the differences grow large enough, they can damage components. Electric utilities have the biggest problems because their switching stations and power plants may have large potential differences. Fiber optics can avoid these problems, as long as the nonconductive glass fiber is enclosed in a nonconductive cable. Many sellers of computer data links recommend that fibers be used for interbuilding connections to avoid noise and ground-loop problems.

Nonconductive cables can also alleviate a serious concern with outdoor cables—lightning strikes, which can cause power and voltage surges large enough to fry electronics on either end.

> Nonconductive fiber-optic cables are immune to ground loops and resistant to lightning surges.

Eliminating Spark Hazards

In some cases, transmitting signals electrically can be downright dangerous. Even modest electric potentials can generate small sparks, especially at switches. Those sparks ordinarily pose no hazard, but they can be extremely dangerous in an oil refinery or chemical plant where the air contains potentially explosive vapors. One tiny spark could make one very large and deadly boom. Again, nonconductive fiber-optic cables can avoid this hazard.

> Sparks from electrical wires can be dangerous in explosive atmospheres.

Ease of Installation

Increasing transmission capacity of wire cables generally makes them thicker and more rigid. Shielding against EMI or eavesdropping has similar effects. Such thick cables can be difficult to install in existing buildings where they must go through walls and cable ducts. Fiber cables are easier to install because they're smaller and more flexible. This helps reduce installation costs, especially in cases where only fiber cables are small and flexible enough to fit through existing ductwork in buildings or underground.

> Fiber cables can be much easier to install than metal cables.

The small size, light weight, and flexibility of fiber-optic cables also make them easier to use in temporary or portable installations.

Point-to-Point Fiber Links

I mentioned earlier that fibers are used both for point-to-point data links and in networks. The two applications differ in important ways, and I'll talk about them separately.

Point-to-point fiber data links work on the same principles as point-to-point telecommunications, but they operate over shorter distances and at lower data rates. Some operate at standard rates in the digital telephone hierarchy, usually 1.5 and 45 Mbit/s, but most operate at rates used in data communications. Their design is less standardized than networks, but most are made to interface with standard computer equipment. Fiber data links do not compete with copper link for general applications, but they are used in cases where fiber performance offers important benefits, such as those described earlier.

> ● Point-to-point fiber links usually work at lower data rates and shorter distances than point-to-point telecommunications.

An array of point-to-point links can function as a network if the nodes attached to them can route the signals. This forms what is called a switching fabric. For example, the central node of a star local area network could route signals between pairs of other nodes. You could even think of a personal computer as having similar capabilities, such as sending data from a disk to a printer.

Technology

Point-to-point data communications rarely push the limits of fiber-optic technology in distance or data rate, and cost is usually a major design constraint. This leads to widespread use of multimode fibers at 850 or 1300 nm, and sometimes to the use of all-plastic fiber. LED sources are used as long as they meet performance requirements.

> ● Point-to-point fiber links rarely push technical limits.

Many commercial point-to-point fiber links accept input and deliver output in standard electrical formats. Some come with integral electrical connectors that plug into standard sockets on computers or peripheral devices. Some companies call such products modems or fiber-optic modems because their function is analogous to that of electronic modems, which convert digital data for transmission in different form over telephone lines. Normally, transmitter and receiver are packaged together. (Note, however, that such fiber-optic modems can't send data over wire telephone lines.) Most fiber-optic modems are made to replace electrical cables and be functionally transparent or invisible to the user.

In practice, many fiber-optic data links are semi-customized products assembled from standard components. The transmitter and receiver are standard components, adaptable for use with various connector types. Specifications indicate allowable transmission loss through different fibers. Often there is a minimum acceptable loss (to prevent receiver overload), as well as a maximum loss. Typical loss budgets are 10 to 15 dB for LED sources. Suppliers cut the cable to length, mount the desired components, and plug the proper pieces together to produce a system.

Multiplexed Transmission

Many fiber-optic data links take advantage of the high capacity of fiber optics to replace many parallel metal cables. The transmitter multiplexes the input signals to generate the optical output, and the receiver demultiplexes the signals. Individual channels may be transmitting at standard telephone speeds or computer data transmission rates. These systems are called fiber-optic multiplexers.

● Multiplexed fiber links can replace many parallel cables.

Applications in Point-to-Point Transmission

Most fiber data links are installed to meet requirements that cannot be met as economically (or, often, at all) with wires or coaxial cables. For example, multiplexing many signals onto a single-fiber cable saves the cost, installation, and handling problems of many parallel cables between two points. Other common rationales for installing fibers include the following:

● Most point-to-point fiber links are installed to meet special requirements.

● Need to transmit at higher speeds or over longer distances than is possible with copper wires, either at present or expected in the future.

● Interference problems, from electrical power lines, radio-frequency signals, or other sources that blocked data transmission.

● Differences in potential between two points, which could cause problems if signals were transmitted electrically.

● Saving installation costs by using routes where metal cables could not fit or would not work (e.g., alongside power lines).

● Meeting stringent security requirements of military or security agencies, or of financial institutions, concerned that no one can tap data transmission.

Many such applications are fairly routine. Both security concerns and electromagnetic interference are common in otherwise ordinary-seeming office environments. Space constraints, interference problems, and high data rates lead to the use of fiber links inside high-performance computers and telephone switching systems (which are essentially special-purpose computers). Heavy machinery in factories generates severe EMI, which can affect computer control systems. Other applications can be unusual. They include the following:

● Transmission of data from a screen room used in electromagnetic testing to the outside world. The screening that shields the room from intense electric and magnetic fields generated in the laboratory prevents electronic transmission.

● Transmission of data from underground nuclear tests to the surface. Cables must hang down deep, narrow shafts to instruments near the bomb blast. Because each test costs tens of millions of dollars, it is crucial to collect as much data as possible—putting a premium on high-speed transmission. (The cables near the bomb do not survive the tests; they collect and transmit data for the milliseconds between the detonation and their destruction.)

Fiber Optic Data Networks

●
Fibers play different
roles in different
standardized
networks.

Fiber optics have found growing use in computer data networks as data rates have increased. The higher the data rate and the larger the area covered by the network, the more likely fibers are to be used. Thus fibers are more likely to be used in high-speed wide area networks or metropolitan area networks than in slower local area networks.

Countless network architectures have been developed and tested, but only a few have gained wide acceptance. These common networks use standardized designs and have standardized interfaces with devices attached to them. Fiber optics play a variety of roles in these networks. They are optional accessories to extend the performance of Ethernet, for example, but are a basic element of the higher-speed Fiber Distributed Data Interface. I will look first at how fibers can enhance copper network performance, then at networks that explicitly incorporate fiber.

Fiber Optics in Ethernet

●
The 10 Mbit/s
Ethernet LAN was
designed for coax,
but fiber can be
added.

The first LAN to gain much acceptance was the Ethernet standard developed by Digital Equipment, Intel, and Xerox, and codified as IEEE (Institute of Electrical and Electronics Engineers) standard 802.3. The standard Ethernet distributes digital data packets of variable length at 10 Mbit/s to transceivers dispersed along a coaxial cable bus, as shown in Figure 20.5. Separate cables up to 50 m long, containing four twisted wire pairs, run from the transceivers to individual devices (e.g., personal computers, file servers, or printers). The network can serve up to 1,024 terminals.

FIGURE 20.5.

*Basic elements of
Ethernet.*

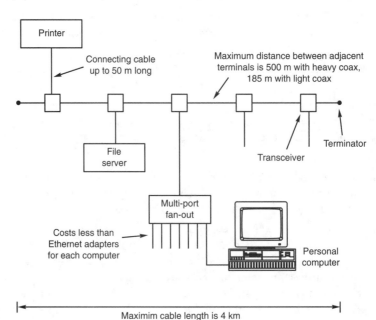

The network has no overall controller; control functions are handled by individual transceivers. If a terminal is ready to send a signal, its transceiver checks if another signal is going along the coaxial cable. Transmission is delayed if another signal is present. If not, the terminal begins transmitting and continues until it finishes or detects a collision—the transmission of data at the same time by a second terminal. Such collisions happen because it takes time—several nanoseconds a meter—for signals to travel along the coax. If the delay is 6 ns/m, a collision would occur if two terminals 300 m apart on the coax started sending within 1.8 μs of each other. The terminal stops transmitting if it detects a collision, and waits a random interval before trying again.

Data signals contain an address header specifying the terminal to which they are directed. The signals pass through all transceivers, but only the addressed one sends the signals to the terminal attached to it.

There are some important variations on the basic Ethernet design. The standard heavy coaxial cable allows transceivers to be up to 500 m apart, but it is expensive. Substituting a lighter grade of coax limits transceiver spacing to no more than 185 m, but this "thin" Ethernet is adequate for many purposes. Another alternative is using twisted wire pairs, which can carry signals no more than 100 m; typically wire-pair Ethernets are arranged in a star configuration, with cables radiating outward from a hub, which can be connected to other hubs by standard Ethernet coax.

Optical fibers can stretch transmission distances beyond the limit imposed by loss of the coaxial cable to limits imposed by delays in signal propagation. Typically, a point-to-point fiber link connects two coaxial segments of the Ethernet, allowing pairs of transceivers to be separated by up to 5 km (the distance corresponding to round-trip time of 45 μs). This allows a single Ethernet to link terminals in different buildings, which can be difficult with the 500 m limit of standard Ethernet coax and especially with the 185 m limit of thin coax. It also allows connection of a few remote terminals to a central Ethernet. Because transmission distances are short, multimode fibers are typical, although single-mode fibers can also be used.

> ●
> Optical fibers can stretch Ethernet beyond the normal limit of coax.

Other fiber-optic variations on Ethernet are possible, but not as common. All-fiber LANs based on Ethernet-like concepts have been built, largely for special-purpose applications.

Token-Ring Networks

Another widely used LAN is the IEEE 802.5 Token-Ring network, developed by IBM. The basic design is a ring of nodes, connected by twisted-wire pairs, which operates at 4 or 16 Mbit/s. Terminal devices are connected to the nodes.

> ●
> Fibers can extend transmission distances in token-ring networks.

The token-ring network gets its name from the way it controls transmission. Instead of having terminals contend for space to send signals, as in Ethernet, this system uses an authorization code called a token. If no message is being sent, the token is sent around the loop. When a node with a message to send receives the token, it holds the token and sends the message, which includes a code identifying its destination. All other nodes ignore the message, which is canceled when it completes its path around the ring. Then the token is passed around the ring again. This scheme is more efficient than contention in heavily used networks.

The distance between nodes is limited by power, and is typically limited to about 100 m with wire. (Higher-power repeaters can extend this distance to about 300 m.) Multimode fiber links can extend node spacing to about 3 km. The fiber links are particularly valuable for interbuilding transmission because they avoid ground loops and potential variations, and some suppliers recommend them.

Fiber Distributed Data Interface (FDDI)

●
The 100 Mbit/s
FDDI standard LAN
is based on fiber
optics.

The Fiber Distributed Data Interface (FDDI) network standard, which operates at 100 Mbit/s, is the next big step above Ethernet and token-ring networks. The FDDI standard calls for the ring topology shown in Figure 20.6, with two rings that can transmit signals in opposite directions to a series of nodes. It also specifies concentrator-type terminals that allow stars and/or branching trees to be added to the main FDDI backbone ring. Normally one ring carries signals while the other is kept in reserve in case of component or cable failure. Data transmission in each direction is 100 Mbit/s. FDDI uses the same token-passing scheme as the IEEE 802.5 Token-Ring network to control transmission around the loop.

FIGURE 20.6.

*FDDI's dual
fiber ring.*

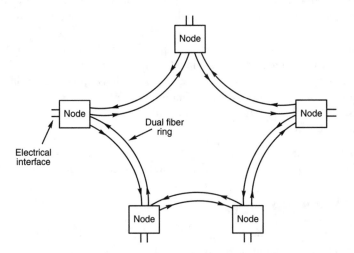

FDDI uses a 4 of 5 transmission code, which adds one extra bit for every four data bits. This means the actual data rate is 125 Mbaud for 100 Mbit/s of user data. This coding scheme balances transmission between on and off bits to enhance operating efficiency.

●
Copper wires can
carry FDDI signals
short distances.

The standard was developed around fiber-optic transmission, but copper wires can be used to carry 100 Mbit/s signals short distances, and wired versions of FDDI have been developed. (They are sometimes called CDDI, with the C from copper substituted for the F from fiber.) Efforts are underway to standardize such networks.

FDDI is nominally a local area network, but in practice it often serves as a backbone network, linking LANs that operate at slower speeds, such as Ethernet and Token-Ring networks. In that case the nodes are gateways to other networks. Devices called concentrators are attached to FDDI nodes to combine signals from many terminals, or to collect signals from a LAN for transmission to the FDDI network. Higher data rates and the growth of video- and graphics-intensive applications may lead to more use of FDDI as a local area network.

NODES

Each FDDI node includes a transmitter, receiver, and possibly an optional optical bypass switch, as well as an electronic interface to the terminal, as shown in Figure 20.7, for one fiber in the ring. In normal operation, the receiver/transmitter pair acts as a repeater that monitors the signal it receives. The receiver detects and amplifies the signal and passes it along to be decoded. If the message is not addressed to that terminal, the signal will be regenerated and passed along to the next terminal, but not processed by the terminal. If the signal is addressed to that terminal, it will be passed on to the terminal (via the electronic connections on top) as well as regenerated and transmitted to the next terminal. This approach allows transmission of signals through up to 2 km of fiber between nodes. The maximum length of the entire ring, constrained by default settings of recovery timers, is 200 km, passing through a thousand nodes.

> FDDI nodes include transmitter, receiver, bypass switch, and a terminal interface.

The dual counter-rotating ring topology of FDDI allows the network to continue operating in the event of a failure, such as a cut cable or a failed node. When the failure is detected, the FDDI network bypasses the problem area, doing a "wrap-around" that converts the network into a single ring until the problem is fixed. An optional optical bypass switch can be used as an extra network survival tool for nodes that may be turned off (such as individual computer terminals). More typically, terminals that may be turned off are connected to the network via an FDDI concentrator, avoiding the need to bypass turned-off nodes.

FIBERS

The FDDI standard specifies some details of the transmission equipment:

> FDDI uses multimode fiber and 1300 nm LEDs.

- Multimode graded-index fiber with 62.5 μm core and 125 μm cladding is recommended, but 50/125 and 100/140 fiber can also be used. Modal bandwidth should be at least 500 MHz-km at 1300 nm. Attenuation between nodes must not exceed 11 dB (or 7 dB for the low-cost FDDI standard).

- 1300 nm transmission, to take advantage of low fiber loss and dispersion and allow use of inexpensive LED sources.

- *pin* photodiode detectors because of their lower cost and better reliability than 1300 nm APDs. Minimum power needed at the detector for a bit-error rate of 1 in 2.5×10^{10} bits must be no higher than -27 dBm.

- The FDDI standard also provides for the use of single-mode fiber with four combinations of laser transmitters and receivers, depending on transmission distance.

FIGURE 20.7.

A node in an FDDI network.

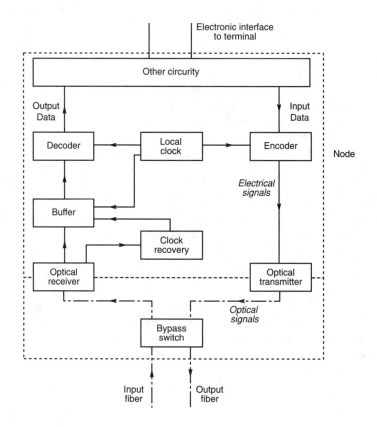

Fiber Channel

Fiber Channel provides data rates to 100 Mb/s.

Standards are in development for a higher-speed wide area network called Fiber Channel (planners use the British spelling, Fibre). It will carry signals at the rates shown in Table 20.2. Each 8-bit byte of data is encoded as a 10-bit word, and the signaling rate includes an overhead of 6.25%, so the rates do not appear to match at first glance. The standard allows transmission over 10 km on single-mode fiber, 2 km on multimode fiber, and up to 50 m (depending on data rate) on coax. Fiber transmission can be at 780 or 1300 nm through multimode fiber, or at 1300 nm through single-mode fiber.

Table 20.2. Fiber channel data rates.

Raw Signal (Megabaud)	Minimum User Data Rate (Megabytes/second)
132.6	12.5
256.6	25
531.2	50
1062.5	100

Fiber Channel will offer different "grades" of service, depending on whether delivery of data packets is guaranteed by the system. Variations will be able to serve as a switched network similar to the telephone system, as a hub-based system broadcasting signals to all attached nodes, or as a ring system.

Other Fiber Networks

Other fiber-optic networks are in development. Most are proprietary or in the research stage, but a few may emerge as standards. Many concepts are still evolving, such as basing data networks on the asynchronous transfer mode (ATM) described in Chapter 15. Developers are also considering a next-generation variation on FDDI, which would carry a larger variety of services at higher speeds. If history is any guide, most of these ideas will fall by the wayside, but a few may emerge eventually as important new systems.

What Have You Learned?

1. Computers may exchange data over point-to-point links between pairs of devices or over local area networks (LANs), which link many nodes or devices. Wide area networks link many LANs and may connect directly with other devices. Networks allow pairs of devices attached to the network to communicate directly with each other.

2. Local area networks come in star, ring, and bus configurations.

3. The vast bulk of data communications is over wires because wire systems are cheaper than fiber equipment.

4. Fibers are used in applications where wires will not meet speed and/or distance requirements. Fibers may also be installed when new networks are built to provide capacity for future upgrades. Continuing increases in data-transmission requirements are leading to wider use of fiber data links and networks.

5. Fibers are immune to electromagnetic interference, so they are used where EMI would degrade transmission on wires. Fibers also do not radiate electromagnetic fields, so they are used where data security is critical. The small size and flexibility of fiber cables can simplify installations.

6. Fiber links are recommended between buildings because they are not affected by small variations in the ground voltage level at separate locations, which can cause problems over electric wiring.

7. The nonconductive nature of fibers avoids spark hazards and damage to electronic equipment from power surges.

8. Multiplexing on fiber cables can replace many parallel wire links.

9. Standards for the Ethernet LAN call for coaxial-cable transmission, but fibers can be used to extend the network over larger areas than otherwise possible.

10. Fibers can extend transmission distances on Token-Ring networks beyond the usual limits of twisted wire pairs.

11. The FDDI LAN is based on fiber-optic transmission of 100 Mbit/s over a ring network. FDDI is often used as a wide area network to connect smaller LANs. FDDI nodes convert optical signals to electronic form to drive terminal equipment and optical transmitters.

12. The Fiber Channel standard provides data rates to 100 megabytes per second (1,062 Mbit/s).

What's Next?

In Chapter 21, I will look at some interesting fiber-optic communications outside the normal world of telecommunications and data transmission—military and vehicle systems.

Quiz for Chapter 20

1. Point-to-point data transmission involves

    a. transmission of signal from a central computer to remote nodes.

    b. interchange of data between pairs of devices.

    c. communication between any two devices connected by a common transmission network.

    d. the connection of two devices through a switched network like the telephone system.

2. Which of the following is a local area network (LAN)?

    a. A system that interconnects many nodes by making all signals pass through a central node.

    b. A ring network with a transmission medium that passes through all nodes.

    c. A common transmission medium or data bus to which all nodes are connected but which does not form a complete ring.

    d. All of the above.

    e. None of the above.

3. What makes optical fibers immune to EMI?

   a. They transmit signals as light rather than electric current.

   b. They are too small for magnetic fields to induce currents in them.

   c. Magnetic fields cannot penetrate the glass of the fiber.

   d. They are readily shielded by outer conductors in cable.

4. The most important drawback of optical fibers for point-to-point data transmission is

   a. that connections are difficult to make.

   b. the higher costs of fiber equipment.

   c. that it cannot provide electrical grounding.

   d. that they do not operate properly at low data rates.

   e. that it is difficult to upgrade.

5. Why would you install a fiber-optic network in a new building rather than a less-expensive wire system if both could meet current requirements?

   a. To avoid ground loops.

   b. Fiber is simpler to install in a new building.

   c. To provide future upgrade capability at a cost much lower than retrofitting later.

   d. To get a big kickback from the fiber supplier.

6. Why are fiber-optic cables recommended for connections between buildings?

   a. To avoid ground loops.

   b. Fiber is simpler to install between new buildings.

   c. To provide future upgrade capability.

   d. To avoid eavesdropping by industrial spies.

   e. To get a big kickback from the fiber supplier.

7. When would optical fibers be used in an Ethernet-type LAN?

   a. Never, the standard calls for coaxial cable.

   b. To extend transmission distance to reach remote terminals.

   c. Routinely, the standard calls for optical fiber.

   d. When transmission speeds exceed 50 Mbit/s.

   e. When the stockroom is out of coaxial cable.

8. The FDDI standard calls for nodes to be

   a. attached to a data bus in the form of a transmissive star coupler.

   b. attached to a pair of fiber rings and a terminal device.

   c. attached to a network of star couplers that detect collisions and transmit only one signal to the next level.

   d. attached to fibers by passive T couplers.

9. Which type of light source and fiber are chosen for FDDI networks?

   a. Single-mode fiber and 1500 nm lasers.

   b. Plastic fiber and 660 nm LEDs.

   c. Multimode fiber and 1300 nm lasers.

   d. Multimode fiber and 1300 nm LEDs.

   e. Multimode fiber and 850 nm lasers or LEDs.

10. What is the maximum data rate in the Fiber Channel standard?

   a. 1 Mb/s.

   b. 10 Mb/s.

   c. 100 Mb/s.

   d. 1 Gb/s.

   e. None of the above.

21

Mobile and Military Fiber Communications

About This Chapter

In the past four chapters, I have examined a wide range of fiber-optic applications in voice, data, and video communications. Some important fiber-optic communication applications do not fall neatly into these categories. One broad area is mobile communications, including cables to moving vehicles (from surface robots to guided missiles), and cables installed in various types of vehicles (planes, ships, cars, and spacecraft). A second—often overlapping—category is military communications. I'll talk first about mobile communications, including many military systems. Then I'll cover other military communications that are either portable or fixed in place permanently. This chapter will show you why fibers are used, and how these special technologies differ from those in more conventional communication systems.

Remotely Controlled Robotic Vehicles

When we think of remotely controlled vehicles, most of us think first of radio-controlled toys that zip across the floor until the batteries run down. Radio controls are cheap and simple, but limited. You can command your radio-controlled car to go faster, slower, forward, backward, or turn right or left—but not much more.

Control of advanced robotic vehicles is a far more demanding job. The operator needs video transmission from a camera in the robot to see the local environment. Other environmental sensing information may also be needed, such as temperature and pressure readings. Signals must flow in the opposite direction so the operator can control the vehicle. Fiber-optic cables carry signals in both directions for a variety of remotely controlled vehicles, often using wavelength-division multiplexing so two signals can travel in opposite directions through a single fiber at different wavelengths. Although care must be taken to protect the cables, fiber-optic cables can work in places where radio signals cannot, including underwater and in electromagnetically noisy environments. Small single-fiber cables can also be made quite rugged and better able to survive being run over than heavier metal cables. Other fiber advantages include their ability to carry high-bandwidth signals over greater distances, and their light weight.

> ●
>
> Fiber-optic cables carry signals to control robotic vehicles.

Remotely controlled robots are not widely used today, but they are attractive for a variety of applications. Their prime advantage is that they can go into places unsafe for humans. Robots can probe the radioactive parts of nuclear reactors, to make measurements or repairs, or to dissemble old reactors at the ends of their operating lifetimes. Robots can descend deep into the ocean or walk across the surface of the moon or Mars. Robots can be scouts for armies, and they can even deliver weapons to their target (we call them guided missiles). Let's look at how some of this technology works.

Fiber-Optic Guided Missiles

Guided missiles are, in a sense, simple robots with rather deadly missions—to deliver weapons to their targets. The Pentagon has developed a system that uses a ruggedized optical fiber to carry control signals to and from a short-range missile on the battlefield, letting a soldier guide it to its target from a safe hiding place. Called FOG-M, for fiber-optic guided missile, the program gives a good idea of how remote control through optical fibers works.

> ●
>
> A fiber can send images from a television camera in a missile back to a soldier guiding the missile to its target.

The idea of FOG-M, shown in Figure 21.1, is to send images from a video camera in the missile to a soldier on the ground, who guides the missile to its target. The missile is launched toward the target with a single uncabled optical fiber trailing from it to the launcher. Looking at the video image, the soldier operates controls that direct the missile to its target. The soldier can follow the missile right to impact.

Military agencies like FOG-M because the soldier operating it keeps safe under cover. This is an advantage over laser-guided bombs, where the laser operator must be in line with the target to project a laser spot onto it. Missiles can be guided by wire, but the wires can't carry video images, which allow more accurate guidance. Only an optical fiber has the combination of small size, light weight, strength, low attenuation, and bandwidth needed to transmit video signals over the 10 or 20 km needed for missile guidance.

> ●
>
> Fiber unwinds from a special reel on the missile.

The basic hardware for FOG-M is shown in Figure 21.2. The missile contains a video camera, a fiber-optic video transmitter, a low-bandwidth receiver, and a special reel of fiber. One end of the fiber is mounted on the launcher, so it remains behind when the missile is fired. As the missile speeds toward its target, the fiber rapidly unwinds from the reel, forming a long arc over the battlefield. The reel is a critical component, because the fiber will tangle or break if not unwound at the right rate. The camera transmits a video image to a soldier at the launcher, who sends control signals back through the fiber to guide the missile to keep the target in the proper place in his field of view.

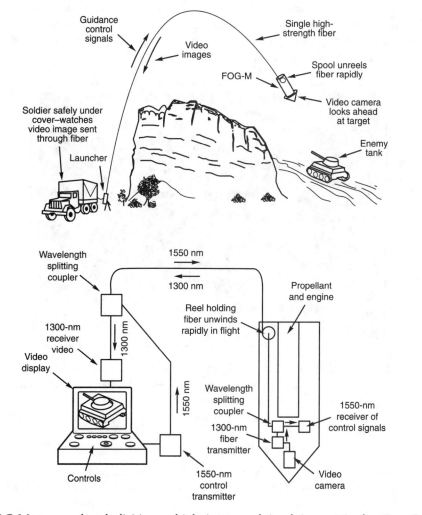

FIGURE 21.1.

A soldier guiding a FOG-M missile to its target.

FIGURE 21.2.

FOG-M components.

FOG-M uses wavelength-division multiplexing to send signals in opposite directions. In this drawing, video signals from the missile are sent at 1300 nm through fibers with low dispersion at that wavelength. Low-bandwidth control signals are sent at 1550 nm in the opposite direction.

Robotic Vehicles on Land

Fiber cables can control robotic land vehicles for scouting or weapon delivery.

The same principles can be applied to robotic vehicles for use on land. Much of the work is military. The Marine Corps has tested unmanned ground vehicles, linked to a controller through a fiber-optic cable, like the one shown in Figure 21.3. In addition to video, the controller receives signals from sensors on the vehicle. Those inputs let the controller drive the vehicle, and even fire machine guns mounted on it. Fiber cables are immune to electromagnetic interference and electromagnetic pulse effects that could block radio or wire communications, vital considerations on the battlefield.

FIGURE 21.3.

Remotely operated vehicle is controlled by fiber cable. (Courtesy of AT&T)

Ruggedization of the cable is critical for military systems because the vehicle is virtually certain to run over it. In the Marine Corps experiments, a ruggedized 2.5-millimeter cable carried signals to and from a modified "HMM-WV"—the high-mobility multipurpose wheeled vehicle designed as a successor to the jeep. The Army is working on a robotic tank, called ROBAT for Robotic Obstacle Breaching Assault Tank, which may require an even more rugged cable 4 mm in diameter.

The high bandwidth of fiber cables makes it possible to consider futuristic visions of using virtual reality techniques to operate unmanned ground vehicles. Sensors on the vehicle would scan the area, serving as the operator's "eyes," while other sensors would listen for sounds and "feel" the terrain. The sights, sounds, and feel would be conveyed to the operator—far from the vehicle—using screens, speakers, and perhaps a moving chair. The goal would be to make the operator feel as if he or she were driving the vehicles, but without being exposed to the dangers faced by the vehicle.

Remote-controlled robots could also serve many nonmilitary purposes in environments hostile or dangerous to humans. Robots could inspect the "hot" interiors of nuclear power plants or perform needed repairs inside the reactor. The robots could be left inside the reactor permanently if they became contaminated. Eventually, specialized robots could be used to dismantle old reactors, without exposing people to the highly radioactive materials inside. Likewise, remotely controlled robots could clean up hazardous wastes. A fiber-optic cable was used to control a multilegged robot built to climb into a hazardous Antarctic volcano and collect data. As with military vehicles, the immunity of fiber-optic cables to EMI and EMP effects is a big plus.

●
Remotely controlled robots could clean up hazardous wastes and dismantle old nuclear reactors.

Submersible Robots

Radio links can substitute for cables in some land applications, but most radio waves don't penetrate far into water. Cables or acoustic signaling is required to maintain contact with submerged vessels. Although manned submersibles can operate without a continuous link to the surface, only cables can provide the transmission capacity needed to remotely operate a submerged vessel. Fiber cables are preferred because of their high bandwidth and durability.

●
Hybrid fiber-electrical cables carry signals and power to remotely operated submersibles.

Hybrid fiber and electrical cables carry the signals and power needed to steer and accelerate submerged vessels, as well as bring video signals and telemetry to the surface, where shipboard operators can monitor them.

Fiber cables also allow operators of a manned submersible to control unmanned robotic vehicles that can be sent into small spaces or dangerous areas. The most famous example came when Robert Ballard's team from the Woods Hole Oceanographic Institution discovered the sunken wreck of the *Titanic* in 1985. The scientists discovered the wreck with *Alvin,* a submersible that carried three people. However, they did not dare to explore the inside of the deteriorating wreck. Instead, they used a 250-pound robot, tethered to *Alvin* with a fiber-optic cable that carried control signals. It was this fiber-controlled robot that photographed details of the dark interior of the wreck.

Fibers in Aircraft

It was not too long ago that most aircraft were controlled by hydraulic systems. When the pilot moved a lever, it would cause hydraulic fluid to move a control surface (e.g., a wing flap), much as hydraulic brakes work in an automobile. Newer planes have fly-by-wire electronic controls that send electronic signals to motors that move control surfaces. Modern aircraft—particularly military planes—also use many electronic systems and sensors, adding to signal transmission requirements. These include radars, navigation and guidance systems, and—in military planes—weapons systems with automatic targeting capabilities and electronic countermeasure equipment.

●
Fibers can serve as the control networks for aircraft.

Like many other users faced with increasing communication requirements, the aerospace industry and the military began investigating fiber optics. In fact, the Pentagon began looking seriously at airborne fiber-optic transmission in the 1970s. An added motivation was development of composite nonmetallic materials for aircraft skins. Such materials are stronger per pound than metals, but unlike a metal fuselage they cannot shield the inside of the plane from electromagnetic interference and potential enemy countermeasures. Using fiber could overcome such problems, as well as reduce cable weight.

New military planes use some fiber links, but fibers are not yet standardized for aircraft.

It takes years for new technology to work its way into military hardware, and complete military specifications have yet to be completed for aircraft fiber optics. Some optical fiber is used in the B-1 bomber, which was originally designed in the 1970s when fiber was new and largely untested. Other planes also use some short lengths of fiber. A 6 m (20 ft) length of fiber cable carries sensor data in the Marine Corps' AV-8B harrier attack aircraft. Future aircraft might benefit from more extensive use of fiber. Calculations indicate that the B-1's weight could have been reduced by as much as a ton if fiber had replaced all its wire cables. Such weight reductions could mean greater range, lower fuel requirements, or higher load capacity.

Few design details have been released on Stealth aircraft, but they probably make much more extensive use of fibers. Stealth technology depends on reducing radar visibility, and that, in turn, requires minimizing the use of metal. The lack of a metal fuselage would leave advanced electronic systems vulnerable to electromagnetic interference and enemy countermeasures, pushing a shift to fiber. Fiber transmission would also help keep the wiring in Stealth planes from radiating its own EMI, which an enemy might spot.

The short distances involved in systems installed on single aircraft make multimode fibers the logical choice for on-board fiber systems. Standard military specifications call for radiation-hardened fiber, a heritage of the cold war. Most military specifications call for radiation hardening of all electronic and communication equipment.

Fibers are attractive for use in large satellites.

Light weight and immunity to electromagnetic interference are also crucial for spacecraft, making fiber-optic systems attractive, especially for large satellites or manned space stations. NASA has been working with fibers for years; it installed one of the first large systems of underground fiber-optic cables at the Kennedy Space Flight Center in Florida in the 1970s.

Shipboard Fiber Systems

Big ships have massive communication requirements.

The communication requirements of a big ship may rival those of an office building. Ships have their own telephone networks to keep officers and crew in contact. They have a variety of sensing and weapon systems, as well as radars, sonar systems, and radio links with military communication systems. Modern ships have computer rooms, both to control weapon systems on board and to analyze their situation. Computers have become so important that the Navy has begun installing local area networks that are essentially militarized versions of FDDI. Figure 21.4 shows how optical fibers can link some of these systems together.

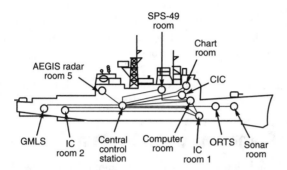

SPS-49
room

Chart
room

AEGIS radar
room 5

CIC

GMLS   IC
room 2

Central
control
station

Computer
room

IC
room 1

ORTS   Sonar
room

FIGURE 21.4.

Cabling in a large ship. (Courtesy of AT&T)

The U.S.S. Washington, launched in July 1992 from Newport News Shipyard, makes extensive use of fibers. Fibers provide the backbone of the ship's telephone system. Fiber links reach all the way to the desks of officers. Fibers collect signals from cameras on the exterior of the ship and carry information to and from many other systems.

Weight is not as critical on ships as it is on aircraft, but scrapping metal cable systems can save space. Fibers are also immune to electromagnetic interference, which can be a problem when mechanical and electronic systems are crammed together on a ship. EMI immunity is also important to avoid potential enemy countermeasures designed to knock out electronic systems.

As in aircraft, the usual choice of fiber is radiation-hardened 62.5/125 fiber, although the Navy is looking at single-mode fiber. The Navy systems use special water-blocking cables, designed to prevent water from flowing through empty parts of the cable itself.

Military Systems

The Pentagon has developed a host of fiber-optic systems with cryptic code numbers and acronyms. Many systems are developmental, in the early stages of establishing feasibility. Some systems are fixed strategic-communication networks that operate much like civilian telecommunication lines. Others are large programs that use a small amount of fiber. Some are classified. It is impossible to describe the whole range of military systems, but I will give a few examples of how and why fibers are used.

Radar Remoting

Radar is invaluable in spotting and tracking enemy aircraft, but it also has an important drawback. Microwave emissions from radar dishes are a target for enemy radiation-seeking missiles. Military planners figure they are bound to lose a few radar dishes in a battle, so they want to put the dishes far from ground control centers. That lets the

●

Fibers carry signals
from remote radars
to control centers.

control center—and the soldiers operating it—survive if a radiation-seeking missile takes out a radar dish.

That was difficult when only metal cables were available. Military radar generates an analog intermediate-frequency output at 70 MHz. That frequency is high enough to carry the signal but too high to go far through metal cable. The only way to locate the radar dish a comfortable distance from the control center was to install repeaters, which created other problems. However, single-mode fiber optics can easily transmit the 70 MHz radar signals a few kilometers without repeaters, so fiber cables have become standard—and radar dishes have been safely moved to remote locations.

Battlefield Communication Systems

● Portable battlefield communication systems use fiber-optic networks.

Military field command centers require extensive communication capabilities. That means that the Army must have portable communication systems ready to lay down quickly at temporary field headquarters. The need for portability means the systems should be small and lightweight. For battlefield use, they must also be rugged. Cables may be laid in the dirt, or strung from trees or buildings. Soldiers can step on them, and vehicles can drive over them. To address these problems, the Army Communications and Electronics Command developed a network that uses ruggedized fiber cables to connect truck-mounted communication facilities.

The fiber cables replace 26-pair metal cables for distances under 1 km and twin coaxial cables for 1-60 km. These cables work, but they have plenty of problems. They are heavy and bulky and take a long time to install, so they aren't very portable. They are vulnerable to lightning and EMI. Worse yet, the bulky metal cables are easy to damage. Wires in the 26-pair cable tend to break as it is installed and picked up, rendering many circuits inoperable. Fibers avoid these problems.

● Substituting fibers for 26-pair cables greatly simplifies logistics.

What does this mean in a practical sense? It takes a few trucks to carry enough 26-pair cable for a regional command center. A couple of hefty soldiers are needed to carry a reel full of cable, which is not a very long cable because of its thickness. By switching to fiber, the Army can cut back to a single truck and lay cable much faster, either by hand or from a helicopter. I vividly recall one early film that highlighted the advantages of fiber by showing two hefty male soldiers struggling with a reel of metal cable while a lightly built female soldier strolled by unreeling a much longer fiber cable.

Two-fiber cables are also replacing 26-pair cables in other portable communication systems. They use radiation-hardened 50/125 fibers, with 850 nm LED transmitters and silicon *pin* photodiode receivers. Equipped with Kevlar strength members and a polyethylene jacket, the two-fiber cable is designed to survive being run over many times. Its ruggedness was demonstrated by laying the cable across the entrance to a parking lot, where it survived being run over by 30,000 cars. A 1 km reel weighs 25 kg (55 lbs), but the cable is typically supplied in 300 meter (1,000 foot) lengths, equipped with special military connectors.

Automotive Fiber Optics

Once upon a time, automotive engineering took a nap reminiscent of Rip Van Winkle's, and automotive innovation was confined mostly to sculpting tail fins. The industry finally woke in the early 1970s, kicked awake, perhaps a bit brutally, by government and public concern over pollution, safety, and fuel economy. After some angry and sometimes befuddled protests, automotive engineers found that electronics could solve many of their problems and enhance their products in many ways.

They also discovered that automobiles are not friendly environments for electronics. Semiconductor electronics are quite vulnerable to electromagnetic interference because they need only a few volts to switch states, and EMI could make strange things happen on the road. Radio signals bouncing from the metal deck of a Chicago bridge confused the control systems of early electronic brakes being tested on buses, even though the transmitter was half a mile away. Delicate electronics were vulnerable to extremes of heat and cold; cars must survive in conditions ranging from a sun-scorched Miami parking lot in midsummer to a frozen Anchorage snowdrift in mid-winter.

The proliferation of automotive electronics, like those cited in Table 21.1, have complicated matters. Many of these systems use microprocessors, even for such simple-seeming tasks as control of intermittent windshield wipers. The more electronic systems involved, the more complex the wiring harnesses needed to deliver power and control signals. Because many electronic systems are optional, automakers must stock many different harnesses in their plants and make sure the proper one is installed in each car. Servicing them can be even more of a problem, as you quickly learn if you try to find and fix a defective electrical connection somewhere in an automobile.

● Automobiles use many electronic systems but have been very slow to adapt optical fibers.

Table 21.1. Some electrical and electronic systems available in cars.

Speedometer	Odometer	Light-on warning
Battery guard	Interior lights	Illuminated entry
Dimming control	Exterior lights	Fuel monitor
Transmission control	Clock/calendar	Alarm
Climate diagnostic panel	Heater	Air conditioner
Rear-window defogger	Airbag	Cellular phone
Automatic latch release	Cruise control	Windshield wipers
Instrument panel	Cigarette lighter	Horn
Electronic fuel injection	Coolant fan control	Hazard flashers
Seat-belt warning buzzer	Windshield washer	Power windows
Power door locks	Radio/tape player/CD player	Radar detector
Automatic seat belts		

Frustrated automotive engineers turned to fiber optics as one possible solution. Fibers can't carry electrical power, but they can carry control signals. They thought that those control signals could be used to switch electrical power off and on to accessories such as power windows. That would let them send a signal from the dashboard to roll down the window, without passing current to drive the motor through the dashboard switch—one of the things that makes wiring harnesses very complex. The control signals don't require much bandwidth, so they could be multiplexed together on a single fiber network with modest bandwidth of 1 Mbit/s. Engineers have demonstrated such systems, and they work—but you won't find them in 1993 cars.

● Fiber optics can carry control signals, but not power.

Fiber-Optic Limitations

What happened to automotive fiber optics is a mixture of technical and organizational problems. The giants of the American auto industry have long been slow to change established technology—especially technology critical for safety. You can drive your car safely without a working radio or power door lock, but not if the steering, fuel injection, or windshield wipers don't work. They want to make sure that new technology can stand up to the rigors of automotive use. They don't want to take big risks, or spend extra money.

● Automakers want to use plastic fibers, but their temperature range is limited.

They also try to make sure that cars are easy to assemble and can be fixed by trained mechanics. They are not about to equip every auto shop in the country with fusion splicers for single-mode fibers. They want instead to use large-core all-plastic fibers that could be cut with a razor blade and spliced together (if necessary) with an inexpensive mount. Plastic fibers can handle the data rate over the required distance, but automotive engineers are not yet convinced that they can withstand the extreme high temperatures in engine compartments.

Bulb-Outage Indicators

Plastic fibers do suffice for some less-demanding auto applications. Some cars use fibers to alert the driver that a bulb has burned out. A plastic fiber placed near a head lamp or other bulb collects some light if it is on, and delivers it to the dashboard. If the light is on, the driver sees a bright spot; if the light is out, the driver does not see the indicator. Plastic fibers can also carry light from bulbs hidden behind the dashboard to a display visible to the driver for other purposes, such as warning lights.

● Plastic fibers are used for illumination and bulb-outage indicators in present cars.

These fiber applications work for two reasons. One is that the fibers are routed to avoid the hottest areas in the car. The other is that they are noncritical. Many of us have driven many a mile with burned-out indicator bulbs, but we wouldn't get very far if the electronic ignition failed.

What Have You Learned?

1. Optical fibers can carry signals to control robotic vehicles on land, in air, or in the water. Fibers' advantages are their small size, light weight, immunity to EMI, and high bandwidth. Ruggedization of cables is critical for vehicle applications.

2. A fiber can transmit images from a television camera in a missile back to a soldier guiding the missile to its target. Fiber unwinds from a special reel on the missile.

3. Remotely controlled robots could serve as scouts or deliver weapons in the battlefield. In civilian applications, remotely controlled robots could clean up hazardous wastes and dismantle old nuclear reactors.

4. Fibers can be used for signal transmission in aircraft because of EMI immunity, small size, and light weight. Fibers are also attractive for transmission in large satellites.

5. Fiber-optic communication networks are being installed on military ships, which have large communication needs.

6. Because fibers can transmit signals farther without repeaters than other cables, they can be used to put radar dishes far from military battlefield control centers, reducing risks to soldiers.

7. Fiber optics have replaced bulky 26-pair wire cables for portable battlefield communication networks.

8. The automobile industry has tested plastic fibers for carrying control signals and other information between the many electronic systems in automobiles. However, engineers worry about whether plastic fibers can withstand the extreme environments in cars. Some cars do use fibers for bulb-outage indicators.

What's Next?

In Chapters 22 and 23, I will look at the many uses of fiber optics outside of communications.

Quiz for Chapter 21

1. What kinds of remotely operated vehicles cannot be controlled by operators through fiber optics?

    a. Guided missiles.

    b. Submersibles.

    c. Battlefield scouting systems.

    d. Guns mounted on vehicles.

    e. Satellites.

2. What signals are transmitted from fiber-guided missiles to the operator?

    a. Video images of the target.

    b. Control commands.

    c. Data on temperature and pressure.

    d. Data on fiber attenuation.

3. How are signals transmitted to and from a fiber-guided missile?

    a. Separately through two fibers in a single cable.

    b. Bidirectionally through one fiber by time-division multiplexing.

    c. Bidirectionally through one fiber by wavelength-division multiplexing.

    d. Signals are transmitted only one way.

    e. From the missile through the fiber; to the missile via radio.

4. Which of the following attributes of fiber optics are important for remote control of land vehicles?

    a. Secure data transmission.

    b. Lightweight, durable cable.

    c. EMI immunity.

    d. b and c.

    e. a, b, and c.

5. Which of the following reasons do not influence the use of fibers for signal transmission in aircraft?

    a. Optical fibers are immune to EMI.

    b. Optical fibers are lighter than wires.

    c. Aircraft lack adequate power supplies for wire-based communications.

    d. Military aircraft must be hardened against enemy electronic countermeasures.

    e. Fiber optics can help reduce aircraft visibility to radar.

6. Why are fiber-optic cables used to connect radar dishes to battlefield control centers?

    a. Because they carry radar signals far enough that the dishes can be placed a few kilometers from the control centers.

    b. Because they are immune to electromagnetic eavesdropping.

    c. Because they are nonconductive.

    d. Because the control centers already use fiber optics.

    e. Because a top military official's brother-in-law sells fiber-optic cable.

7. What have fiber-optic cables replaced in portable battlefield communication systems?

    a. Obsolete plastic fibers that have become brittle with age.

    b. 26-pair cable and coax.

    c. Telephone cables.

    d. Microwave transmission.

    e. Nothing—without fibers such systems were not practical.

8. What type of fiber systems are used in portable battlefield networks?

    a. Plastic fibers transmitting in the red.

    b. Multimode fibers transmitting 850 nm.

    c. Multimode fibers transmitting 1300 nm.

    d. Single-mode fibers transmitting 1300 nm.

    e. Single-mode dispersion-shifted fibers transmitting 1550 nm.

9. What are the major attractions of fiber-optic cables for portable battlefield communications?

    a. Small cables require fewer trucks to transport.

    b. Fiber cable is more durable than metal cables.

    c. Light weight eases the logistics of laying cable.

    d. Cable is radiation-hardened and immune to EMI.

    e. All of the above.

10. What problem has limited the use of fiber optics in automotive control systems?

    a. Optical fibers cannot conduct electric power.

    b. The high cost of single-mode connectors.

    c. Temperature limitations of plastic fibers.

    d. Pollution regulations.

    e. Automotive engineers don't like anything electronic.

Fiber-Optic Sensors

About This Chapter

So far, I have concentrated almost entirely on the ways optical fibers are used for communications. However, fiber optics also have other important uses. This chapter will show how fibers can be used as sensors of many different types. Fiber sensors can measure strengths of composite materials, serve as gyroscopes to measure direction and rotation, or sense acoustic waves at the bottom of the ocean. All this depends on changes in how fibers transmit light under certain conditions.

Fiber Sensing Concepts

The label "fiber sensors" covers a broad range of devices that work in different ways. The simplest use optical fibers merely as a probe, to detect changes in light outside the fiber. The fiber may collect light from a given point, to see if an object (e.g., a part on an assembly line) is present or not. The fiber may also collect light from some other type of optical sensor that responds to its environment in a way that changes the light reaching the fiber. For example, a prism in a tank of liquid may reflect light back into a fiber probe if the liquid level drops below the prism.

Other fiber sensors detect changes in light traveling through the fiber itself, when the outside world affects the fiber. That may seem a bit strange at first, because you saw earlier that the outside world has little effect on the light inside fibers. However, you can sometimes magnify those weak effects by the way you mount or use fibers. You can also use optical effects like interference to detect very tiny effects over long lengths of fiber. This lets you use fiber sensors to detect changes in temperature, pressure, magnetic fields, and other quantities.

● Fibers can serve as probes or as sensors themselves.

I will look first at simple fiber sensors, where the fiber serves as a probe that does not respond directly to the environment. Then I will cover mechanisms that can be used for sensing. Finally I will describe some important types of fiber sensors and their applications.

Fiber-Optic Probes

Fiber-optic probes collect light from remote points, often sampling light that was delivered there by fibers. They come in two broad families that perform different sensing functions. The simpler ones look to see if light is present or absent at the point they observe. Others collect light from remote optical sensors, bringing it back to a place where it can be analyzed.

Simple Probes

● Fiber probes can detect objects when they block or reflect light.

Figure 22.1 shows a simple fiber-optic probe checking for parts on an assembly line. One optical fiber delivers light from an external source, and a second fiber collects light emerging from the first, as long as nothing gets between the two. When a part passes between them on the assembly line, it blocks the light. Thus light off indicates that a part is on the assembly line, and light on indicates that no part is passing by.

FIGURE 22.1.

Fiber-optic probe checks for parts on an assembly line.

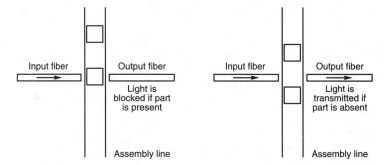

This concept can be used in many ways and is not particularly new. One early example was reading holes in the punched cards used to input data to early mainframe computers. The card passed an array of fibers at a fixed speed, and detectors monitored light transmission as a function of time. Passage of a hole let light reach the detector, until it was blocked by the end of the hole. The light is turned off at the end of the hole, when part of the card blocks it. The technique was simple and effective, but punched cards are now museum pieces.

More refined variations are possible, such as measuring the size of parts to make sure they meet tolerances. An array of fibers could be mounted beside the production line, so passing parts blocked the light to some of them. The parts would pass inspection if all the fibers above the maximum height received light and all those below the minimum height did not. Parts that were too small or too tall would be rejected when light reached fibers that were supposed to be dark or did not reach fibers that were supposed to be illuminated.

Optical Remote Sensing

Fiber probes can also collect light from other types of optical sensors. In this case, the fibers function like wires attached to an electronic sensor. The optical sensor (which is not a fiber) responds in some way to the environment, changing the light that reaches the fiber probe. The fiber carries that light to a detector, which senses the change.

One example is the liquid-level sensor shown in Figure 22.2, which senses when the gasoline in tank trucks reaches a certain level. Many tank trucks are filled from the bottom so vapor left in the tank can be collected to control pollution, and the liquid level must be sensed to prevent overfilling. One fiber delivers light to a prism mounted at the proper level. If there is no liquid in the tank, the light from the fiber experiences total internal reflection at the base of the prism and is directed back into the collecting fiber. If the bottom of the prism is covered with gasoline, total internal reflection cannot occur at the angle that light strikes the prism's bottom face, and no more light is reflected back into the fiber. When the light signal stops, the control system shuts off the gas pump.

Fibers can collect light from other optical sensors.

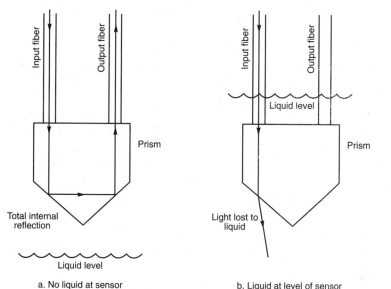

FIGURE 22.2.

A liquid-level sensor.

Another example senses temperature changes by observing the response of a phosphor in a glass blob at the end of a fiber. Ultraviolet light transmitted by the fiber stimulates fluorescence from the phosphor at several wavelengths. The ratio of fluorescence at the different wavelengths changes with temperature. The fiber collects the fluorescent light and delivers it to an optical analyzer that compares intensities at different wavelengths and thus measures the temperature.

Fiber Sensing Mechanisms

Some effects can change how fibers transmit light.

Outside influences can directly affect how optical fibers transmit light. They work in a variety of ways. Many make small changes in refractive index of the fiber, which can affect transmission over long distances. Others affect light of different polarizations differently. Some cause tiny bends in the fiber, which let light leak out, while others change the effective distance that light travels in ways that can be detected optically.

Countless fiber sensors have been demonstrated over the last decade, and most have gone little further than the laboratory. I can't cover them all, but I will look at important mechanisms of fiber sensing and how they work. The categories I describe next overlap to some extent, because particular sensors may rely on more than one effect. After looking at these fundamentals, I'll look at a few important examples of fiber sensing.

Microbending

Microbending losses can be used to sense pressure.

Back in Chapter 4, I mentioned that light can leak out of optical fibers if they are bent sharply. That is undesirable for communications, and cable manufacturers and installers go to considerable effort to avoid it. However, sensors can be based on measurement of the losses caused by controlled microbending.

Figure 22.3 shows the basic idea of a microbending sensor. A fiber is mounted between a pair of plates containing parallel grooves. When pressure pushes the plates together, it causes microbends in the fiber, which let some light leak out, reducing optical power transmitted by the fiber. This allows measurement not just of static pressure but also of instantaneous changes in pressure (i.e., sound waves). The goal is not better microphones for ordinary use but very special microphones, acoustic sensors to rest on the ocean bottom keeping watch for submarines. Microbending sensors can also be used to measure other quantities that can be translated into pressure, including temperature.

Microbending sensors don't have to sit between pairs of grooved plates. They require only an environment that causes the fiber to bend under certain conditions. This could happen, for example, if the fiber were embedded in a material that shrank when cooled.

Crack Sensors

Cracks in fibers cause losses, which could detect structural flaws.

Microbend sensors are not supposed to strain fibers to the breaking point. However, severe strains can break fibers, and in some cases that can be useful for sensing. Suppose, for example, that you want to verify the structural integrity of a dam. You might embed optical fibers in the concrete matrix and watch for sudden changes in light transmission. If internal stresses caused the concrete to crack inside the dam, it could snap the embedded fiber, causing light transmission to drop. This could be a vital early warning of structural problems.

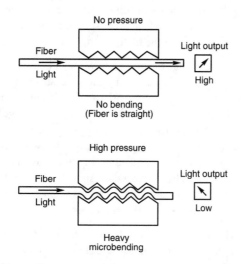

FIGURE 22.3.

Increasing pressure on the plates causes microbending losses useful for sensing.

Refractive Index Change

Changes in temperature and pressure cause small changes in the refractive index of glass. (You can visualize them as due to changes in density, but it isn't quite that simple.) Higher pressure generally raises the index, whereas higher temperature causes it to decrease. These index changes can be used in a variety of sensors.

● Changes in fiber refractive index can be used for sensing.

One example is temperature sensing with a fiber in which the refractive indexes of core and cladding vary with temperature in different ways. Suppose that at 0° C the core index is 1.50, the cladding index is 1.49, and the core index decreases by 0.0005 per degree while the cladding index decreases by 0.0004 per degree. At 100° C, the two refractive indexes would be equal (1.45). At that point, the fiber would stop guiding light, so output light intensity would drop to essentially zero. The example is an artificial one, but the concept has been demonstrated in a sensor that can measure temperature within a few degrees.

Index changes can also be measured in other ways. Remember that the refractive index is the ratio of the speed of light in a vacuum to that in the medium, so changing the refractive index changes the time that light takes to pass through a fiber. You can't measure that change by looking at one fiber alone, but you can by comparing a pair of fibers, one passing through the region being measured, and the other in a controlled environment. Such measurements depend on optical interference.

Interferometric (Phase) Sensing

Interferometric sensing is a general way to measure changes in the effective length of a fiber. The effective length depends not only on refractive index of the fiber but also on its physical length, because strain along its length can stretch a fiber. The result is a change in the phase of the light emerging from the fiber.

- Interferometric sensors measure changes in effective transmission distance, caused by temperature, strain, and other factors.

To understand what a phase change means and how it can be measured, consider a simplified case in which a fiber transmits coherent light waves. Stretching the fiber by half a wavelength of light does not change the intensity of light emerging from it, but it does change the phase of the light wave by 180°. You can detect the change by mixing the light from the stretched fiber with light from an unchanged fiber that was originally in phase with that from the stretched fiber. Before the stretching, the two waves were in phase, so their amplitudes added to give a bright spot. After the stretching, their light waves are out of phase and interfere destructively to give a dark spot. Further stretching will shift the phase of light in the first fiber further until the two waves are in phase (meaning the stretching has shifted phase by 360°), producing a bright spot.

- Interferometric measurements are complex but very sensitive.

Although interferometric measurements can be cumbersome, they are very sensitive and can be used in many different sensing applications. For example, if a fiber is coated with metal or wrapped around a special cylinder, a magnetic field applies pressure to it in a way proportional to field strength. Measuring the change in effective length makes this a magnetic-field sensor. Tying a weight to the end of a fiber stretches the fiber when the object accelerates, making an accelerometer.

Polarization Sensors

- Changes in polarization can be used for sensing.

Standard single-mode fibers transmit light without regard to polarization, but as you saw in Chapter 4, some special fibers maintain input polarization or transmit only one polarization. These fibers have different refractive indexes for light of different polarizations, and outside factors may affect these indexes differently. That could lead to a measurable change in polarization.

Another example of polarization sensing is measurement of current by observing changes in polarization caused by the magnetic field generated by the current. A phenomenon called Faraday rotation rotates the plane of polarized light in certain materials by an amount proportional to the magnetic field. This rotation can be seen by winding a suitable fiber around a conductor and measuring the change in the plane of polarization when the current is on.

Fiber Grating Sensors

A new approach to fiber sensors is to create patterns of refractive-index differentials in the material of the fiber core. This is done by directing ultraviolet laser pulses into the fiber from the side. Optics focus the ultraviolet beams to produce an interference pattern, with alternating light and dark zones. At the peaks, the ultraviolet light is intense enough to cause optical damage at sites occupied by germanium atoms in the fiber, changing the refractive index of the glass. This creates a permanent periodic pattern in the fiber; it is called a grating because it works like a diffraction grating, although it is actually more like a series of layers stacked along the length of the fiber.

Fiber gratings selectively transmit some wavelengths and reflect others, depending on the spacing of the pattern and the refractive index of the glass. Changes in the refractive index of the glass, caused by environmental effects such as changes in temperature or pressure, can change the wavelengths of peak reflection and/or transmission. So can changes in the spacing, caused by strain along the length of the fiber. Measurement of these wavelength changes can be the basis of a sensor. Fiber gratings appear particularly useful when embedded in materials called smart structures or smart skins, as I will describe later.

The Sagnac Effect

There are too many fiber sensing concepts to cover in the brief space allocated here, but one more deserves mention—the Sagnac Effect, which can detect rotation. The basic idea is to send light traveling in opposite directions through a loop of single-mode fiber. If the ring is perfectly still, the light going in opposite directions will travel exactly the same distance in one complete cycle. However, if the ring is rotating around its axis, it will move slightly in one direction or the other by the time the light gets back to its starting point—so one beam will have traveled further than the other. This is the basis of the fiber-optic rotation sensor, or gyroscope, described in the next section and shown in Figure 22.4.

Fiber-Optic Gyroscopes

The fiber-optic gyroscope is probably the most successful fiber sensor so far. It relies on optical processes to sense rotation around the axis of a ring of fiber. Rotation sensing is vital for aircraft and missiles, which have traditionally used gimbaled mechanical gyroscopes as references. Fiber gyroscopes (and laser gyroscopes that serve a similar purpose but operate on different principles) offer a number of advantages, including no moving parts, greater reliability, and no need for a warm-up period to start the gyro.

Figure 22.4 shows the workings of a fiber gyro. Light from a single source is split into two beams directed into opposite ends of a loop of single-mode fiber. In actual sensors, the fiber is wound many times around a cylinder, but the drawing shows only one turn. Light takes a finite time to travel around a fiber loop with radius r, and in that time the loop can rotate an angle θ, which in practice is very small. This rotation moves the starting point a

> Fiber gratings are sensitive to changes in refractive index, making them useful sensors.

> A fiber-optic gyroscope measures rotation interferometrically.

distance Δ. Light going in the direction of the rotation must travel a distance 2πr + Δ to get back to its starting point, but light traveling in the opposite direction travels only 2πr - Δ. That slight difference in distance will mean a difference in a phase when the two beams are superimposed. This difference, the Sagnac Effect described previously, can be detected by interferometry if the beams travel through a suitable single-mode fiber.

FIGURE 22.4.

Workings of a fiber-optic gyroscope.

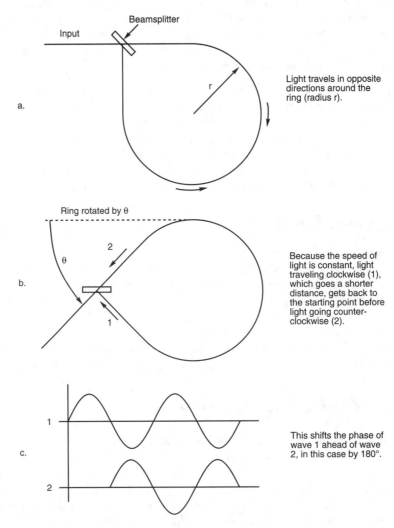

a. Light travels in opposite directions around the ring (radius r).

b. Because the speed of light is constant, light traveling clockwise (1), which goes a shorter distance, gets back to the starting point before light going counter-clockwise (2).

c. This shifts the phase of wave 1 ahead of wave 2, in this case by 180°.

Fiber gyros are not as accurate as laser gyros, which are used on new civilian and military aircraft. However, fiber gyros are good and inexpensive, making them suitable for applications such as guiding missiles (which you don't want to load with lots of expensive equipment), and short-range aircraft. Without any moving parts, they should pose fewer operational problems than mechanical gyros. Some developers have proposed building navigation systems for automobiles, based on fiber gyros and signals from global positioning system satellites. They aren't there yet, but fiber gyro technology is still advancing.

Smart Skins and Structures

Fiber sensors can be embedded in composites and other materials such as concrete to create "smart structures" or "smart skins." The goal is to create a structural element (including the skins of aircraft) equipped to monitor internal conditions. The initial emphasis is on verifying that components meet initial structural requirements, but developers envision eventually using the fiber sensors throughout the life of the component. Figure 22.5 shows a simple example of how fibers could be embedded between layers of a composite material; in this case, they are encased in an epoxy layer.

⬤
Fiber sensors are
embedded in
composites to make
smart structures
and skins.

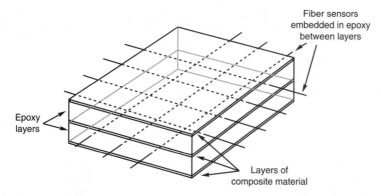

Fiber sensors
embedded in epoxy
between layers

Epoxy
layers

Layers of
composite material

FIGURE 22.5.

*Sensing fibers in a
smart skin are
embedded in an epoxy
matrix between layers
of a composite
material.*

The initial use of the fiber sensors is to monitor fabrication and curing of the composite. The fiber sensors could monitor temperature to be sure curing conditions met requirements. They could also monitor strain, to verify that the component was not stressed excessively. Detection of cracked fibers would indicate serious stress problems. Later, the fiber sensors could provide data on stresses and strains that occur after the composite component is mounted in final position. Eventually this information could be used by operating engineers, but currently its main use is in studying properties of structures and aircraft.

Once a smart-skin or smart-structure system is in operation, engineers could use the fiber sensors for periodic checks of performance and structural integrity. For example, fibers embedded in aircraft wings could be plugged into monitoring equipment in the service bay, to make sure they suffered no invisible internal cracks that could cause catastrophic failure. Dams and bridges likewise could be monitored with fibers.

Some military planners think the ultimate step would be to plug the fiber sensors into a real-time control system designed to optimize performance. The performance limits of aircraft materials and structures are not known precisely, so engineers err on the side of safety and avoid pushing too far. Real-time fiber monitors could tell computers how well components were withstanding stresses in operation. Ultimately, perhaps, the computers could use the sensor data to apply corrections in real-time that would push the performance envelope further, without endangering pilots or aircraft.

Acoustic Sensors

Fiber sensor arrays could detect passing submarines.

The U.S. Navy supported much early fiber-sensor research in its efforts to improve the technology for detecting submerged nuclear submarines. The Navy program led to many other kinds of fiber sensors and developed some highly sensitive arrays of fiber acoustic sensors for underwater use. However, the Cold War ended before submarine acoustic sensors were ready to be deployed, and that technology has not come into widespread use.

System Development and Implications

Fiber sensors have not come into wide use for a variety of reasons.

Development of fiber sensors has continued for many years, and in some ways the technology seems successful. Although many early proposals never came to fruition, that is inevitable with any new technology. Research and development programs continue to yield encouraging results. Yet few fiber sensors have found widespread use. The obvious question is why?

The answer is complex. In part, most fiber sensors do not have the compelling advantages that fibers enjoy for high-speed communications. Existing technologies can do most of the jobs that fiber sensors can, without the cost and complications of having to provide special interfaces between fiber sensors and electronic computers and control systems. Fiber sensor costs remain high because the technology has never moved beyond the development stage—and without compelling advantages, production will remain too low to benefit from the economies of scale.

Some technical problems remain unsolved. One is isolating fiber response to different quantities. If both increased temperature and higher pressure change refractive index, how do you isolate which causes what? Another is the need to interface optical sensors with electronic controls and computer systems. Costs remain high for most optical sensors.

Fiber gyros are an exception. They received a healthy helping of the military funding that has fueled much fiber sensor development—but they also work enough better than mechanical gyros to earn a place in many military systems. Other fiber sensors are promising for military applications, but like acoustic sensors some are stalled by the defense cutbacks that followed the end of the Cold War. Smart skins and structures are promising for a range of civilian and military applications, but the technology is still emerging.

The best chance for fiber sensors may depend on the spread of fiber-optic networks and other optical information processing systems. Once the fibers are in place, they may be capable of collecting information from sensors and feeding it to receivers that can interpret it. Optical signal processing and computing would help further.

Fiber probes would be the simplest to implement, but other types can be developed. Suppose, for instance, a sensor compares intensity of three fluorescent wavelengths to measure temperature. Light at the three wavelengths could be collected by a single fiber, separated at a wavelength-division demultiplexer, and fed to three separate detectors. Output from the three detectors could be compared electronically to generate a digitized temperature reading. Further in the future, output of the wavelength-division demultiplexer might instead go to bistable optical devices, designed to switch to certain logical states (and turn on other controls) if the temperature goes above or below the desired range. The technology is young, and further possibilities are many.

What Have You Learned?

1. Fibers can serve as probes that collect light for sensing. Fibers can also function as sensors themselves.

2. Fiber probes can detect objects that block or reflect light. This lets them measure shapes, count parts, and do other simple tasks.

3. Fiber optics can collect light from remote optical sensors so it can be measured.

4. Certain effects can alter the way fibers transmit light. These are detected by fiber sensors.

5. Placing a fiber in a position where pressure can cause microbending allows the fiber to sense pressure; the more pressure, the more light lost from the fiber.

6. Temperature and pressure can change the refractive index of the glass in a fiber, and this change can be used to measure changes in temperature and pressure.

7. Interferometric sensors measure changes in effective transmission distance, caused by temperature, strain, and other factors. Interferometric measurements are complex but very sensitive.

8. Changes in polarization inside the fiber can be used to detect environmental changes.

9. Fiber gratings are zones in a fiber core where the refractive index is alternately high and low. They are formed by illuminating the fiber with ultraviolet light. They can be used as sensors because their optical characteristics are very sensitive to the refractive index and the spacing between the layers.

10. Loops of fiber can measure rotation by sensing differences in the time light takes to travel in opposite directions around the loop. Such fiber gyroscopes can be used in guidance systems.

11. Fiber sensors can be embedded in composite materials to make smart structures and smart skins.

12. Fiber sensors have not come into wide use, because few offer compelling advantages over other technologies.

What's Next?

In Chapter 23, I will look at other noncommunication applications of fiber optics in a wide variety of fields.

Quiz for Chapter 22

1. How do fiber-optic probes work?

   a. They detect the presence or absence of light at a point.

   b. They detect the pressure of objects placed on top of them.

   c. Changes in temperature make them expand or contract.

   d. None of the above.

2. Which of the following is an example of a fiber collecting light from a remote optical sensor?

   a. Fiber-optic gyroscope.

   b. Liquid-level sensor based on total internal reflection from a prism.

   c. Acoustic sensor based on microbending.

   d. Fiber grating used as a pressure sensor.

   e. Smart skins.

3. How can microbending effects be sensed?

   a. By observing tension along the length of the fiber.

   b. By monitoring changes in light transmitted by the fiber.

   c. By looking for changes in data rate of a signal transmitted through the fiber.

   d. By measuring light emitted by the fiber.

4. Which of the following can change the refractive index of a fiber?

   a. Temperature changes.

   b. Pressure changes.

   c. Sound waves.

   d. All of the above.

   e. None of the above.

5. Which sort of change in a fiber sensor can be measured by interferometry?

   a. Changes in the wavelength of light.

   b. Changes in intensity of light.

   c. Changes in effective length caused by strain.

   d. Changes in optical absorption.

6. An example of an interferometric sensor is

    a. a punched card reader.

    b. a microbending sensor of acoustic waves.

    c. a magnetic-field sensor that detects strain changes in a metal-coated fiber.

    d. a sensor that measures the height of parts on a production line.

7. How do changes in fiber gratings affect light transmission?

    a. Microbending causes increased attenuation.

    b. They alter wavelengths transmitted and reflected.

    c. They change polarization.

    d. They modulate light with a digital code.

8. The Sagnac effect detects what?

    a. Rotation of a ring of fiber around its axis.

    b. Changes in the color of light.

    c. Changes in temperature.

    d. Changes in pressure.

    e. Several different quantities, depending on the type of fiber used.

9. How do fiber-optic gyroscopes detect rotation?

    a. By changes in the wavelength of light in the fiber.

    b. By interferometrically measuring changes in light going in opposite directions around the fiber loop.

    c. By inertial changes in the fiber loop.

    d. By observing microbending losses.

10. What can fiber sensors measure when embedded in a smart structure?

    a. Curing conditions of a composite material.

    b. Internal strain in a composite material.

    c. Structural integrity of a completed component.

    d. Stresses on a component during use.

    e. All of the above.

Noncommunication Fiber Optics

About This Chapter

Communications and sensing were latecomers in the world of fiber optics; the early developers of optical fibers had other things in mind. You have seen some of these other uses of fibers in passing, but now it's time to take a closer look at applications including light piping, imaging, inspection, and medical treatment. Many require fibers quite different from those used for communications, or bundled together in a particular way. If you are going to work regularly with fiber optics, you should know about such applications, although many are limited to specialized fields. The major applications—classical fiber-optic light piping and imaging, and medicine—differ in important ways, so I will look at each separately. First, however, I will look at light guiding in short, bundled fibers.

Basics of Fiber Bundles

Single fibers are used in communications and sensing, but most early fiber-optic applications used many fibers bundled together. One reason was that the larger area of the bundle could collect more light, especially from the large-area light sources available before semiconductor lasers could operate reliably at room temperature. Another was that bundles could transmit images projected onto one end to the other end—if all fibers in the bundle were in the same relative position on both ends. These imaging bundles are sometimes called coherent bundles (because they transmit elements of an image in the proper position, not because they carry coherent light).

Fiber bundles are made from many individual fibers; they may be rigid or flexible.

If you want to visualize a fiber bundle, you can start with a handful of uncooked spaghetti all the same length. The fibers are like the strands of spaghetti, but much thinner. If you've been careful with your spaghetti, most of the strands are in the same position on both ends of the bundle. That's harder to do with optical fibers, but it lets you transmit images through the bundle. Alternatively, the fibers may be in random positions on each end (like the spaghetti would be after you cooked it). Random fibers can deliver light from one end of the bundle to the other, but not images. The difference between randomly arranged fibers and imaging bundles is a fundamental one in fiber bundles.

The second basic dichotomy is between rigid and flexible bundles. Your uncooked handful of spaghetti is a rigid bundle, which makes it easier for you to handle. It's much harder to hold cooked spaghetti—or loose fibers—together, but a flexible fiber bundle is better for many applications that require looking inside or illuminating hard-to-reach places. Note that you can have flexible and rigid random bundles, and flexible and rigid imaging bundles, although in practice rigid random bundles are rare.

Making Fiber Bundles

Many fibers are fused together for imaging bundles.

Randomly arranged fibers can be bundled together like spaghetti, but imaging bundles take considerably more care. They are made in a series of phases. First, a step-index fiber is drawn from a preform using a process like the one I described in Chapter 4, but the fiber is larger—about 2.5 millimeters in diameter—and easier to handle. A group of these fibers—typically 37 to 169—are bundled together, heated, and drawn again, to make a rigid "multifiber" about 2 mm in diameter, as shown in Figure 23.1. Then a number of multifibers (typically 61 to 271) are packed together, heated, and drawn again to produce a rigid fiber bundle, containing many thousands of fibers. Each of the fibers is about 3-20 micrometers in diameter.

The same basic process can be used to make both rigid and flexible bundles, with a few important differences. Look carefully at Figure 23.1, and you will note that the large core is surrounded by two rings of cladding. One is the conventional cladding used with all fibers. Composition of the other depends on the type of bundle being made.

For rigid bundles, that outer layer is an absorptive glass that keeps light from leaking between cores in the bundle. Too much such leakage could degrade the image.

Flexible bundles can be made by leaching acid-soluble glass from special rigid bundles.

For flexible bundles, that outer layer is a glass that is soluble in acid. Manufacturers cover the ends of the rigid rod, then dip it in an acid that dissolves away that leachable layer in the middle of the rod, leaving a flexible bundle of many thin fibers, which are arranged so their ends are aligned for imaging. (Another way to make a flexible bundle is to loop a single thin fiber many times in a circle, glue together one part of that loop, then cut through the middle of the glued portion. Because the fibers were cut apart, they are in the same position on both rigid ends, but loose in the middle.)

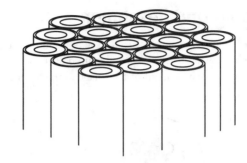

FIGURE 23.1.
*Stages in making a
fiber bundle.*
(Courtesy of Schott
Fiber Optics Inc.)

Step 1
MONOFIBER (drawn
to ~2.5mm dia)

Step 2
MULTIFIBER (37 to 169 MONOFIBERS
drawn to ~ 2.0 mm dia)

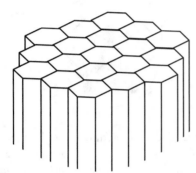

Step 3
MULTI-MULTIFIBER (61 to 271
MULTIFIBERS, 5000 to 20,000 pixels,
drawn to ~6-18 μm fiber dia)

Individual fibers in a flexible coherent bundle can be small, but not quite as small as in a fused bundle. Some performance limits of flexible bundles are comparable to those of rigid bundles (e.g., packing fraction and resolution). When flexible bundles are used, an added concern is breakage of individual fibers, which does not occur in fused bundles. Each fiber break prevents light transmission from one spot on the input face. The loss of a single fiber is not critical, but as more fibers break, the transmitted light level drops and resolution can decline as well. Eventually breakage reaches a point where the image-transmitting bundle is no longer usable. Because of the breakage problem, plastic fibers are often used in flexible bundles.

Randomly aligned bundles are made by collecting many fibers into a bundle, much like collecting strands of spaghetti. This would be very difficult if the fibers were as thin as those in imaging bundles, but such fine fibers are not needed because the resolution does not matter; random bundles serve purely as "light pipes." Typically, random bundles are made of fibers with diameters in the 100-micrometer range, which are flexible enough to bend freely with minimum fiber breakage.

● Randomly aligned
bundles serve as
"light pipes."

Imaging and Resolution

So far, I've glossed over the crucial issue of resolution in an imaging fiber bundle. Figure 23.2 shows how an image is carried from one end of the bundle to the other. Each fiber core carries its own segment of the image to the other end, maintaining their alignment.

FIGURE 23.2.

Image transmission through a fiber bundle.

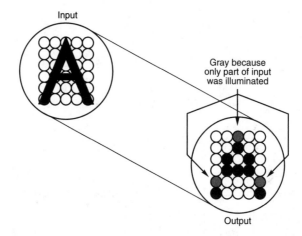

To visualize what happens, imagine that each fiber core captures a chunk of the image and delivers it to the other end of the bundle. This process averages out any details that fall within a single core. For example, if the input to a single core is half black and half white, the output will be gray. Thus, the fiber cores must be small to see much detail. For a static fiber bundle, the resolution is about half a line pair per fiber core, meaning two fiber core widths are needed to measure a line pair. Numerically, that means 10 μm fiber cores could resolve 50 line pairs per millimeter (1 line pair per 20 μm). Imaging bundles have fiber cores as small as 3 μm. Resolution is significantly higher—about 0.8 line pair per fiber core diameter—if the fiber bundle is moving with respect to the object.

● **Bundle resolution depends on the core sizes of the fibers it contains.**

Before you wonder too much about the quality of fiber-bundle images, you should realize that if fiber cores were a common size of 10 μm, the letter A in Figure 23.2 would be only 60 μm high, less than 1/16 of a millimeter tall. That's many times smaller than the finest of fine prints used in legal documents. You have to look very hard, and may need a strong magnifying lens, to see the individual fiber spots on a good imaging bundle.

Cladding Effects

● **Light that falls into fiber claddings in bundles is lost, but typically 90% falls onto fiber cores.**

The cores conduct light in fiber bundles, but they are surrounded by cladding layers. Bundles are made with thin cladding layers, but some light must fall onto the cladding rather than the core. The fate of that light depends on the bundle design. Rigid bundles have an outer cladding layer that absorbs light so that little can pass between fiber cores. Light that leaks out of the cores of individual fibers in flexible bundles cannot easily enter other fibers. However, neither type can completely prevent any light from leaking between fibers.

Most light entering the cladding is lost, which can limit transmission efficiency. This makes the fraction of the surface made up by fiber cores an important factor in a bundle's light-collection efficiency. That is, the collection efficiency depends (in part) on the packing fraction, defined as

PACKING FRACTION = (TOTAL CORE AREA)/(TOTAL SURFACE AREA)

A typical value is around 90%.

Transmission Characteristics

Short-distance fibers, particularly those used in bundles, do not have as low attenuation as communication fibers, because bundles carry light no more than a matter of meters. Typical attenuation of bundled fiber is around 1 dB/m, over a thousand times higher than that of communication fibers at 1300 nm.

Likewise, operating wavelengths differ. Visible light is needed for imaging and illumination, and even for other applications the short distances make it unnecessary to operate at wavelengths where fibers are most transparent. Glass fiber bundles are typically usable at wavelengths of 400-2200 nm, and special types made from glass with good ultraviolet transmission are usable at somewhat shorter wavelengths. Plastic fibers are usable at visible wavelengths, 400-700 nm. Some special-purpose bundles are made of other materials, but they are not widely used.

Also, different considerations affect numerical aperture in bundled and communication fibers. Transmission distances are short and pulse dispersions are irrelevant in bundled fibers, but light collection efficiency is important. To give large acceptance angles, bundled fibers tend to have higher NAs than communication fibers (from 0.35 to 1.1). Virtually all bundled fibers are step-index multimode types, which are easy to make and offer the desired large numerical aperture.

Optics of Bundled Fibers

The basic principles of fiber optics are the same if fibers are separate or bundled. However, some implicit assumptions always hide behind any discussion of basics. In describing communication fibers earlier, I made some assumptions that don't always work for bundles or other noncommunication fibers. It's time to go back and face some complications that I earlier simplified away.

●
Typical attenuation of bundled fiber is around 1 dB/m.

●
Bundled fibers are step-index multi-mode types with large NA.

●
Some simplifying assumptions valid for single communication fibers are not valid for bundles.

Light Rays in Optical Fibers

Light rays entering a fiber at one angle emerge in a diverging ring.

Looking at individual light rays, as in Figure 23.3, gives a slightly different view consistent with what you learned earlier. If a light ray enters the fiber at an angle θ within the fiber's acceptance angle, it will emerge at roughly the same angle to the fiber axis, although not necessarily in the same direction. "Roughly" is the operative word because the ray will emerge in a ring of angles centered on θ because of imperfections in the fiber, effects of fiber length, and other factors.

FIGURE 23.3.

Light rays emerge from a fiber in a diverging ring.

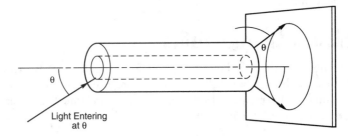

Light Entering
at θ

This does not conflict with what you learned earlier, because then you looked at light guiding only collectively, for a large number of rays. If you had broken down the cone of light entering the fiber into individual rays, you would have seen the same effect. At that point, there was no need to do so.

Step-index fibers with constant-size cores do not focus light.

One other thing should be pointed out: step-index fibers with constant-diameter cores do not focus light. (As you will see later, graded-index fibers can focus light passing along their length.) All light emerges from a step-index fiber at roughly the same angle that it entered, not at a changed angle, as would happen if it did focus light. As long as the fiber's sides and ends are straight and perpendicular to each other, a step-index fiber or a fiber bundle can no more focus light than a pane of flat window glass can. This has one important practical consequence that you'll discover the first time you look through an imaging bundle. You have to put the distant end up very close to what you want to see, or the image will become blurred. You see the image on the close end because light travels straight through each fiber, but the bundle does not rearrange light rays that got into the wrong fibers (because the far end was not close enough to the object).

If the fiber's output end is cut at an angle not perpendicular to its axis, light entering at an angle θ still emerges in a cone, but the center of the cone is at an angle to the fiber axis. If the slant angle (from the perpendicular) is a small value φ, the angle β by which the rays are offset is approximately

$$\beta = \phi(n-1)$$

where n is the refractive index of the fiber core, as long as φ is small.

Tapered Fibers

I assumed earlier that fiber cores are straight and uniform, but they could also be tapered (although not over long distances). Figure 23.4 shows what happens to a light ray entering such a fiber at an angle θ_1. If the ray meets criteria for total internal reflection, it is confined in the core. However, it meets the core-cladding boundary at different angles on each bounce so each total internal reflection is at different angles from the axis. The result is that it emerges from the fiber at a different angle, θ_2. If input core diameter is d_1 and output core diameter is d_2, the relationship between input and output angles is

$$d_1 \sin \theta_1 = d_2 \sin \theta_2$$

The same relationship holds for the fiber's outer diameter as long as core and outer diameter change by the same factor d_2/d_1.

●
Tapered fibers magnify or demagnify objects seen through them. Tapered fibers are used in bundles.

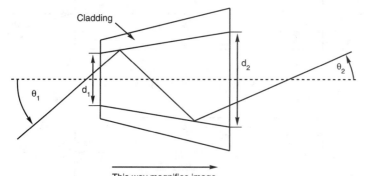

Cladding

d_2

d_1

θ_1

θ_2

This way magnifies image

FIGURE 23.4.

Light passing from the narrow to the broad end of a tapered fiber.

As a numerical example, suppose the input angle was 30° and the taper expanded diameter by a factor of 2. The output angle would be about 14.5° (the inverse sine of 0.25). Thus, light exiting the broad end of a taper would emerge at a smaller angle to the axis than at which it entered (and, conversely, light entering the broad end would leave the narrow end at a broader angle). This effect lets fiber tapers magnify or demagnify objects. Images are magnified if light goes from the narrow end to the broad end, and are shrunk if light goes from the broad end to the narrow end.

In practice, single fibers are not used as tapers. Instead, fiber tapers are rigid bundles of fibers, each expanding or shrinking by the same factor. You can also use this viewpoint to see why a taper magnifies. All the light from each core on the small end is spread out over the larger core area on the larger end—making the image larger (although not increasing resolution measured as the number of lines across the image). Essentially, the taper makes each spot larger. The effect is somewhat like a lens, but not exactly the same.

Focusing with Graded-Index Fibers

Graded-index fibers can focus light in certain cases.

Unlike step-index fibers, graded-index fibers can focus light in certain cases. This does not make graded-index fibers useful for image transmission or other fiber-bundle applications, but as described later in this chapter, segments of graded-index fibers can function as components in some optical systems.

In Chapter 4, you saw that light follows a sinusoidal path through graded-index fiber. When you looked at how a cone of light was transmitted through a long fiber, you saw output as a cone of the same angle. Now look instead at the path of an individual ray through a short segment of graded-index fiber, shown in Figure 23.5, and compare that with the path of a light ray in step-index fiber.

FIGURE 23.5.

Rays in graded-index and step-index fibers.

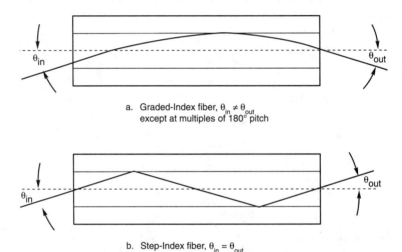

a. Graded-Index fiber, $\theta_{in} \neq \theta_{out}$
except at multiples of 180° pitch

b. Step-Index fiber, $\theta_{in} = \theta_{out}$

There is an important but subtle difference. Total internal reflection from a step-index boundary keeps light rays at the same angle to the fiber axis all along the fiber. However, graded-index fibers refract light rays, so the angle of the ray to the axis is constantly changing as the ray follows a sinusoidal path. If you cut the fiber after the light ray has gone through 180° or 360° of the sinusoid, the light emerges at the same angle that it entered. However, if the distance the light ray travels is not an integral multiple of 180° of the sinusoid, it emerges at a different angle. This property allows segments of graded-index fiber to focus light.

Pitch is a critical parameter of graded-index lenses.

In design of graded-index fiber lenses (usually sold under the trade name Selfoc), the key parameter is the fraction of a full sinusoidal cycle that light goes through before emerging. That fraction is called the pitch. A 0.23-pitch lens, for instance, has gone through 0.23 of a cycle, or $0.23 \times 360° = 82.8°$. The value of the pitch depends on factors including refractive-index gradient, index of the fiber, core diameter, and wavelength of light.

Although the lenses are segments of fiber, they are short by fiber-optic standards, just a few millimeters long. Thus, they can be considered as rod lenses as well as fiber lenses. These tiny lenses are used in a variety of optical systems.

Light Piping

The simplest application of optical fibers of any type is light piping. The term describes the process—piping light from one place to another. The light can be carried by one or many fibers without regard to how the fibers are arranged, as long as they collect light and deliver it to the same places. Thus, alignment of individual fibers need not be the same at the two ends of a fiber bundle. It doesn't matter if a fiber is at the center of the input end and at the outer edge of the output end.

Illumination

Most light-piping is for illumination, the delivery of light to some desired location. Why bother with optical fibers to do a light bulb's job? A flexible bundle of optical fibers can efficiently concentrate light in a small area or deliver light around corners to places it could not otherwise reach (e.g., inside machinery). A bundle can illuminate places where light bulbs cannot be used (e.g., areas where explosive vapors are being used). And fiber bundles can be divided to deliver light from one bulb to many places.

● A fiber bundle can illuminate small or hard-to-reach areas.

Such illumination fibers are also used in indicators (e.g., the automobile bulb-outage indicator discussed in Chapter 19). And illumination fibers—in the form of fiber-optic lamps and displays—gave many of us our first real view of fiber technology.

Optical Power Delivery

Some light-piping fibers can carry enough optical power to do more than illuminate. They carry laser beams used for medical treatment or industrial materials-working.

● Single large-core fibers can carry optical power for medicine and materials-working.

Conventional laser systems use lenses or mirrors to focus their beams on the desired point. However, these beam delivery systems are bulky and can be cumbersome for fine tasks, such as delicate surgery. Fiber systems for beam delivery are much easier to manipulate. Some are designed for surgeons to use with their hands. Others are built for robotic control in factories.

Power delivery requires single fibers with large cores or bundled fibers. Single fibers are best for many applications if the input fiber can be concentrated into a single core. Power transmission capacities are surprisingly high for single large-core fibers with losses measured in decibels per kilometer, which can easily carry tens of watts over a few meters.

Signs

●

Fiber arrays can create illuminated signs.

If all the fibers in an illumination bundle wind up in the same place, they illuminate a single spot. If they are routed to different places, they can form a patterned image, such as the fiber-optic display shown in Figure 23.6. All the fibers are brought together in one place to collect light from a bulb, then they are splayed out to create the desired pattern. Diffusing lenses at the ends of the fiber can spread light out to make large spots.

FIGURE 23.6.

A fiber-optic sign.

Fused and Imaging Fibers

Like incoherent bundles for illumination, coherent fiber bundles are not the standard means of image transmission or projection. Conventional lenses and optical systems generally are much less costly, and typically project better images. However, lenses and conventional optics cannot be used everywhere. Imaging fiber bundles can probe inaccessible areas, from the interior of the human body to the interiors of machines. Fiber bundles fit into places where lenses can't, and offer other capabilities, such as directing light from a streetlight or display in a particular direction. I'll take a look at a sampling of these applications.

Faceplates

Image transmission does not have to be over a long distance. Indeed, one of the most common uses of fiber-optic image transmission uses coherent bundles less than 1 inch long. That is the fiber-optic faceplate, a rigid array of fibers fused together and cut into thin sheets that concentrate light in a particular direction.

You've probably seen fiber-optic faceplates without realizing it. Inexpensive glass types are put in front of traffic signals to direct their light to the spot where traffic planners expect to find people in cars looking for the traffic light. You don't see any light from the side of the street, but the light comes into view when you move into the right position. You usually find them where two or more lights are close together, and traffic planners apparently think that drivers could be confused if they could see too many of them. The idea sounds good, but if the lights are misaligned or the faceplates have too narrow a viewing angle, the lights can be very hard to see. (Pedestrians, bicyclists, and truck drivers can easily miss lights targeted to standard automobiles.)

Faceplates are also used to enhance images generated in cathode ray tubes, by concentrating light into a narrow range of viewing angles. The faceplate becomes the surface of the CRT. The goal is not to replace the glass front of an ordinary television picture tube. These fiber-optic faceplates are much smaller and are used with small, high-resolution displays or image-intensification tubes, demanding applications that can justify the added expense. They can help flatten the curved field of a cathode-ray tube, correct for distortion, and enhance the display's effective brightness by suppressing ambient light. Most are used for military systems. Faceplates may not sound as exciting as other uses of coherent fiber bundles, but they are among the most widespread uses of the technology.

● Fiber-optic faceplates concentrate light from a traffic light or display in a particular direction.

Image Manipulation, Splitting, and Combining

Coherent fiber bundles can do more than just transmit images; they can also manipulate them. Twisting a coherent bundle by 180° inverts the image. You can do the same with lenses, but a fiber-optic image inverter does not require as long a working distance, which is of critical importance in some military systems. (Some image inverters are less than 1 inch long.)

Another type of image manipulation possible with fused fiber optics is the image combiner/splitter shown in Figure 23.7. This is made by laying down a series of fiber-optic ribbons, alternating them as if shuffling a deck of cards. One ribbon goes from the single input to output 1, the next from the input to output 2, the next to output 1, and so on. Put a single image into the input, and you get two identical (but fainter) output images. Put separate images into the two outputs, and you get one combined image.

● Coherent fiber bundles can manipulate images.

Similar ideas could be used in other image manipulators or in devices to perform operations on optical signals. However, before you rush out for a patent application on your own bright idea, you must face the ugly reality of cost. Manufacture of the fiber-optic image combiner in Figure 23.7 requires time and exacting precision, making it too expensive for most uses. Image inverters are used in some systems, but only where less-costly lens systems won't do the job.

FIGURE 23.7.

A fiber-optic image combiner and duplicator. (Courtesy of Galileo Electro-Optics Corp.)

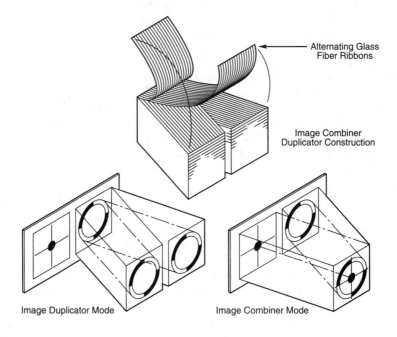

Tapers

Earlier you saw how tapered fiber cores could bend light rays. The same principle could be used for imaging with tapered bundles. Controlling the degree of tapering across the area of the taper can vary the degree of magnification. For example, the center of the image can be magnified more than the outer part, or vice versa. However, the practical problems are similar to those with other fiber-optic image manipulators—lenses cost much less and have higher resolution, so they are used for most applications.

Medical Endoscopes

Imaging fiber bundles called endoscopes let physicians look inside body cavities without surgery.

The most important use of imaging fiber bundles is in medicine, to let physicians look inside the body without surgery. These special-purpose coherent fiber bundles, up to a couple of meters long, are called endoscopes. Endoscopes may be rigid or flexible, but flexible types are preferred for many purposes because they are easier to insert and manipulate through body orifices.

Suppose a physician wants to examine a patient's bronchial tubes. He passes an endoscope down the patient's throat toward the lungs. Some fibers in the bundle transmit light from a lamp to illuminate the airway. Others return light so the physician can see conditions inside. Similar instruments are used to examine the colon.

In some cases, the physician can do more than just look. With a laser that emits a wavelength transmitted by the fiber, the physician can perform laser surgery through the endoscope. For example, an endoscope could be aimed at bronchial cancer, then laser pulses fired to vaporize the cancer cells. The physician could look through the endoscope to check progress after each pulse. (Care must be taken to keep the bright laser pulses out of the physician's eye.) Only a small fraction of endoscopes are used with laser surgery, but the possibilities of avoiding some major surgery are exciting. That can save the patients not only money but the risks associated with major surgery.

⬤ Some endoscopes can be used for internal laser surgery.

Optical fibers can also deliver laser energy inside the body for other procedures. One idea that has been in development for several years is threading single fibers through arteries to deliver laser energy to a blocked region. The technique is promising for some cases, particularly for blood vessels that are easily accessed and relatively straight, but problems can arise in other cases. You can't just blast away at all clogged arteries with a fiber-optic version of a plumber's snake, because a misguided laser beam can remove arterial wall instead of arterial plaque.

Graded-Index Fiber Lenses

The graded-index fiber segments described earlier in this chapter have uses quite different from imaging fiber bundles and faceplates. Their applications are as lenses.

⬤ Graded-index fiber segments can be used as optical components.

Some uses are in fiber optics. A graded-index fiber microlens can focus output from an LED or diode laser so it could be coupled efficiently into a fiber. However, most transmitters have light sources butted up against the collecting fiber, without intermediate components. In addition, many fiber pigtails could serve the same function without the need of any other discrete component.

Most applications of graded-index fiber microlenses are in optical systems such as photocopiers. One example is the use of a linear array of fiber-optic microlenses to focus light reflected from a small area of a page being copied onto individual sensors in a linear array that detects reflected light.

What Have You Learned?

1. Rigid or flexible bundles of optical fibers can transmit images if the fibers that make them up are properly aligned at the ends (coherent). Rigid bundles are made of fibers fused together; flexible bundles contain separate fibers bonded at the ends. Resolution is limited by the size of the fiber cores, typically around 10 μm.

2. Bundles of fibers in which the ends are not aligned with one another serve as "light pipes" to illuminate hard-to-reach places. The bundle can be broken up on one end to form an image or display (e.g., a WALK sign).

3. Light that falls into fiber claddings in bundles is lost, but typically 90% falls onto fiber cores.

4. Imaging and other short-distance fibers generally have much higher attenuation than communication fibers. Bundled fibers are step-index multimode types with large NA.

5. Step-index multimode fibers do not focus light, but segments of graded-index fiber do focus light and can serve as lenses. Tapered fibers magnify or demagnify objects seen through them; they are used in bundles.

6. Thin fiber-optic faceplates are used in directional traffic lights and to concentrate light from certain displays in a particular direction. They are used with certain high-performance imaging tubes, but not for ordinary cathode-ray tubes.

7. Coherent fiber bundles can invert, split, and combine images.

8. Endoscopy is the use of coherent fiber bundles to view inside the body without surgery.

9. Large-core fibers can deliver laser power for medicine or materials-working.

Quiz for Chapter 23

1. Which of the following statements is false?

   a. Coherent fiber bundles can transmit images.

   b. Coherent fiber bundles can focus light.

   c. Graded-index fiber segments can focus light.

   d. Imaging fiber bundles contain step-index multi-mode fibers.

2. A graded-index fiber lens has a pitch of 0.45. How much of a sinusoidal oscillation cycle do light rays experience in passing through it?

   a. 27°.

   b. 45°.

   c. 81°.

   d. 162°.

   e. 180°.

3. What does it mean to say that a fiber bundle has a packing fraction of 90%?

   a. 90% of the fibers are intact.

   b. 90% of the input surface is made up of optical fibers.

   c. 90% of the input surface is made up of fiber core.

   d. 90% of the input surface is made up of fiber cladding.

   e. It transmits 90% of the incident light through its entire length.

4. You want to resolve an image with 8 line pairs per millimeter. In theory, what is the largest fiber core size that you could use in a stationary coherent bundle?

    a. 8 μm.

    b. 50 μm.

    c. 62.5 μm.

    d. 100 μm.

    e. 125 μm.

5. Endoscopes used in medicine to view inside the body are

    a. flexible fiber-optic bundles.

    b. sometimes able to transmit laser beams to treat disease.

    c. rigid fiber-optic bundles.

    d. all of the above.

    e. none of the above.

6. Fiber-optic faceplates are

    a. specialized sensors that detect temperature variations across a surface.

    b. thin rigid fiber bundles used to make traffic signals visible only from a limited viewing angle.

    c. assemblies of graded-index fiber lenses that focus light in photocopiers.

    d. used on most television sets.

7. Average attenuation of bundled fibers is

    a. 0.5 dB/km.

    b. 1-5 dB/km.

    c. 10-100 dB/km.

    d. around 1 dB/m.

8. What types of fibers are used in imaging bundles?

    a. Step-index multimode.

    b. Graded-index multimode.

    c. Step-index single-mode.

    d. All of the above.

9. The practical use of fiber-optic bundles to manipulate images is limited by what?

    a. Poor resolution.

    b. High attenuation.

    c. Fragility.

    d. High cost.

10. Which of the following is the most important advantage of random fiber bundles for illumination?

    a. Flexibility.

    b. Low cost.

    c. Disposability.

    d. Durability.

Laser Safety

With the exercise of reasonable common sense, fiber-optic systems are inherently safe. However, special rules have been developed to cover the safe use of lasers. Their most visible impact is in warning labels printed on many laser data sheets and packages to meet requirements imposed by the Center for Devices and Radiological Health, which is part of the Department of Health and Human Services.

Why are there so many warnings when a couple of milliwatts from a semiconductor laser are not about to burn holes through anything? Because the eye can focus the invisible infrared beam onto a tiny spot on the retina, the light-sensitive layer at the back of the eye. The light intensity in the small spot on the retina formed by focusing a 1 mW laser beam is similar to that produced on the retina by looking directly at the sun, and is thus potentially dangerous.

Just as with the sun, a momentary glance into a laser beam will not blind you, but staring into it for a long period could leave a blind spot. The beam from an unfocused semiconductor laser or optical fiber spreads out so rapidly (by laser standards) that looking at it from a few feet away should pose no threat. (That is not true for gas or crystalline solid-state lasers, which have very narrow beams.) However, you should never aim the output of a semiconductor laser directly into your eye. The same applies to the output of an optical fiber connected to a semiconductor laser. Although federal regulations do not apply to LEDs, their output could have the same effect if it was as powerful and as tightly focused as a laser beam (e.g., if it emerged from a fiber right into the eye). LED intensities are rarely that great, but caution is wise.

Glossary

Acceptance Angle   The angle over which the core of an optical fiber accepts incoming light; usually measured from the fiber axis. Related to numerical aperture (NA).

All-Dielectric Cable   Cable made entirely of dielectric (insulating) materials without any metal conductors, armor, or strength members.

Analog   A signal that varies continuously (e.g., sound waves). Analog signals have frequency and bandwidth measured in hertz.

Ångstrom (Å)   A unit of length, 0.1 nm or 10^{-10} m, often used to measure wavelength but not part of the SI system of units. Often written Angstrom because the special symbol is not available.

Armor   A protective layer, usually metal, wrapped around a cable.

ATM (Asynchronous Transfer Mode)   A digital transmission switching format, with cells containing 5 bytes of header information followed by 48 data bytes. Part of the B-ISDN standard.

Attenuation   Reduction of signal magnitude, or loss, normally measured in decibels. Fiber attenuation is normally measured per unit length in decibels per kilometer.

Attenuator   An optical element that reduces the intensity of light passing through it (i.e., attenuates it).

Avalanche Photodiode (APD)   A semiconductor photodetector with integral detection and amplification stages. Electrons generated at a p/n junction are accelerated in a region where they free an avalanche of other electrons. APDs can detect faint signals but require higher voltages than other semiconductor electronics.

Average Power   The average level of power in a signal that varies with time.

Axis   The center of an optical fiber.

Backbone System   A transmission network that carries high-speed telecommunications between regions (e.g., a nationwide long-distance telephone system). Sometimes used to describe the part of a local area network that carries signals between branching points.

Backscattering   Scattering of light in the direction opposite to that in which it was originally traveling.

Bandwidth   The highest frequency that can be transmitted in analog operation. Also (especially for digital systems), the information-carrying capacity of a system.

Baud   Strictly speaking, the number of signal-level transitions per second in digital data. For some common coding schemes, this equals bits per second, but this is not true for more complex coding, where it is often misused. Telecommunication specialists prefer bits per second, which is less ambiguous.

Beamsplitter   A device that divides incident light into two separate beams.

Bel   A relative measurement, denoting a factor of ten change. Rarely used in practice; most measurements are in decibels (0.1 bel).

BER   See *bit-error rate.*

Bidirectional   Operating in both directions. Bidirectional couplers operate the same way regardless of the direction light passes through them. Bidirectional transmission sends signals in both directions, sometimes through the same fiber.

Bifringent   Having a refractive index that differs for light of different polarizations.

B-ISDN   See *Broadband-Integrated Services Digital Network.*

Bistable Optics   Optical devices with two stable transmission states.

Bit-Error Rate (BER)   The fraction of bits transmitted incorrectly.

Broadband   In general, covering a wide range of frequencies. The broadband label is sometimes used for a network that carries many different services or for video transmission.

Broadband-Integrated Services Digital Network (B-ISDN)   A family of standards for digital telecommunications integrating voice, video, data, and other services. Includes ATM and SONET.

Broadcast Transmission   Sending the same signal to many different places, like a television broadcasting station. Broadcast transmission can be over optical fibers if the same signal is delivered to many subscribers.

Bundle (of fibers)   A rigid or flexible group of fibers assembled in a unit. Coherent fiber bundles have fibers arranged in the same way on each end and can transmit images.

Bypass   A circuit that carries telephone signals from a subscriber to another point without the use of local telephone company circuits.

Byte   Eight bits of digital data. (Sometimes parity and check bits are included, so one "byte" may include ten bits, but only eight of them are data.)

CATV   An acronym for cable television, derived from Community Antenna TeleVision.

CCITT   International Consultative Commission on Telephone and Telegraph, an arm of the International Telecommunications Union, which sets standards.

Cells   Blocks of data transmitted in Asynchronous Transfer Mode.

Central Office   A telephone company facility for switching signals among local telephone circuits; connects to subscriber telephones. Also called a switching office.

Chromatic Dispersion   Pulse spreading arising from differences in the speed that light of different wavelengths travels through materials. Measured in picoseconds (of dispersion) per kilometer (of fiber length) per nanometer (of source bandwidth), it is the sum of waveguide and material dispersion.

Cladding   The layer of glass or other transparent material surrounding the light-carrying core of an optical fiber. It has a lower refractive index than the core, and thus confines light in the core. Coatings may be applied over the cladding.

Coherent Bundle (of fibers)   Fibers packaged together in a bundle so they retain a fixed arrangement at the two ends and can transmit an image.

Coherent Communications   In fiber optics, a communication system where the output of a local laser oscillator is mixed with the received signal, and the difference frequency is detected and amplified.

Compression   Reducing the number of bits needed to encode a digital signal, typically by eliminating long strings of identical bits or bits that do not change in successive sampling intervals (e.g., video frames).

Connector   A device mounted on the end of a fiber-optic cable, light source, receiver, or housing that mates to a similar device to couple light into and out of optical fibers. A connector joins two fiber ends or one fiber end and a light source or detector.

Core   The central part of an optical fiber that carries light.

Coupler   A device that connects three or more fiber ends, dividing one input between two or more outputs or combining two or more inputs into one output.

Coupling   Transfer of light into or out of an optical fiber. (Note that coupling does not require a coupler.)

Critical Angle   The angle at which light undergoes total internal reflection.

Cut-Back Measurements   Measurement of optical loss made by cutting a fiber to compare loss of a short segment with loss of a longer one.

Cut-Off Wavelength   The longest wavelength at which a single-mode fiber can transmit two modes, or (equivalently) the shortest wavelength at which a single-mode fiber carries only one mode.

Cycles per Second   The frequency of a wave, or number of oscillations it makes per second. One cycle per second equals one hertz.

Dark Current   The noise current generated by a photodiode in the dark.

Data Link   A fiber-optic transmitter, cable, and receiver that transmit digital data between two points.

dBm   Decibels below 1 mW.

dBμ   Decibels below 1 μW.

Decibel (dB)   A logarithmic comparison of power levels, defined as ten times the base-ten logarithm of the ratio of the two power levels. One-tenth of a bel.

Demultiplexer   A device that separates a multiplexed signal into its original components; the inverse of a multiplexer.

Detector   A device that generates an electrical signal when illuminated by light. The most common in fiber optics are photodiodes.

Dielectric   Nonconductive.

Digital   Encoded as a signal in discrete levels, typically binary 1s and 0s.

Diode   An electronic device that lets current flow in only one direction. Semiconductor diodes used in fiber optics contain a junction between regions of different doping. They include light emitters (LEDs and laser diodes) and detectors (photodiodes).

Diode Laser   A semiconductor diode in which the injection of current carriers produces laser light by amplifying photons produced when holes and electrons recombine at the junction between p- and n-doped regions.

Directional Coupler   A coupler in which light is transmitted differently when it goes in different directions.

Dispersion   The spreading out of light pulses as they travel in an optical fiber, proportional to length.

Dispersion Compensation   The use of fiber segments with different dispersions to reduce total dispersion.

Dispersion-Shifted Fiber   Optical fiber designed to have its nominal zero-dispersion point at 1550 nm rather than the 1300 nm in step-index single-mode fiber.

DS*x*   A transmission rate in the North American Digital Telephone hierarchy; see Table 17.1. Also called T carrier.

Duplex   A duplex cable contains two fibers; a duplex connector links two pairs of fibers.

Edge-Emitting Diode   An LED that emits light from its edge, producing more directional output than LEDs that emit from their top surface.

Electromagnetic Interference (EMI)   Noise generated when stray electromagnetic fields induce currents in electrical conductors.

Electromagnetic Radiation   Waves made up of oscillating electrical and magnetic fields perpendicular to one another and traveling at the speed of light. Can also be viewed as photons or quanta of energy. Electromagnetic radiation includes radio waves, microwaves, infrared, visible light, ultraviolet radiation, X rays, and gamma rays.

EMI   See *electromagnetic interference.*

Endoscope   A fiber-optic bundle used for imaging and viewing inside the human body.

Erbium-Doped Fiber Amplifier   Optical fibers doped with the rare earth element erbium, which can amplify light in the 1550 nm region when pumped by an external light source.

Evanescent Wave   Light guided in the inner part of an optical fiber's cladding rather than in the core.

Excess Loss   Loss of a passive coupler above that inherent in dividing light among the output ports.

External Modulation   Modulation of a light source by an external device.

Extrinsic Loss   Splice losses arising from the splicing process itself.

Eye Pattern   A pattern displayed when an oscilloscope is driven by a receiver output and triggered by the signal source that drove the transmitter. The more open the eye, the better the signal quality.

Faceplate   A rigid array of short fibers fused together to direct light onto a traffic signal or imaging tube.

FDDI   See *Fiber Distributed Data Interface.*

Ferrule   A tube within a connector with a central hole that contains and aligns a fiber.

Fiber Distributed Data Interface (FDDI)   A standard for a 100 Mbit/s fiber-optic local area network.

Fiber Grating   An optical fiber in which the refractive index of the core varies periodically along its length, scattering light in a way similar to a diffraction grating, and transmitting or reflecting certain wavelengths selectively. Fiber gratings are sensitive to outside effects, making them good sensors.

Fiber-Optic Gyroscope   A coil of optical fiber that can detect rotation about its axis.

Fiber To The Curb (FTTC)   Fiber-optic service to a node that is connected by wires to several nearby homes, typically on a block.

FITL   Fiber In The Loop.

Fluoride Glasses   Materials that have the amorphous structure of glass but are made of fluoride compounds (e.g., zirconium fluoride) rather than oxide compounds (e.g., silica).

FOG-M   Fiber-Optic Guided Missile.

Frames   Blocks of data transmitted in the SONET format; also individual images shown in sequence on video screens. (One video frame has the full number of lines for a standard format.)

Frequency Division Multiplexing   Combining analog signals by assigning each a different carrier frequency and merging them in a single signal with a broad range of frequencies.

Fused Fibers   A bundle of fibers fused together so they maintain a fixed alignment with respect to each other in a rigid rod.

GaAlAs   Gallium aluminum arsenide.

GaAs   Gallium arsenide.

Gallium Aluminum Arsenide (GaAlAs)   A semiconductor compound used in LEDs, diode lasers, and certain detectors.

Gallium Arsenide (GaAs)   A semiconductor compound used in LEDs, diode lasers, detectors, and electronic components.

Graded-Index Fiber   A fiber in which the refractive index changes gradually with distance from the fiber axis, rather than abruptly at the core-cladding interface.

Graded-Index Fiber Lens   A short segment of graded-index fiber that focuses light passing along it.

Hard-Clad Silica Fiber   A fiber with a hard plastic cladding surrounding a step-index silica core. (Other plastic-clad silica fibers have a soft plastic cladding.)

HDTV   High-definition (or high-resolution) television; television with about double the resolution of present systems.

Head End   The central facility where signals are combined and distributed in a cable television system.

Hertz   Frequency in cycles per second.

Hierarchy   A set of transmission speeds arranged to multiplex successively higher numbers of circuits.

Hydrogen Losses   Increases in fiber attenuation that occur when hydrogen diffuses into the glass matrix and absorbs some light.

Index of Refraction   The ratio of the speed of light in a vacuum to the speed of light in a material, usually abbreviated n.

Index-Matching Gel   A gel or fluid with refractive index close to glass that reduces refractive-index discontinuities that can cause reflective losses.

Indium Gallium Arsenide (InGaAs)   A semiconductor material used in lasers, LEDs, and detectors.

Indium Gallium Arsenide Phosphide (InGaAsP)   A semiconductor material used in lasers, LEDs, and detectors.

Infrared   Wavelengths longer than 700 nm and shorter than about 1 mm. We cannot see infrared radiation but can feel it as heat. Transmission of glass optical fibers is best in the infrared at wavelengths of 1100-1600 nm.

Infrared Fiber   Colloquially, optical fibers with best transmission at wavelengths of 2 μm or longer, made of materials other than silica glass.

InGaAs   Indium gallium arsenide.

InGaAsP   Indium gallium arsenide phosphide. Properties depend on composition, which is sometimes written $In_{1-x}Ga_xAs_{1-y}P_y$.

Injection Laser   Another name for a semiconductor or diode laser.

Integrated Optics   Optical devices that perform two or more functions and are integrated on a single substrate; analogous to integrated electronic circuits.

Integrated Services Digital Network (ISDN)   A digital standard calling for 144 kbit/s transmission, corresponding to two 64 kbit/s digital voice channels and one 16 kbit/s data channel.

Intensity   Power per unit solid angle.

Interferometric Sensors   Fiber-optic sensors that rely on interferometric detection.

Intrinsic Losses   Splice losses arising from differences in the fibers being spliced.

Irradiance   Power per unit area.

ISDN   See *Integrated Services Digital Network.*

Junction Laser   A semiconductor diode laser.

Kevlar   A strong synthetic material used in cable strength members; the name is a trademark of the Dupont Company.

LAN   See *local area network.*

Large-Core Fiber   Usually, a fiber with a core of 200 μm or more.

Laser   From *Light Amplification by Stimulated Emission of Radiation,* one of the wide range of devices that generates light by that principle. Laser light is directional, covers a narrow range of wavelengths, and is more coherent than ordinary light. Semiconductor diode lasers are the standard light sources in fiber-optic systems.

LED   See *light-emitting diode*.

Light   Strictly speaking, electromagnetic radiation visible to the human eye at 400 to 700 nm. Commonly, the term is applied to electromagnetic radiation with properties similar to visible light, including the invisible near-infrared radiation in most fiber-optic communication systems.

Light Piping   Use of optical fibers to illuminate.

Light-Emitting Diode (LED)   A semiconductor diode that emits incoherent light at the junction between p- and n-doped materials.

Lightguide   An optical fiber or fiber bundle.

Lightwave   As an adjective, a synonym for optical, often (but not always) meaning fiber-optic.

Local Area Network (LAN)   A network that transmits data among many nodes in a small area (e.g., a building or campus).

Local Loop   The part of the telephone network extending from the central (switching) office to the subscriber.

Longitudinal Modes   Oscillation modes of a laser along the length of its cavity. Each longitudinal mode contains only a very narrow range of wavelengths, so a laser emitting a single longitudinal mode has a very narrow bandwidth. Distinct from transverse modes.

Loose Tube   A protective tube loosely surrounding a cabled fiber, often filled with gel.

Loss   Attenuation of optical signal, normally measured in decibels.

Loss Budget   An accounting of overall attenuation in a system.

Margin   Allowance for attenuation in addition to that explicitly accounted for in system design.

Mass Splicing   Simultaneous splicing of many fibers in a cable.

Material Dispersion   Pulse dispersion caused by variation of a material's refractive index with wavelength.

Mbit/s   Megabits (million bits) per second.

Mechanical Splice   A splice in which fibers are joined mechanically (e.g., glued or crimped in place) but not fused together.

Microbending   Tiny bends in a fiber that allow light to leak out and increase loss.

Micrometer   One-millionth of a meter, abbreviated μm.

Modal Dispersion   Dispersion arising from differences in the times that different modes take to travel through multimode fiber.

Mode   An electromagnetic field distribution that satisfies theoretical requirements for propagation in a waveguide or oscillation in a cavity (e.g., a laser). Light has modes in a fiber or laser.

Mode Field Diameter   The diameter of the one mode of light propagating in a single-mode fiber, typically slightly larger than core diameter.

Mode Stripper   A device that removes high-order modes in a multimode fiber to give standard measurement conditions.

Multimode   Transmits or emits multiple modes of light.

Multiplexer   A device that combines two or more signals into a single output.

n Region   A semiconductor doped to have an excess of electrons as current carriers.

NA   See *numerical aperture.*

Nanometer   A unit of length, 10^{-9} m. It is part of the SI system and has largely replaced the non-SI Angstrom (0.1 nm) in technical literature.

Nanosecond   One-billionth of a second, 10^{-9} second.

National Electrical Code   A wiring code that specifies safety standards for copper and fiber-optic cable.

Near Infrared   The part of the infrared near the visible spectrum, typically 700 to 1500 or 2000 nm; it is not rigidly defined.

Network   A system of cables or other connections that links many terminals, all of which can communicate with each other through the system.

No Return to Zero (NRZ)   A digital code in which the signal level is low for a 0 bit and high for a 1 bit, and does not return to 0 between successive 1 bits.

Noise Equivalent Power (NEP)   The optical input power to a detector needed to generate an electrical signal equal to the inherent electrical noise.

Normal (angle)   Perpendicular to a surface.

NRZ   See *no return to zero.*

NTSC   The analog video broadcast standard used in North America, set by the National Television System Committee.

Numerical Aperture (NA)   The sine of half the angle over which a fiber can accept light. Strictly speaking, this is multiplied by the refractive index of the medium containing the light, but that equals 1 for air, the normal medium from which NA is measured.

OC-*x* Optical Carrier, a carrier rate specified in the SONET standard; see Table 17.1.

Optical Amplifier A device that amplifies an input optical signal without converting it into electrical form. The best developed are optical fibers doped with the rare earth erbium.

Optical Loss Test Set An optical power meter and light source calibrated for use together.

Optical Time-Domain Reflectometer (OTDR) An instrument that measures transmission characteristics by sending a short pulse of light down a fiber and observing backscattered light.

Optical Waveguide Technically, any structure that can guide light. Sometimes used as a synonym for optical fiber, it can also apply to planar light waveguides.

p Region Part of a semiconductor doped with electron acceptors in which holes (vacancies in the valence electron level) are the dominant current carriers.

Packing Fraction The fraction of the surface area of a fiber-optic bundle that is fiber core.

PCS Fiber See *plastic-clad silica fiber.*

Peak Power Highest instantaneous power level in a pulse.

Phase The position of a wave in its oscillation cycle.

Photodetector A light detector.

Photodiode A diode that can produce an electrical signal proportional to light falling upon it.

Photonic A term coined for devices that work using photons, analogous to "electronic" for devices working with electrons.

Photons Quanta of electromagnetic radiation. Light can be viewed as either a wave or a series of photons.

***pin* Photodiode** A semiconductor detector with an intrinsic (i) region separating the p- and n-doped regions. It has fast linear response and is used in fiber-optic receivers.

Planar Waveguide A waveguide fabricated in a flat material such as a thin film.

Plastic-Clad Silica (PCS) Fiber A step-index multimode fiber in which a silica core is surrounded by a lower-index plastic cladding.

Plenum Cable Cable made of fire-retardant material that meets electrical code requirements (UL 910) for low smoke generation and installation in air spaces.

Point-to-Point Transmission Carrying a signal between two end points, without branching to other points.

Polarization   Alignment of the electric and magnetic fields that make up an electromagnetic wave; normally refers to the electric field. If all light waves have the same alignment, the light is polarized.

Polarization-Maintaining Fiber   Fiber that maintains the polarization of light that enters it.

Preform   A cylindrical rod of specially prepared and purified glass from which an optical fiber is drawn.

Pulse Dispersion   The spreading out of pulses as they travel along an optical fiber.

Quantum Efficiency   The fraction of photons that strike a detector which produce electron-hole pairs in the output current.

Quaternary   A semiconductor compound made of four elements (e.g., InGaAsP).

Radiation-Hardened   Insensitive to the effects of nuclear radiation, usually for military applications.

Radiometer   An instrument, distinct from a photometer, to measure power (watts) of electromagnetic radiation.

Rays   Straight lines that represent the path taken by light.

Receiver   A device that detects an optical signal and converts it into an electrical form usable by other devices.

Recombination   Combination of an electron and a hole in a semiconductor that releases energy, sometimes leading to light emission.

Refraction   The bending of light as it passes between materials of different refractive index.

Refractive Index   Ratio of the speed of light in a vacuum to the speed of light in a material; abbreviated n.

Refractive-Index Gradient   The change in refractive index with distance from the axis of an optical fiber.

Regenerator   A receiver-transmitter pair that detects a weak signal, cleans it up, then sends the regenerated signal through another length of fiber.

Repeater   Often a receiver-transmitter pair that detects, cleans up, and amplifies a weak signal for retransmission through another length of optical fiber. Sometimes a repeater contains multiple regenerators, one for each fiber in a cable.

Responsivity   The ratio of detector output to input, usually measured in units of amperes per watt (or microamperes per microwatt).

Return to Zero (RZ)   A digital coding scheme where signal level is low for a 0 bit and high for a 1 bit during the first half of a bit interval, then in either case returns to zero for the second half of the bit interval.

Ribbon Cables   Cables in which many fibers are embedded in a plastic material in parallel, forming a flat ribbon-like structure.

Ring Architecture   A network scheme in which a transmission line forms a complete ring. If the ring is broken, signals can still be sent among the terminals.

Rise Time   The time it takes output to rise from low levels to peak value. Typically measured as the time to rise from 10% to 90% of maximum output.

RZ   See *return to zero.*

SDH   See *Synchronous Digital Hierarchy.*

Selfoc Lens   A trade name used by the Nippon Sheet Glass Company for a graded-index fiber lens; a segment of graded-index fiber made to serve as a lens.

Semiconductor Laser   A laser in which injection of current into a semiconductor diode produces light by recombination of holes and electrons at the junction between p- and n-doped regions.

Sheath   An outer protective layer of a fiber-optic cable.

SI Units   The standard international system of metric units.

Signal-to-Noise Ratio   The ratio of signal to noise, measured in decibels; an indication of signal quality in analog systems.

Silica Glass   Glass made mostly of silicon dioxide, SiO_2, used in conventional optical fibers.

Simplex   Single element (e.g., a simplex connector is a single-fiber connector).

Single-Frequency Laser   A laser that emits a range of wavelengths small enough to be considered a single frequency.

Single Mode   Containing only one mode. When dealing with lasers, beware of ambiguities because of the difference between transverse and longitudinal modes. A laser operating in a single transverse mode typically does not operate in a single longitudinal mode.

Single-Polarization Fibers   Optical fibers capable of carrying light in only one polarization.

Smart Structures (or Smart Skins)   Materials containing sensors (fiber-optic or other types) to measure their properties during fabrication and use.

Soliton   An optical pulse that regenerates original shape at certain points as it travels along an optical fiber. Solitons can be combined with optical amplifiers to carry signals very long distances.

SONET (Synchronous Optical Network)   A standard for fiber-optic transmission that is part of B-ISDN. Data rates are listed in Table 17.1.

Splice A permanent junction between two fiber ends.

Splitting Ratio The ratio of power emerging from two output ports of a coupler.

Star Coupler A coupler with more than three ports.

Step-Index Multimode Fiber A step-index fiber with a core large enough to carry light in multiple modes.

Step-Index Single-Mode Fiber A step-index fiber with a small core capable of carrying light in only one mode.

Submarine Cable A cable designed to be laid underwater.

Subscriber Loop The part of the telephone network from a central office to individual subscribers.

Supertrunk A cable that carries several video channels between facilities of a cable television company.

Surface-Emitting Diode An LED that emits light from its flat surface rather than its side. Simple and inexpensive, with emission spread over a wide angle.

Switched Network A network that routes signals to their destinations by switching circuits.

Synchronous Digital Hierarchy (SDH) The international version of SONET, the Synchronous Optical Network standard. The biggest difference is in the names of the transmission rates in Table 17.1.

Synchronous Optical Network See *SONET*.

T Carrier A system operating at one of the standard levels in the North American Digital Hierarchy; rates are listed in Table 17.1

T Coupler A coupler with three ports.

TDM See *time-division multiplexing*.

Ternary A semiconductor compound made of three elements (e.g., GaAlAs).

III-V (3-5) Semiconductor A semiconductor compound made of one or more elements from the IIIA column of the periodic table (Al, Ga, and In) and one or more elements from the VA column (N, P, As, or Sb). Used in LEDs, diode lasers, and detectors.

Threshold Current The minimum current needed to sustain laser action in a diode laser.

Tight Buffer A material tightly surrounding a fiber in a cable, holding it rigidly in place.

Time-Division Multiplexing (TDM) Digital multiplexing by taking one pulse at a time from separate signals and combining them in a single bit stream.

Total Internal Reflection   Total reflection of light back into a material when it strikes the interface with a material having a lower refractive index at an angle below a critical value.

Transceiver   A combination of transmitter and receiver providing both output and input interfaces with a device.

Transverse Modes   Modes across the width of a waveguide (e.g., a fiber or laser). Distinct from longitudinal modes, which are along the length.

Tree   A network architecture in which transmission routes branch out from a central point.

Trunk Line   A transmission line running between telephone switching offices.

Ultraviolet   Electromagnetic waves invisible to the human eye, with wavelengths about 10-400 nm.

Videoconferencing   Conducting conferences via a video telecommunications system.

Videophone   A telephone-like service with a picture as well as sound.

Visible Light   Electromagnetic radiation visible to the human eye at wavelengths of 400-700 nm.

Voice Circuit   A circuit capable of carrying one telephone conversation or its equivalent; the standard subunit in which telecommunication capacity is counted. The U.S. analog equivalent is 4 kHz. The digital equivalent is 56 kbit/s in North America and 64 kbit/s in Europe.

Waveguide   A structure that guides electromagnetic waves along its length. An optical fiber is an optical waveguide.

Waveguide Couplers   A coupler in which light is transferred between planar waveguides.

Waveguide Dispersion   The part of chromatic dispersion arising from the different speeds light travels in the core and cladding of a single-mode fiber (i.e., from the fiber's waveguide structure).

Wavelength   The distance an electromagnetic wave travels in the time it takes to oscillate through a complete cycle. Wavelengths of light are measured in nanometers (10^{-9} m) or micrometers (10^{-6} m).

Wavelength-Division Multiplexing (WDM)   Multiplexing of signals by transmitting them at different wavelengths through the same fiber.

Y Coupler   A variation on the T coupler in which input light is split between two channels (typically planar waveguide) that branch out like a Y from the input.

Zero-Dispersion Wavelength   Wavelength at which net chromatic dispersion of an optical fiber is nominally zero. Arises when waveguide dispersion cancels out material dispersion.

Index

UNDERSTANDING

D

E

F

G

N

Q–R

U–V